Herbert Spencer

The Principles of Biology

Vol. II

Herbert Spencer

The Principles of Biology
Vol. II

ISBN/EAN: 9783744747257

Printed in Europe, USA, Canada, Australia, Japan

Cover: Foto ©Suzi / pixelio.de

More available books at **www.hansebooks.com**

THE PRINCIPLES OF SOCIOLOGY

BY
HERBERT SPENCER

VOL. II—2

NEW YORK
D. APPLETON AND COMPANY
1896

Authorized Edition.

CONTENTS.

Part V.—POLITICAL INSTITUTIONS
(CONTINUED).

CHAP.		PAGE
X.	MINISTRIES	442
XI.	LOCAL GOVERNING AGENCIES	451
XII.	MILITARY SYSTEMS	473
XIII.	JUDICIAL AND EXECUTIVE SYSTEMS	492
XIV.	LAWS	513
XV.	PROPERTY	538
XVI.	REVENUE	557
XVII.	THE MILITANT TYPE OF SOCIETY	568
XVIII.	THE INDUSTRIAL TYPE OF SOCIETY	603
XIX.	POLITICAL RETROSPECT AND PROSPECT	643

Part VI.—ECCLESIASTICAL INSTITUTIONS.

I.	THE RELIGIOUS IDEA	671
II.	MEDICINE-MEN AND PRIESTS	705
III.	PRIESTLY DUTIES OF DESCENDANTS	712
IV.	ELDEST MALE DESCENDANTS AS QUASI-PRIESTS	715
V.	THE RULER AS PRIEST	722
VI.	THE RISE OF A PRIESTHOOD	729
VII.	POLYTHEISTIC AND MONOTHEISTIC PRIESTHOODS	737

CHAP.		PAGE
VIII.	ECCLESIASTICAL HIERARCHIES	749
IX.	AN ECCLESIASTICAL SYSTEM AS A SOCIAL BOND .	763
X.	THE MILITARY FUNCTIONS OF PRIESTS . . .	775
XI.	THE CIVIL FUNCTIONS OF PRIESTS . . .	786
XII.	CHURCH AND STATE	793
XIII.	NONCONFORMITY	802
XIV.	THE MORAL INFLUENCES OF PRIESTHOODS .	808
XV.	ECCLESIASTICAL RETROSPECT AND PROSPECT .	818
XVI.	RELIGIOUS RETROSPECT AND PROSPECT . . .	827

PART V.

POLITICAL INSTITUTIONS
(CONTINUED).

CHAPTER X.

MINISTRIES.

§ 504. Men chosen by the ruler to help him, we meet with in early stages of social evolution—men whose positions and duties are then vague and variable. At the outset there is nothing to determine the selection of helpers save considerations of safety, or convenience, or liking. Hence we find ministers of quite different origins.

Relationship leads to the choice in some places and times; as with the Bachassins, among whom the chief's brother conveys his orders and sees them executed; as of old in Japan, where the Emperor's son was prime minister and the daimios had cadets of their families as counsellors; as in ancient Egypt where "the principal officers of the Court or administration appear to have been at the earliest period the relatives" of the king. Though in some cases family-jealousy excludes kinsmen from these places of authority, in other cases family-feeling and trust, and the belief that the desire for family-predominance will ensure loyalty, lead to the employment of brothers, cousins, nephews, &c.

More general appears to be the unobtrusive growth of personal attendants, or household servants, into servants of State. Those who are constantly in contact with the ruler have opportunities of aiding or hindering intercourse with him, of biassing him by their statements, and of helping or impeding the execution of his commands; and they thus gain power, and tend to become advising and executive agents. From the earliest times onwards we meet with illustrations. In ancient Egypt—

"The office of fan-bearer to the king was a highly honourable post, which none but the royal princes, or the sons of the first nobility, were permitted to hold. These constituted a principal part of his staff; and in the field they either attended on the monarch to receive his orders, or were despatched to have the command of a division."

In Assyria the attendants who thus rose to power were not relatives, but were habitually eunuchs; and the like happened in Persia. "In the later times, the eunuchs acquired a vast political authority, and appear to have then filled all the chief offices of state. They were the king's advisers in the palace, and his generals in the field." Kindred illustrations are furnished by the West. Shown among the primitive Germans, the tendency for officers of the king's household to become political officers, was conspicuous in the Merovingian period: the seneschal, the marshal, the chamberlain, grew into public functionaries. Down to the later feudal period in France, the public and household administrations of the king were still undistinguished. So was it in old English times. According to Kemble, the four great officers of the Court and Household were the Hrægc Thegn (servant of the wardrobe); the Steallere and Horsthegn (first, Master of the Horse, then General of the Household Troops, then Constable or Grand Marshal); the Discthegn (or thane of the table—afterwards Seneschal); the Butler (perhaps Byrele or Scenca). The like held under the conquering Normans; and it holds in a measure down to the present time.

Besides relatives and servants, friends are naturally in some cases fixed on by the ruler to get him information, give him advice, and carry out his orders. Among ancient examples the Hebrews furnish one. Remarking that in the small kingdoms around Israel in earlier times, it was customary for the ruler to have a single friend to aid him, Ewald points out that under David, with a larger State and a more complex administration, " the different departments are necessarily more subdivided, and new offices of 'friends' or ministers of the king assume a sort of independent importance." Like needs produced kindred effects in the first days of the Roman empire. Duruy writes:—

"Augustus, who called himself a plain Roman citizen, could not, like a king, have ministers, but only friends who aided him with their experience. ... The multitude of questions ... induced him afterwards to distribute the chief affairs regularly among his friends. ... This council was gradually organized."

And then in later days and other regions, we see that out of the group known as "friends of the king" there are often some, or there is one, in whom confidence is reposed and to whom power is deputed. In Russia the relation of Lefort to Peter the Great, in Spain that of Albuquerque to Don Pedro, and among ourselves that of Gaveston to Edward II., sufficiently illustrate the genesis of ministerial power out of the power gained by personal friendship and consequent trust. And then with instances of this kind are to be joined instances showing how attachment between the sexes comes into play. Such facts as that after Albuquerque fell, all offices about the court were filled by relations of the king's mistress; that in France under Louis XV. "the only visible government was that by women" from Mme. de Prie to Mme. du Barry; and that in Russia during the reign of Catherine II., her successive lovers acquired political power, and became some of them prime ministers and practically autocrats; will serve adequately to recall a tendency habitually displayed.

Regarded as able to help the ruler supernaturally as well as naturally, the priest is apt to become his chosen ally and agent. The Tahitians may be named as having a prime minister who is also chief priest. In Africa, among the Eggarahs (Inland Negroes), a priest "officiates as minister of war." How political power of priests results from their supposed influence with the gods, is well shown by the case of Mizteca (part of Mexico).

"The high-priests were highly respected by the caziques, who did nothing without their advice; they commanded armies, and ruled the state, reproved vice, and when there was no amendment, threatened famine, plague, war, and the anger of the gods."

Other places in ancient America—Guatemala, Vera Paz, &c., furnish kindred facts; as do historic peoples from the earliest times downwards. In ancient Egypt the king's advisers mostly belonged to the priestly caste. Under the Roman emperors ecclesiastics became ministers and secret counsellors. In mediæval days Dominican and Franciscan monks held the highest political offices. And in later times the connexion was shown by the ministerial power of cardinals, or, as in Russia, of patriarchs. This acquisition of leading political functions by functionaries of the church, has in some cases special causes in addition to the general cause. A royal chaplain (uniting the character of personal attendant with that of priest) stands in a relation to the king which almost necessitates acquisition of great influence. Moreover, being fitted by culture for secretarial work, he falls naturally into certain State-duties; as he did into those of chancellor among ourselves in early days.

Recognizing the fact that at the outset, these administrative agents, whatever further characters they have, are usually also soldiers, and are included in the primitive consultative body, of which they become specialized parts, we may say of them generally, that they are relatives, friends, attendants, priests, brought into close relations with the ruler, out of whom he is obliged by stress of business to choose assistants; and that at first vague and irregular, their appointments and functions gradually acquire definiteness.

§ 505. Amid much that is too indefinite for generalization, a few tolerably constant traits of ministers, and traits of ministries, may be briefly indicated.

That a trusted agent commonly acquires power over his principal, is a fact everywhere observable. Even in a gentleman's household a head servant of long standing not unfrequently gains such influence, that his master is in various matters guided by him—almost controlled by him. With chief officers of State it has often-been the same; and especially where hereditary succession is well established. A ruler who, young, or idle, or pleasure-seeking, performs his duties by proxy, or who, through personal liking or entire trust, is led to transfer his authority, presently becomes so ill informed concerning affairs, or so unused to modes of procedure, as to be almost powerless in the hands of his agent.

Where hereditary succession pervades the society and fixes its organization, there is sometimes shown a tendency to inheritance, not of the rulership only, but also of these offices which grow into deputy-rulerships. Under the Norman dukes before the Conquest, the places of seneschal, cup-bearer, constable, and chamberlain, were "hereditary grand serjeanties." In England in Henry II.'s time, succession to the posts of high-steward, constable, chamberlain, and butler, followed from father to son in the houses of Leicester, Miles, Vere, and Albini. So was it with the Scotch in King David's reign: "the offices of great steward and high constable had become hereditary in the families of Stewart and De Morevil." And then in Japan the principle of inheritance of ministerial position had so established itself as to insure ministerial supremacy. In these cases there come into play influences and methods like those which conduce to hereditary kingship. When, as during the later feudal period in France, we see efforts made to fix in certain lines of descent, the chief offices of State (efforts which, in that case, sometimes succeeded and sometimes failed), we are shown that ministers use the facilities which their places give them, to establish succession to these places in their own families, in the same way that early kings do. Just as, during the stage of elective kingship, the king is apt to use the advantages derived from his position to secure the throne for his son, by getting him chosen during his own life, and thus to initiate hereditary succession; so the minister who has been allowed to acquire great power, is prompted to employ it for the purpose of establishing a monopoly of his office among his own descendants. Generally his desire is effectually antagonized by that of the ruler; but where, as in Japan, seclusion of the ruler impedes his hold on affairs, this desire of the minister takes effect.

Since there ever tend to arise these struggles between a king and one or more of those who serve him—since his efforts to maintain his authority are sometimes so far defeated that he is obliged to accept assistants who are hereditary; there results a jealousy of those whose interests are at variance with his own, and an endeavour to protect himself by excluding them from office. There comes a motive for choosing as ministers men who, having no children, cannot found houses which, growing powerful, may compete for supremacy; and hence in certain times the preference for celibate priests. Or, from allied motives, men neither clerical nor military are selected; as in France, where in the 15th and 17th centuries, members of the *bourgeois* class came to be preferred. A policy like that shown in the befriending of towns as a set-off against feudal chiefs, prompted the official employment of citizens instead of nobles. Under other conditions, again, there is a jealousy of ecclesiastics and an exclusion of them from power. For generations before the time of Peter the Great, the head of the church in Russia was " considered the second person in the empire; he was consulted on all State-affairs, until at length, their [his] spiritual pride outrunning all decorum, venturing upon, and even attempting to control the sovereign power, it was resolved by Peter the Great to abolish the patriarchate altogether." Between Louis XIV. and the Pope, there was a conflict for supremacy over the French church; and on more occasions than one, certain of the clergy encouraged "the absolutist pretensions of the Roman Pontiffs:" the result being that such prelates as held office were those who subordinated clerical to political aims, and that by Louis XIV., after 1661, " no churchman was allowed to touch the great engine of State-government" Among ourselves may be traced, if less clearly, the working of kindred tendencies. During the 15th century, "clergymen were secretaries of government, the privy seals, cabinet councillors, treasurers of the crown, ambassadors, commissioners to open parliament, and to Scotland; presidents of the king's council, supervisors of the royal works, chancellors, keepers of the records, the masters of the rolls, &c.;" but with antagonism to the Church came partial, and in later days complete, disappearance of the clerical element from the administration. Under Henry VIII. the King's secretary, and afterwards the chancellor, ceased to be ecclesiastics; while of the council of sixteen executors appointed to govern during the minority of his son, three only were in holy orders. And though, during a subsequent temporary revival of papal influence, there was a re-acquirement of ministerial position by priests, they afterwards again ceased to be chosen.

Whether a ruler is able to prevent high offices of State from being held by men whose ambitions and interests he fears, depends, however, upon his acquirement of adequate predominance. A class which, being powerful, is excluded as therefore dangerous, being still more powerful, cannot be excluded; and is apt either to monopolize administrative functions or practically to dictate the choice of ministers. In ancient Egypt, where the priesthood was pre-eminent in influence, the administration was chiefly officered by its members, with the result that at one time there was usurpation of the kingship by priests; and the days during which the Catholic church was most powerful throughout Europe, were the days during which high political posts were very generally held by prelates. In other cases supremacy of the military class is shown; as in Japan, where soldiers have habitually been the ministers and practically usurpers; as in feudal England, when Henry III. was obliged by the barons to accept Hugh Le Despenser as chief justiciary, and other nominees as officers of his household; or as when, in the East, down to our own time, changes of ministry are insisted on by the soldiery. Naturally in respect of these administrative offices, as in respect of all other places of power, there arises a conflict between the chiefs of the warrior class, who are the agents of the terrestrial ruler, and the chiefs of the clerical class, who profess to be agents of the celestial ruler; and the predominance of the one or the other class, is in many cases implied by the extent to which it fills the chief offices of State.

Such facts show us that where there has not yet been established any regular process for making the chief advisers and agents of the ruler into authorized exponents of public opinion, there nevertheless occurs an irregular process by which some congruity is maintained between the actions of these deputy rulers and the will of the community; or, at any rate, the will of that part which can express its will.

§ 506. Were elaboration desirable, and collection of the needful data less difficult, a good deal might here be added respecting the development of ministries.

Of course it could, in multitudinous cases, be shown how, beginning as simple, they become compound—the solitary assistant to the chief, helping him in all ways, developing into the numerous great officers of the king, dividing among them duties which have become extensive and involved. Along with this differentiation of a ministry might also be traced the integration of it that takes place under certain conditions: the observable change being from a state in which the departmental officers separately take from the ruler their instructions, to a state in which they form an incorporated body. There might be pursued an inquiry respecting the conditions under which this incorporated body gains power and accompanying responsibility; with the probable result of showing that development of an active executive council, and accompanying reduction of the original executive head to an automatic state, characterizes that representative form of government proper to the industrial type. But while results neither definite nor important are likely to be reached, the reaching of such as are promised would necessitate investigation at once tedious and unsatisfactory.

For such ends as are here in view, it suffices to recognize the general facts above set forth. As the political head is at first but a slightly-distinguished member of the group—now a chief whose private life and resources are like those of any other warrior, now a patriarch or a feudal lord who, becoming predominant over other patriarchs or other feudal lords, at first lives like them on revenues derived from private possessions—so the assistants of the political head take their rise from the personal connexions, friends, servants, around him: they are those who stand to him in private relations of blood, or liking, or service. With the extension of territory, the increase of affairs, and the growth of classes having special interests, there come into play influences which differentiate some of those who surround the ruler into public functionaries, distinguished from members of his family and his household. And these influences, joined with special circumstances, determine the kinds of public men who come into power. Where the absoluteness of the political head is little or not at all restrained, he makes arbitrary choice irrespective of rank, occupation, or origin. If, being predominant, there are nevertheless classes of whom he is jealous, exclusion of these becomes his policy; while if his predominance is inadequate, representatives of such classes are forced into office. And this foreshadows the system under which, along with decline of monarchical power, there grows up an incorporated body of ministers having for its recognized function to execute the public will.

CHAPTER XI.

LOCAL GOVERNING AGENCIES.

§ 507. This title is needed because the classes of facts to be here dealt with, cover a wider area than those comprehended under the title "Local Governments."

We have to deal with two kinds of appliances for control, originally one but gradually becoming distinguished. Alike among peoples characterized by the reckoning of kinship through females, and among peoples characterized by descent of property and power through males, the regulative system based on blood-relationship is liable to be involved with, and subordinated by, a regulative system originating from military leadership. Authority established by triumph in war, not unfrequently comes into conflict with authority derived from the law of succession, when this has become partially settled, and initiates a differentiation of political headship from family headship. We have seen that, from primitive stages upwards, the principle of efficiency and the principle of inheritance are both at work in determining men's social positions; and where, as happens in many cases, a war-chief is appointed when the occasion arises, notwithstanding the existence of a chief of acknowledged legitimacy, there is a tendency for transmitted power to be over-ridden by power derived from capacity. From the beginning, then, there is apt to grow up a species of government distinct from family-government; and the aptitude takes effect where many family-groups, becoming united, carry on militant activities. The growth of the family into the gens, of the gens into the phratry, of the phratry into the tribe, implies the multiplication of groups more and more remotely akin, and less and less easily subordinated by the head of some nominally-leading group; and when local aggregation brings interfusion of tribes which, though of the same stock, have lost their common genealogy, the rise of some headship other than the headships of family-groups becomes imminent. Though such political headship, passing through the elective stage, often becomes itself inheritable after the same manner as the original family-headships, yet it constitutes a new kind of headship.

Of the local governing agencies to which family-headships and political headships give origin, as groups become compounded and re-compounded, we will consider first the political, as being most directly related to the central governing agencies hitherto dealt with.

§ 508. According to the relative powers of conqueror and conquered, war establishes various degrees of subordination. Here the payment of tribute and occasional expression of homage, interfere but little with political independence; and there political independence is almost or quite lost. Generally, however, at the outset the victor either finds it necessary to respect the substantial autonomies of the vanquished societies, or finds it his best policy to do this. Hence, before integration has proceeded far, local governments are usually nothing more than those governments of the parts which existed before they were united into a whole.

We find instances of undecided subordination everywhere. In Tahiti "the actual influence of the king over the haughty and despotic district chieftains, was neither powerful nor permanent." Of our own political organization in old English times Kemble writes:—" the whole executive government may be considered as a great aristocratic association, of which the ealdormen were the constituent earls, and the king little more than president." Similarly during early feudal times; as, for example, in France. "Under the first Capetians, we find scarcely any general act of legislation. . . . Everything was local, and all the possessors of fiefs first, and afterwards all the great suzerains, possessed the legislative power within their domains." This is the kind of relation habitually seen during the initial stages of those clustered groups in which one group has acquired power over the rest.

In cases where the successful invader, external to the cluster instead of internal, is powerful enough completely to subjugate all the groups, it still happens that the pre-existing local organizations commonly survive. Ancient American states yield examples. "When the kings of Mexico, Tezcuco, and Tacuba conquered a province, they used to maintain in their authority all the natural chiefs, the highest as well as the lower ones." Concerning certain rulers of Chibcha communities, who became subject to Bogota, we read that the Zipa subdued them, but left them their jurisdiction and left the succession to the caziqueship in their families. And as was pointed out under another head, the victorious Yncas left outstanding the political headships and administrations of the many small societies they consolidated. Such is, in fact, the most convenient policy. As is remarked by Sir Henry Maine, "certain institutions of a primitive people, their corporations and village-communities, will always be preserved by a suzerain-state governing them, on account of the facilities which they afford to civil and fiscal administration;" and the like may be said of the larger regulative structures. Indeed the difficulty of suddenly replacing an old local organization by an entirely new one, is so great that almost of necessity the old one is in large measure retained.

The autonomies of local governments, thus sometimes scarcely at all interfered with and in other cases but partially suppressed, manifest themselves in various ways. The original independence of groups continues to be shown by the right of private war between them. They retain their local gods, their ecclesiastical organizations, their religious festivals. And in time of general war the contingents they severally furnish remain separate. Egyptian nomes, Greek cities, feudal lordships, yield illustrations.

§ 509. The gradual disappearance of local autonomies is a usual outcome of the struggle between the governments of the parts, which try to retain their powers, and the central government, which tries to diminish their powers.

In proportion as his hands are strengthened, chiefly by successful wars, the major political head increases his restraints over the minor political heads; first by stopping private wars among them, then by interfering as arbitrator, then by acquiring an appellate jurisdiction. Where the local rulers have been impoverished by their struggles with one another, or by futile attempts to recover their independence, or by drafts made on their resources for external wars—where, also, followers of the central ruler have grown into a new order of nobles, with gifts of conquered or usurped lands as rewards for services; the way is prepared for administrative agencies centrally appointed. Thus in France, when the monarch became dominant, the seigneurs were gradually deprived of legislative authority. Royal confirmation became requisite to make signorial acts valid; and the crown acquired the exclusive right of granting charters, the exclusive right of ennobling, the exclusive right of coining. Then with decline in the power of the original local rulers came deputies of the king overlooking them: provincial governors holding office at the king's pleasure were nominated. In subsequent periods grew up the administration of intendants and their sub-delegates, acting as agents of the crown; and whatever small local powers remained were exercised under central supervision. English history at various stages yields kindred illustrations. When Mercia was formed out of petty kingdoms, the local kings became ealdormen; and a like change took place afterwards on a larger scale. "From the time of Ecgberht onwards there is a marked distinction between the King and the Ealdorman. The King is a sovereign, the Ealdorman is only a magistrate." Just noting that under Cnut, ealdormen became subordinated by the appointment of earls, and again that under William I. earldoms were filled up afresh, we observe that after the Wars of the Roses had weakened them, the hereditary nobles had their local powers interfered with by those of centrally-appointed lords-lieutenant. Not only provincial governing agencies of a personal kind come to be thus subordinated as the integration furthered by war progresses, but also those of a popular kind. The old English Scirgeréfa, who presided over the Sciregemot, was at first elective, but was afterwards nominated by the king. Under a later *régime* there occurred a kindred change: "9 Edward II. abolished the popular right to election" to the office of sheriff. And similarly, "from the beginning of Edward III.'s reign, the appointment of conservators" of the peace, who were originally elected, "was vested in the crown," " and their title changed to that of justices."

With sufficient distinctness such facts show us that, rapidly where a cluster of small societies is subjugated by an invader, and slowly where one among them acquires an established supremacy, the local rulers lose their directive powers and become executive agents only; discharging whatever duties they retain as the servants of newer local agents. In the course of political integration, the original governing centres of the component parts become relatively automatic in their functions.

§ 510. A further truth to be noted is that there habitually exists a kinship in structure between the general government and the local governments. Several causes conspire to produce this kinship.

Where one of a cluster of groups has acquired power over the rest, either directly by the victories of its ruler over them, or indirectly by his successful leadership of the confederation in war, this kinship becomes a matter of course. For under such conditions the general government is but a development of that which was previously one of the local governments. We have a familiar illustration furnished by old English times in the likeness between the hundred-moot (a small local governing assembly), the shire-moot (constituted in an analogous way, but having military, judicial, and fiscal duties of a wider kind, and headed by a chief originally elected), and the national witanagémot (containing originally the same class-elements, though in different proportions, headed by a king, also at first elected, and discharging like functions on a larger scale). This similarity recurs under another phase. Sir Henry Maine says:—

"It has often, indeed, been noticed that a Feudal Monarchy was an exact counterpart of a Feudal Manor, but the reason of the correspondence is only now beginning to dawn upon us, which is, that both of them were in their origin bodies of assumed kinsmen settled on land and undergoing the same transmutation of ideas through the fact of settlement."

Of France in the early feudal period, Maury says, "the court of every great feudatory was the image, of course slightly reduced, of that of the king;" and the facts he names curiously show that locally, as generally, there was a development of servants into ministerial officers. Kindred evidence comes from other parts of the world—Japan, several African States, sundry Polynesian islands, ancient Mexico, Mediæval India, &c.; where forms of society essentially similar to those of the feudal system exist or have existed.

Where the local autonomy has been almost or quite destroyed, as by a powerful invading race bringing with it another type of organization, we still see the same thing; for its tendency is to modify the institutions locally as it modifies them generally. From early times eastern kingdoms have shown us this; as instance the provincial rulers, or satraps, of the Persians. "While . . . they remained in office they were despotic—they represented the Great King, and were clothed with a portion of his majesty, . . . They wielded the power of life and death." And down to the present day this union of central chief-despot with local sub-despots survives; as is implied by Rawlinson's remark that these ancient satraps had "that full and complete authority which is possessed by Turkish pashas and modern Persian khans or beys—an authority practically uncontrolled." Other ancient societies of quite other types displayed this tendency to assimilate the structures of the incorporated parts to that of the incorporating whole. Grecian history shows us that oligarchic Sparta sought to propagate oligarchy as a form of government in dependent territories, while democratic Athens propagated the democratic form. And, similarly, where Rome conquered and colonized, there followed the Roman municipal system.

This last instance reminds us that as the character of the general government changes, the character of the local government changes too. In the Roman empire that progress towards a more concentrated form of rule which continued militancy brought, spread from centre to periphery. "Under the Republic every town had, like Rome, a popular assembly which was sovereign for making the law and 'creating' magistrates;" but with the change towards oligarchic and personal rule in Rome, popular power in the provinces decreased: "the municipal organization, from being democratic, became aristocratic." In France, as monarchical power approached absoluteness, similar changes were effected in another way. The government seized on municipal offices, "erecting them into hereditary offices, and . . . selling them at the highest price: . . . a permanent mayor and assessors were imposed upon all the municipalities of the kingdom, which ceased to be elective;" and then these magistrates began to assume royal airs—spoke of the sanctity of their magistracy, the veneration of the people, &c. Our own history interestingly shows simultaneous movements now towards freer, and now towards less free, forms, locally and generally. When, under King John, the central government was liberalized, towns acquired the power to elect their own magistrates. Conversely when, at the Restoration, monarchical power increased, there was a framing of the "municipalities on a more oligarchical model." And then comes the familiar case of the kindred liberalizations of the central government and the local governments which have occurred in our own time.

§ 511. From those local governing agencies which have acquired a political character, we turn now to those which have retained the primitive family character. Though with the massing of groups, political organization and rule become separate from, and predominant over, family-organization and rule, locally as well as generally, yet family-organization and rule do not disappear; but in some cases retaining their orginal nature, in some cases give origin to other local organizations of a governmental kind. Let us first note how wide-spread is the presence of the family-cluster, considered as a component of the political society.

Among the uncivilized Bedouins we see it existing separately: "every large family with its relations constituting a small tribe by itself." But, says Palgrave, "though the clan and the family form the basis and are the ultimate expression of the civilized Arab society, they do not, as is the case among the Bedouins, sum it up altogether." That is, political union has left outstanding the family-organization, but has added something to it. And it was thus with Semitic societies of early days, as those of the Hebrews. Everywhere it has been thus with the Aryans.

'The [Irish] Sept is a body of kinsmen whose progenitor is no longer living, but whose descent from him is a reality. . . . An association of this sort is well known to the law of India as the Joint Undivided Family. . . . The family thus formed by the continuance of several generations in union, is identical in outline with a group very familiar to the students of the older Roman law—the Agnatic Kindred."

Not only where descent in the male line has been established, but also where the system of descent through females continues, this development of the family into gens, phratry, and tribe, is found. It was so with such ancient American peoples, as those of Yucatan, where, within each town, tribal divisions were maintained; and, according to Mr. Morgan and Major Powell, it is still so with such American tribes as the Iroquois and the Wyandottes.

After its inclusion in a political aggregate, as before its inclusion, the family-group evolves a government *quasi*-political in nature. According to the type of race and the system of descent, this family-government may be, as among ancient Semites and Ayrans, an unqualified patriarchal despotism; or it may be, as among the Hindoos at present, a personal rule arising by selection of a head from the leading family of the group (a selection usually falling on the eldest); or it may be, as in American tribes like those mentioned, the government of an elected council of the gens, which elects its chief. That is to say, the triune structure which tends to arise in any incorporated assembly, is traceable in the compound family-group, as in the political group: the respective components of it being variously developed according to the nature of the people and the conditions.

The government of each aggregate of kinsmen repeats, on a small scale, functions like those of the government of the political aggregate. As the entire society revenges itself on other such societies for injury to its members, so does the family-cluster revenge itself on other family-clusters included in the same society. This fact is too familiar to need illustration; but it may be pointed out that even now, in parts of Europe where the family-organization survives, the family vendettas persist. "L'Albanais vous dira froidement . . . Akeni-Dgiak? avez-vous du sang à venger dans votre famille;" and then, asking the name of your tribe, he puts his hand on his pistol. With this obligation to take vengeance goes, of course, reciprocal responsibility. The family in all its branches is liable as a whole, and in each part, for the injuries done by its members to members of other families; just as the entire society is held liable by other entire societies. This responsibility holds not alone for lives taken by members of the family-group, but also for damages they do to property, and for pecuniary claims.

"Dans les districts Albanais libres, les dettes sont contractées à terme. En cas de non-paiement, on a recours aux chefs de la tribu du débiteur, et si ceux-ci refusent de faire droit, on arrête le premier venu qui appartient à cette tribu, et on l'accable de mauvais traitements jusqu'à ce qu'il s'entende avec le véritable débiteur, ou qu'il paie lui-même ses dettes, risque à se pouvoir ensuite devant les anciens de sa tribu ou de poursuivre par les armes celui qui lui a valu ce dommage."

And of the old English mægth we read that "if any one was imprisoned for theft, witchcraft, &c., his kindred must pay the fine . . . and must become surety for his good conduct on his release."

While, within the political aggregate, each compound family-group thus stood towards other such included groups in quasi-political relations, its government exercised internal control. In the gens as constituted among the American peoples above named, there is administration of affairs by its council. The gentile divisions among historic peoples were ruled by their patriarchs; as are still those of the Hindoos by their chosen elders. And then besides this judicial organization within the assemblage of kindred, there is the religious organization, arising from worship of a common ancestor, which entails periodic joint observances.

Thus the evidence shows us that while the massing together of groups by war, has, for its concomitant, development of a political organization which dominates over the organizations of communities of kindred, yet these communities of kindred long survive, and partially retain their autonomies and their constitutions.

§ 512. Social progress, however, transforms them in sundry ways—differentiating them into groups which gradually lose their family-characters. One cause is change from the wandering life to the settled life, with the implied establishment of definite relations to the land, and the resulting multiplication and interfusion.

To show that this process and its consequences are general, I may name the calpulli of the ancient Mexicans, which "means a district inhabited by a family . . . of ancient origin;" whose members hold estates which "belong not to each inhabitant, but to the calpulli;" who have chiefs chosen out of the tribe; and who "meet for dealing with the common interests, and regulating the apportionment of taxes, and also what concerns the festivals." And then I may name as being remote in place, time, and race, the still-existing Russian mir, or village-commune; which is constituted by descendants of the same family-group of nomads who became settled; which is "a judicial corporation . . . proprietor of the soil, of which individual members have but the usufruct or temporary enjoyment;" which is governed by "the heads of families, assembled in council under the presidency of the *starosta* or mayor, whom they have elected." Just noting these allied examples, we may deal more especially with the Teutonic mark, which was "formed by a primitive settlement of a family or kindred," when, as said by Cæsar of the Suevi, the land was divided among "gentes et cognationes hominum." In the words of Kemble, marks were—

"Great family-unions, comprising households of various degrees of wealth, rank, and authority; some in direct descent from the common ancestors, or from the hero of the particular tribe; others, more distantly connected . . . ; some, admitted into communion by marriage, others by adoption, others by emancipation; but all recognizing a brotherhood, a kinsmanship or *sibsceaft*; all standing together as one unit in respect of other similar communities; all governed by the same judges and led by the same captains; all sharing in the same religious rites; and all known to themselves and to their neighbours by one general name."

To which add that, in common with family-groups as already described, the cluster of kindred constituting the mark had, like both smaller and larger clusters, a joint obligation to defend and avenge its members, and a joint responsibility for their actions.

And now we are prepared for observing sundry influences which conspire to change the grouping of kindred into political grouping, locally as well as generally. In the first place, there is that admission of strangers into the family, gens, or tribe, which we have before recognized as a normal process, from savage life upwards. Livingstone, remarking of the Bakwains that "the government is patriarchal," describes each chief man as having his hut encircled by the huts of his wives, relatives, and dependents, forming a kotla: "a poor man attaches himself to the kotla of a rich one and is considered a child of the latter." Here we see being done informally, that which was formally done in the Roman household and the Teutonic mark. In proportion as the adopted strangers increase, and in proportion also as the cluster becomes diluted by incorporating with itself emancipated dependents, the links among its members become weakened and its character altered. In the second place, when, by concentration and multiplication, different clusters of kindred placed side by side, become interspersed, and there ceases to be a direct connexion between locality and kinship, the family or gentile bonds are further weakened. And then there eventually results, both for military and fiscal purposes, the need for a grouping based on locality instead of on relationship. An early illustration is furnished by the Kleisthenian revolution in Attica, which made a division of the territory into demes, replacing for public purposes tribal divisions by topographical divisions, the inhabitants of each of which had local administrative powers and public responsibilities.

We are here brought to the vexed question about the origin of tythings and hundreds. It was pointed out that the ancient Peruvians had civil as well as military divisions into tens and hundreds, with their respective officers. In China, where there is pushed to an extreme the principle of making groups responsible for their members, the clan-divisions are not acknowledged by the government, but only the tythings and hundreds: the implication being that these last were results of political organization as distinguished from family-organization. In parts of Japan, too, "there is a sort of subordinate system of wards, and heads of tens and hundreds, in the *Otonos* of towns and villages, severally and collectively responsible for each other's good conduct." We have seen that in Rome, the groupings into hundreds and tens, civil as well as military, became political substitutes for the gentile groupings. Under the Frankish law, "the tything-man is *Decanus*, the hundred-man *Centenarius;*" and whatever may have been their indigenous names, divisions into tens and hundreds appear to have had (judging from the statements of Tacitus) an independent origin among the Germanic races.

And now remembering that these hundreds and tythings, formed within the marks or other large divisions, still answered in considerable degrees to groups based on kinship (since the heads of families of which they were constituted as local groups, were ordinarily closer akin to one another than to the heads of families similarly grouped in other parts of the mark), we go on to observe that there survived in them, or were re-developed in them, the family-organization, rights, and obligations. I do not mean merely that by their hundred-moots, &c., they had their internal administrations; but I mean chiefly that they became groups which had towards other groups the same joint claims and duties which family-groups had. Responsibility for its members, previously attaching exclusively to the cluster of kindred irrespective of locality, was in a large measure transferred to the local cluster formed but partially of kindred. For this transfer of responsibility an obvious cause arose as the gentes and tribes spread and became mingled. While the family-community was small and closely aggregated, an offence committed by one of its members against another such community could usually be brought home to it bodily, if not to the sinning member; and as a whole it had to take the consequences. But when the family-community, multiplying, began to occupy a wide area, and also became interfused with other family-communities, the transgressor, while often traceable to some one locality within the area, was often not identifiable as of this or that kindred; and the consequences of his act, when they could not be visited on his family, which was not known, were apt to be visited on the inhabitants of the locality, who were known. Hence the genesis of a system of suretyship which is so ancient and so widespread. Here are illustrations:—

"This then is my will, that every man be in surety, both within the towns and without the towns."—Eádg. ii. Supp. § 3.

"And we will that every freeman be brought into a hundred and into a tithing, who desires to be entitled to *lád* or *wer*, in case any one should slay him after he have reached the age of xii years: or let him not otherwise be entitled to any free rights, be he householder, be he follower."— Cnut, ii. § xx.

". . . in all the vills throughout the kingdom, all men are bound to be in a guarantee by tens, so that if one of the ten men offend, the other nine may hold him to right."—Edw. Conf., xx.

Speaking generally of this system of mutual guarantee, as exhibited among the Russians, as well as among the Franks, Koutorga says—

"Tout membre de la société devait entrer dans une décanie, laquelle avait pour mission la défence et la garantie de tous en général et de chacun en particulier; c'est-à-dire que la décanie devait venger le citoyen qui lui appartenait et exiger le wehrgeld, s'il avait été tué; mais en même temps elle se portait caution pour tous les seins."

In brief, then, this form of local governing agency, developing out of, and partially replacing, the primitive family-form, was a natural concomitant of the multiplication and mixture resulting from a settled life.

§ 513. There remains to be dealt with an allied kind of local governing agency—a kind which, appearing to have been once identical with the last, eventually diverged from it.

Kemble concludes that the word "gegyldan" means "those who mutually pay for one another ... the associates of the tithing and the hundred;" and how the two were originally connected, we are shown by the statement that as late as the 10th century in London, the citizens were united into frithgylds, "or associations for the maintenance of the peace, each consisting of ten men; while ten such gylds were gathered into a hundred." Prof. Stubbs writes:—

"The collective responsibility for producing an offender, which had lain originally on the mægth or kindred of the accused, was gradually devolved on the voluntary association of the guild; and the guild superseded by the local responsibility of the tithing."

Here we have to ask whether there are not grounds for concluding that this transfer of responsibility originally took place through development of the family-cluster into the gild, in consequence of the gradual loss of the family-character by incorporation of unrelated members. That we do not get evidence of this in written records, is probably due to the fact that the earlier stages of the change took place before records were common. But we shall see reasons for believing in such earlier stages if we take into account facts furnished by extinct societies and societies less developed than those of Europe.

Of the skilled arts among the Peruvians, Prescott remarks:—"these occupations, like every other calling and office in Peru, always descended from father to son;" and Clavigero says of the Mexicans "that they perpetuated the arts in families to the advantage of the State:" the reason Gomara gives why "the poor taught their sons their own trades," being that "they could do so without expense"—a reason of general application. Heeren's researches into ancient Egyptian usages, have led him to accept the statement of early historians, that "the son was bound to carry on the trade of his father and that alone;" and he cites a papyrus referring to an institution naturally connected with this usage—"the guild or company of curriers or leather-dressers." Then of the Greeks, Hermann tell us that various arts and professions were—

"peculiar to certain families, whose claims to an exclusive exercise of them generally ascended to a fabulous origin. We moreover find 'pupil and son' for many successive generations designated by the same term; and closely connected with the exclusiveness and monopoly of many professions, is the little respect in which they were, in some instances, held by the rest of the people: a circumstance which Greek authors themselves compare with the prejudice of caste prevalent among other nations."

China, as at present existing, yields evidence :—

"The popular associations in cities and towns are chiefly based upon a community of interests, resulting either from a similarity of occupation, when the leading persons of the same calling form themselves into guilds, or from the municipal regulations requiring the householders living in the same street to unite to maintain a police, and keep the peace of their division. Each guild has an assembly-hall, where its members meet to hold the festival of their patron saint."

And, as I learn from the Japanese minister, a kindred state of things once existed in Japan. Children habitually followed the occupations of their parents; in course of generations there resulted clusters of relatives engaged in the same trade; and these clusters developed regulative arrangements within themselves. Whether the fact that in Japan, as in the East generally, the clustering of traders of one kind in the same street, arises from the original clustering of the similarly-occupied kindred, I find no evidence; but since, in early times, mutual protection of the members of a trading kindred, as of other kindred, was needful, this seems probable. Further evidence of like meaning may be disentangled from the involved phenomena of caste in India. In No. CXLII of the *Calcutta Review*, in an interesting essay by Jogendra Chandra Ghosh, caste is regarded as "a natural development of the Indian village-communities;" as "distinguished not only by the autonomy of each guild," "but by the mutual relations between these autonomous guilds;" and as being so internally organized "that caste government does not recognize the finding or the verdict of any court other than what forms part of itself." In answer to my inquiries, the writer of this essay has given me a mass of detailed information, from which I extract the following:—

"A Hindoo joint family signifies (1) that the members all mess together; (2) and live in the same house; (3) that the male members and unmarried girls are descended from a common ancestor; and (4) that the male members put their incomes together. . . . The integral character of the family is destroyed when the joint mess and common purse cease to exist. However, the branches thus disunited continue to observe certain close relations as *gnatis* up to some seven or fourteen generations from the common ancestor. Beyond that limit they are said to be merely of the same *gotra*."

Passing over the detailed constitution of a caste as consisting of many such *gotras*, and of the groups produced by their intermarriages under restrictions of exogamy of the *gotras* and endogamy of the caste—passing over the feasts, sacrificial and other, held among members of the joint family when their groups have separated; I turn to the facts of chief significance. Though, under English rule, inheritance of occupation is no longer so rigorous, yet—

"the principle is universally recognized that every caste is bound to follow a particular occupation and no other. . . . The partition of the land, or the house as well, is governed by the law of equal succession; and as fresh branches set up new houses, they are found all clustered together, with the smallest space between them for roadway. . . . But when, as in *bazaars*, men take up houses for commercial purposes, the clustering is governed either by family and caste-relations, or by common avocations [which imply some caste-kinship] and facility of finding customers."

In which facts we may see pretty clearly that were there none of the complications consequent on the intermarriage regulations, there would simply result groups united by occupation as well as by ancestry, clustering together, and having their internal governments.

Returning from consideration of these facts supplied by other societies, let us now observe how numerous are the reasons for concluding that the gild, familiar to us as a union of similarly-occupied workers, was originally a union of kindred. In the primitive compound family there was worship of the common ancestor; and the periodic sacrificial feasts were occasions on which all the descendants assembled. Describing the origin of gilds, Thierry writes:—

"Dans l'ancienne Scandinavie, ceux qui se réunissaient aux époques solennelles pour sacrifier ensemble terminaient la cérémonie par un festin religieux. Assis autour du feu et de la chaudière du sacrifice, ils buvaient à la ronde et vidaient successivement trois cornes remplies de bière, l'une pour les dieux, l'autre pour les braves du vieux temps, et la troisième pour les parents et les amis dont les tombes, marquées par des monticules de gazon, se voyaient çà et là dans la plaine ; on appelait celle-ci la coupe de l'amitié. Le nom d'amitié (minne) se donnait aussi quelquefois à la réunion de ceux qui offraient en commun le sacrifice, et, d'ordinaire, cette réunion était appelée *ghilde*."

And Brentano, giving a similar account, says—"'Gild' meant originally the sacrificial meal made up of the common contributions; then a sacrificial banquet in general; and lastly a society." Here we find a parallelism with the observances of the Hindoo joint-family, consisting of clusters of relatives carrying on the same occupation, who meet at feasts which were primarily sacrificial to ancestors; and we find a parallelism with the religious observances of such clusters of similarly-occupied relatives as the Asklepiadæ among the Greeks; and we find a parallelism with the gild-feasts of the ancestor-worshipping Chinese, held in honour of the patron saint: all suggesting the origin of those religious services and feasts habitual in early gilds of our own society. To state briefly the further likenesses of nature:—We have, in the primitive compound family, the obligation of blood-revenge for slain relatives; and in early gilds, as in ancient Sleswig, there was blood-revenge for members of the gild. We have, in the compound family, responsibility for transgressions of its members; and gilds were similarly responsible: the wergylds falling in part on them, after murders were compounded for by money. We have, in the compound family, joint claims to sustenance derived from the common property and labour; and in the gild we have the duty of maintaining incapable members. Within the family there was control of private conduct, either by a despotic head or by a council, as there is now within the local clusters of the Hindoo castes; and in like manner the ordinances of gilds extended to the regulation of personal habits. Lastly, this family or caste government, as still shown us in India, includes in its punishments excommunication; and so, too, was there outlawry from the gild.*

It is inferable, then, that the gild was evolved from the family. Continuance of a business, art, or profession, among descendants, is, in early stages, almost inevitable. Acquisition of skill in it by early practice is easy; the cost of teaching is inappreciable; and retention of the "craft" or "mystery" within the family is desirable: there being also the reason that while family-groups are in antagonism, the teaching of one another's members cannot usually be practicable. But in course of time there come into play influences by which the character of the gild as an assemblage of kindred is obscured. Adoption, which, as repeatedly pointed out, is practised by groups of all kinds, needs but to become common to cause this constitutional change. We have seen that among the Greeks, "pupil" and "son" had the same name. At the present time in Japan, an apprentice, standing in the position of son to his master, calls him "father;" and in our own craft-gilds "the apprentice became a member of the family of his master, who instructed him in his trade, and who, like a father, had to watch over his morals, as well as his work." The eventual admission of the apprentice into the gild, when he was a stranger in blood to its members, qualified, in so far, its original nature; and where, through successive generations, the trade was a prosperous

---

* A friend who has read this chapter in proof, points out to me passages in which Brentano draws from these parallelisms a like inference. Referring to the traits of certain fully-developed gilds, he says:—"If we connect them with what historians relate about the family in those days, we may still recognize in them the germ from which, in later times, at a certain stage of civilization, the Gild had necessarily to develop itself . . . the family appears as the pattern and original type, after which all the later Gilds were formed."

one, tempting masters to get more help than their own sons could furnish, this process would slowly bring about predominance of the unrelated members, and an ultimate loss of the family-character. After which it would naturally happen that the growing up of new settlements and towns, bringing together immigrants who followed the same calling but were not of the same blood, would lead to the deliberate formation of gilds after the pattern of those existing in older places: an appearance of artificial origin being the result; just as now, in our colonies, there is an apparently artificial origin of political institutions which yet, as being fashioned like those of the mother-country, where they were slowly evolved, are traceable to a natural origin.

Any one who doubts the transformation indicated, may be reminded of a much greater transformation of allied kind. The gilds of London,—goldsmiths', fishmongers', and the rest,—were originally composed of men carrying on the trades implied by their names; but in each of these companies the inclusion of persons of other trades, or of no trade, has gone to the extent that few if any of the members carry on the trades which their memberships imply. If, then, the process of adoption in this later form, has so changed the gild that, while retaining its identity, it has lost its distinctive trade-character, we are warranted in concluding that still more readily might the earlier process of adoption into the simple family or the compound family practising any craft, eventually change the gild from a cluster of kindred to a cluster formed chiefly of unrelated persons.

§ 514. Involved and obscure as the process has been, the evolution of local governing agencies is thus fairly comprehensible. We divide them into two kinds, which, starting from a common root, have diverged as fast as small societies have been integrated into large ones.

Through successive stages of consolidation, the political heads of the once-separate parts pass from independence to dependence, and end in being provincial agents—first partially-conquered chiefs paying tribute; then fully-conquered chiefs governing under command; then local governors who are appointed by the central governor and hold power under approval: becoming eventually executive officers.

There is habitually a kinship in character between the controlling systems of the parts and the controlling system of the whole (assuming unity of race), consequent on the fact that both are ultimately products of the same individual nature. With a central despotism there goes local despotic rule; with a freer form of the major government there goes a freer form of the minor governments; and a change either way in the one is followed by a kindred change in the other.

While, with the compounding of small societies into large ones, the political ruling agencies which develop locally as well as generally, become separate from, and predominant over, the ruling agencies of family-origin, these last do not disappear; but, surviving in their first forms, also give origin to differentiated forms. The assemblage of kindred long continues to have a qualified semi-political autonomy, with internal government and external obligations and claims. And while family-clusters, losing their definiteness by interfusion, slowly lose their traits as separate independent societies, there descend from them clusters which, in some cases united chiefly by locality and in others chiefly by occupation, inherit their traits, and constitute governing agencies supplementing the purely political ones.

It may be added that these supplementary governing agencies, proper to the militant type of society, dissolve as the industrial type begins to predominate. Defending their members, held responsible for the transgressions of their members, and exercising coercion over their members, they are made needful by, and bear the traits of, a *régime* of chronic antagonisms; and as these die away their *raison d'être* disappears. Moreover, artificially restricting, as they do, the actions of each member, and also making him responsible for other deeds than his own, they are at variance with that increasing assertion of individuality which accompanies developing industrialism.

CHAPTER XII.

MILITARY SYSTEMS.

§ 515. Indirectly, much has already been said concerning the subject now to be dealt with. Originally identical as is the political organization with the military organization, it has been impossible to treat of the first without touching on the second. After exhibiting the facts under one aspect we have here to exhibit another aspect of them; and at the same time to bring into view classes of related facts thus far unobserved. But, first, let us dwell a moment on the alleged original identity.

In rude societies all adult males are warriors; and, consequently, the army is the mobilized community, and the community is the army at rest, as was remarked in § 259.

With this general truth we may join the general truth that the primitive military gathering is also the primitive political gathering. Alike in savage tribes and in communities like those of our rude ancestors, the assemblies which are summoned for purposes of defence and offence, are the assemblies in which public questions at large are decided.

Next stands the fact, so often named, that in the normal course of social evolution, the military head grows into the political head. This double character of leading warrior and civil ruler, early arising, ordinarily continues through long stages; and where, as not unfrequently happens, military headship becomes in a measure separated from political headship, continued warfare is apt to cause a re-identification of them.

As societies become compounded and re-compounded, coincidence of military authority with political authority is shown in detail as well as in general—in the parts as in the whole. The minor war-chiefs are also minor civil rulers in their several localities; and the commanding of their respective groups of soldiers in the field, is of like nature with the governing of their respective groups of dependents at home.

Once more, there is the general fact that the economic organizations of primitive communities, coincide with their military organizations. In savage tribes war and hunting are carried on by the same men; while their wives (and their slaves where they have any) do the drudgery of domestic life. And, similarly, in rude societies that have become settled, the military unit and the economic unit are the same. The soldier is also the landowner.

Such, then, being the primitive identity of the political organization with military organization, we have in this chapter to note the ways in which the two differentiate.

§ 516. We may most conveniently initiate the inquiry by observing the change which, during social evolution, takes place in the incidence of military obligations; and by recognizing the accompanying separation of the fighting body from the rest of the community.

Though there are some tribes in which military service (for aggressive war at any rate) is not compulsory, as the Comanches, Dakotas, Chippewas, whose war-chiefs go about enlisting volunteers for their expeditions; yet habitually where political subordination is established, every man not privately possessed as a chattel is bound to fight when called on. There have been, and are, some societies of considerably-advanced structures in which this state of things continues. In ancient Peru the common men were all either actually in the army or formed a reserve occupied in labour; and in modern Siam the people "are all soldiers, and owe six months' service yearly to their prince." But, usually, social progress is accompanied by a narrowed incidence of military obligation.

When the enslavement of captives is followed by the rearing of their children as slaves, as well as by the consigning of criminals and debtors to slavery—when, as in some cases, there is joined with the slave-class a serf-class composed of subjugated people not detached from their homes; the community becomes divided into two parts, on one of which only does military duty fall. Whereas, in previous stages, the division of the whole society had been into men as fighters and women as workers, the division of workers now begins to include men; and these continue to form an increasing part of the total male population. Though we are told that in Ashantee (where everyone is in fact owned by the king) the slave-population "principally constitutes the military force," and that in Rabbah (among the Fúlahs) the army is composed of slaves liberated "on consideration of their taking up arms;" yet, generally, those in bondage are not liable to military service: the causes being partly distrust of them (as was shown among the Spartans when forced to employ the helots) partly contempt for them as defeated men or the offspring of defeated men, and partly a desire to devolve on others, labours at once necessary and repugnant. Causes aside, however, the evidence proves that the army at this early stage usually coincides with the body of freemen; who are also the body of landowners. This, as before shown in § 458, was the case in Egypt, Greece, Rome, and Germany. How natural is this incidence of military obligation, we see in the facts that in ancient Japan and mediæval India, there were systems of military tenure like that of the middle ages in Europe; and that a kindred connexion had arisen even in societies like those of Tahiti and Samoa.

Extent of estate being a measure of its owner's ability to bear burdens, there grows up a connexion between the amount of land held and the amount of military aid to be rendered. Thus in Greece under Solon, those whose properties yielded less than a certain revenue were exempt from duty as soldiers, save in emergencies. In Rome, with a view to better adjustment of the relation between means and requirements, there was a periodic " revision of the register of landed property, which was at the same time the levy-roll." Throughout the middle ages this principle was acted upon by proportioning the numbers of warriors demanded to the sizes of the fiefs; and again, afterwards, by requiring from parishes their respective contingents.

A dissociation of military duty from land-ownership begins when land ceases to be the only source of wealth. The growth of a class of free workers, accumulating property by trade, is followed by the imposing on them, also, of obligations to fight or to provide fighters. Though, as apparently in the cases of Greece and Rome, the possessions in virtue of which citizens of this order at first become liable, are lands in which they have invested; yet, at later stages, they become liable as possessors of other property. Such, at least, is the interpretation we may give to the practice of making industrial populations furnish their specified numbers of warriors; whether, as during the Roman conquests, it took the shape of requiring " rich and populous" towns to maintain cohorts of infantry or divisions of cavalry, or whether, as with chartered towns in mediæval days, there was a contract with the king as suzerain, to supply him with stated numbers of men duly armed.

Later on, the same cause initiates a further change. As fast as industry increases the relative quantity of transferable property, it becomes more easy to compound for service in war; either by providing a deputy or by paying to the ruler a sum which enables him to provide one. Originally the penalty for non-fulfilment of military obligation was loss of lands; then a heavy fine, which, once accepted, it became more frequently the custom to bear; then an habitual compounding for the special services demanded;

then a levying of dues, such as those called scutages, in place of special compositions. Evidently, industrial growth made this change possible; both by increasing the population from which the required numbers of substitutes could be obtained, and by producing the needful floating capital.

So that whereas in savage and semi-civilized communities of warlike kinds, the incidence of military obligation is such that each free man has to serve personally, and also to provide his own arms and provisions; the progress from this state in which industry does but occupy the intervals between wars, to a state in which war does but occasionally break the habitual industry, brings an increasing dissociation of military obligation from free citizenship: military obligation at the same time tending to become a pecuniary burden levied in proportion to property of whatever kind. Though where there is a conscription, personal service is theoretically due from each on whom the lot falls, yet the ability to buy a substitute brings the obligation back to a pecuniary one. And though we have an instance in our own day of universal military obligation not thus to be compounded for, we see that it is part of a reversion to the condition of predominant militancy.

§ 517. An aspect of this change not yet noted, is the simultaneous decrease in the ratio which the fighting part of the community bears to the rest. With the transition from nomadic habits to settled habits, there begins an economic resistance to militant action, which increases as industrial life develops, and diminishes the relative size of the military body.

Though in tribes of hunters the men are as ready for war at one time as at another, yet in agricultural societies there obviously exists an impediment to unceasing warfare. In the exceptional case of the Spartans, the carrying on of rural industry was not allowed to prevent daily occupation of all freemen in warlike exercises; but, speaking generally, the sowing and reaping of crops hinder the gathering together of freemen for offensive or defensive purposes. Hence in course of time come decreased calls on them. The ancient Suevi divided themselves so as alternately to share war-duties and farm-work: each season the active warriors returned to till the land, while their places were "supplied by the husbandmen of the previous year." Alfred established in England a kindred alternation between military service and cultivation of the soil. In feudal times, again, the same tendency was shown by restrictions on the duration and amount of the armed aid which a feudal tenant and his retainers had to give—now for sixty, for forty, for twenty days, down even to four; now alone, and again with specified numbers of followers; here without limit of distance, and there within the bounds of a county. Doubtless, insubordination often caused resistances to service, and consequent limitations of this kind. But manifestly, absorption of the energies in industry, directly and indirectly antagonized militant action; with the result that separation of the fighting body from the general body of citizens was accompanied by a decrease in its relative mass.

There are two cooperating causes for this decrease of its relative mass, which are of much significance. One is the increasing costliness of the soldier, and of war appliances, which goes along with that social progress made possible by industrial growth. In the savage state each warrior provides his own weapons; and, on war-excursions, depends on himself for sustenance. At a higher stage this ceases to be the case. When chariots of war, and armour, and siege-implements come to be used, there are presupposed sundry specialized and skilled artizan-classes; implying a higher ratio of the industrial part of the community to the militant part. And when, later on, there are introduced fire-arms, artillery, ironclads, torpedoes, and the like, we see that there must co-exist a large and highly-organized body of producers and distributors; alike to furnish the required powers and bear the entailed cost. That is to say, the war-machinery, both living and dead, cannot be raised in efficiency without lowering the ratio it bears to those sustaining structures which give it efficiency.

The other cooperating cause which simultaneously comes into play, is directly due to the compounding and re-compounding of societies. The larger nations become, and the greater the distances over which their military actions range, the more expensive do those actions grow. It is with an army as with a limb, the effort put forth is costly in proportion to the remoteness of the acting parts from the base of operations. Though it is true that a body of victorious invaders may raise some, or the whole, of its supplies from the conquered society, yet before it has effected conquest it cannot do this, but is dependent for maintenance on its own society, of which it then forms an integral part: where it ceases to form an integral part and wanders far away, living on spoils, like Tatar hordes in past ages, we are no longer dealing with social organization and its laws, but with social destruction. Limiting ourselves to societies which, permanently localized, preserve their individualities, it is clear that the larger the integrations formed, the greater is the social strain consequent on the distances at which fighting has to be done; and the greater the amount of industrial population required to bear the strain. Doubtless, improved means of communication may all at once alter the ratio; but this does not conflict with the proposition when qualified by saying—other things equal.

In three ways, therefore, does settled life, and the development of civilization, so increase the economic resistance to militant action, as to cause decrease of the ratio borne by the militant part to the non-militant part.

§ 518. With those changes in the incidence of military obligation which tend to separate the body of soldiers from the body of workers, and with those other changes which tend to diminish its relative size, there go changes which tend to differentiate it in a further way. The first of these to be noted is the parting of military headship from political headship.

We have seen that the commencement of social organization is the growth of the leading warrior into the civil governor. To illustrative facts before named may be added the fact that an old English ruler, as instance Hengist, was originally called "Here-toga"—literally army-leader; and the office developed into that of king only after settlement in Britain. But with establishment of hereditary succession to political headship, there comes into play an influence which tends to make the chief of the State distinct from the chief of the army. That antagonism between the principle of inheritance and the principle of efficiency, everywhere at work, has from the beginning been conspicuous in this relation, because of the imperative need for efficient generalship. Often, as shown in § 473, there is an endeavour to unite the two qualifications; as, for example, in ancient Mexico, where the king, before being crowned, had to fill successfully the position of commander-in-chief. But from quite early stages we find that where hereditary succession has been established, and there does not happen to be inheritance of military capacity along with political supremacy, it is common for headship of the warriors to become a separate post filled by election. Says Waitz, "among the Guaranis the chieftainship generally goes from father to first-born son. The leader in war is, however, elected." In Ancient Nicaragua "the war-chief was elected by the warriors to lead them, on account of his ability and bravery in battle; but the civil or hereditary chief often accompanies the army." Of the New Zealanders we read that "hereditary chiefs were generally the leaders," but not always: others being chosen on account of bravery. And among the Sakarran Dyaks there is a war chief, in addition to the ordinary chief. In the case of the Bedouins the original motive has been defeated in a curious way.

"During a campaign in actual warfare, the authority of the sheikh of the tribe is completely set aside, and the soldiers are wholly under the command of the agyd. . . . The office of agyd is hereditary in a certain family, from father to son; and the Arabs submit to the commands of an agyd, whom they know to be deficient both in bravery and judgment, rather than yield to the orders of their sheikh during the actual expedition; for they say that expeditions headed by the sheikh, are always unsuccessful."

It should be added that in some cases we see coming into play further motives. Forster tells us that in Tahiti the king sometimes resigns the post of commander-in-chief of the fighting force, to one of his chiefs: conscious either of his own unfitness or desirous of avoiding danger. And then in some cases the anxiety of subjects to escape the evils following loss of the political head, leads to this separation; as when, among the Hebrews, "the men of David sware unto him, saying, Thou shalt go no more out with us to battle, that thou quench not the light of Israel;" or as when, in France in 923, the king was besought by the ecclesiastics and nobles who surrounded him, to take no part in the impending fight.

At the same time the ruler, conscious that military command gives great power to its holder, frequently appoints as army-leader his son or other near relative: thus trying to prevent the usurpation so apt to occur (as, to add another instance, it occurred among the Hebrews, whose throne was several times seized by captains of the host). The *Iliad* shows that it was usual for a Greek king to delegate to his heir the duty of commanding his troops. In Merovingian times kings' sons frequently led their fathers' armies; and of the Carolingians we read that while the king commanded the main levy, "over other armies his sons were placed, and to them the business of commanding was afterwards increasingly transferred." It was thus in ancient Japan. When the emperor did not himself command his troops, "this charge was only committed to members of the Imperial house," and "the power thus remained with the sovereign." In ancient Peru there was a like alternative. "The army was put under the direction of some experienced chief of the royal blood, or, more frequently, headed by the Ynca in person."

The widening civil functions of the political head, obviously prompt this delegation of military functions. But while the discharge of both becomes increasingly difficult as the nation enlarges; and while the attempt to discharge both is dangerous; there is also danger in doing either by deputy. At the same time that there is risk in giving supreme command of a distant army to a general, there is also risk in going with the army and leaving the government in the hands of a vicegerent; and the catastrophes from the one or the other cause, which, spite of precautions, have taken place, show us alike that there is, during social evolution, an inevitable tendency to the differentiation of the military headship from the political headship, but that this differentiation can become permanent only under certain conditions.

The general fact would appear to be that while militant activity is great, and the whole society has the organization appropriate to it, the state of equilibrium is one in which the political head continues to be also the militant head; that in proportion as there grows up, along with industrial life, a civil administration distinguishable from the military administration, the political head tends to become increasingly civil in his functions, and to delegate, now occasionally, now generally, his militant functions; that if there is a return to great militant activity, with consequent reversion to militant structure, there is liable to occur a re-establishment of the primitive type of headship, by usurpation on the part of the successful general—either practical usurpation, where the king is too sacred to be displaced, or complete usurpation where he is not too sacred; but that where, along with decreasing militancy, there goes increasing civil life and administration, headship of the army becomes permanently differentiated from political headship, and subordinated to it.

§ 519. While, in the course of social evolution, there has

been going on this separation of the fighting body from the community at large, this diminution in its relative mass, and this establishment of a distinct headship to it, there has been going on an internal organization of it.

The fighting body is at first wholly without structure. Among savages a battle is a number of single combats: the chief, if there is one, being but the warrior of most mark, who fights like the rest. Through long stages this disunited action continues. The *Iliad* tells of little more than the personal encounters of heroes, which were doubtless multiplied in detail by their unmentioned followers; and after the decay of that higher military organization which accompanied Greek and Roman civilization, this chaotic kind of fighting recurred throughout mediæval Europe. During the early feudal period everything turned on the prowess of individuals. War, says Gautier, consisted of "bloody duels;" and even much later the idea of personal action dominated over that of combined action. But along with political progress, the subjection of individuals to their chief is increasingly shown by fulfilling his commands in battle. Action in the field becomes in a higher degree concerted, by the absorption of their wills in his will.

A like change presently shows itself on a larger scale. While the members of each component group have their actions more and more combined, the groups themselves, of which an army is composed, pass from disunited action to united action. When small societies are compounded into a larger one, their joint body of warriors at first consists of the tribal clusters and family-clusters assembled together, but retaining their respective individualities. The head of each Hottentot kraal, " has the command, under the chief of his nation, of the troops furnished out by his kraal." Similarly, the Malagasy " kept their own respective clans, and every clan had its own leader." Among the Chibchas, " each cazique and tribe came with different signs on their tents, fitted out with the mantles by which they distinguished themselves from each other." A.

kindred arrangement existed in early Roman times: the city-army was "distributed into tribes, curiæ, and families." It was so, too, with the Germanic peoples, who, in the field, "arranged themselves, when not otherwise tied, in families and affinities;" or, as is said by Kemble of our ancestors in old English times, "each kindred was drawn up under an officer of its own lineage and appointment, and the several members of the family served together." This organization, or lack of organization, continued throughout the feudal period. In France, in the 14th century, the army was a "horde of independent chiefs, each with his own following, each doing his own will;" and, according to Froissart, the different groups "were so ill-informed" that they did not always know of a discomfiture of the main body.

Besides that increased subordination of local heads to the general head which accompanies political integration, and which must of course precede a more centralized and combined mode of military action, two special causes may be recognized as preparing the way for it.

One of these is unlikeness of kinds in the arms used. Sometimes the cooperating tribes, having habituated themselves to different weapons, come to battle already marked off from one another. In such cases the divisions by weapons correspond with the tribal divisions; as seems to have been to some extent the case with the Hebrews, among whom the men of Benjamin, of Gad, and of Judah, were partially thus distinguished. But, usually, the unlikenesses of arms consequent on unlikenesses of rank, initiate these military divisions which tend to traverse the divisions arising from tribal organization. The army of the ancient Egyptians included bodies of charioteers, of cavalry, and of foot; and the respective accoutrements of the men forming these bodies, differing in their costliness, implied differences of social position. The like may be said of the Assyrians. Similarly, the *Iliad* shows us among the early Greeks a state in which the

contrasts in weapons due to contrasts in wealth, had not yet resulted in differently-armed bodies, such as are formed at later stages with decreasing regard for tribal or local divisions. And it was so in Western Europe during times when each feudal superior led his own knights, and his followers of inferior grades and weapons. Though within each group there were men differing alike in their rank and in their arms, yet what we may call the vertical divisions between groups were not traversed by those horizontal divisions throughout the whole army, which unite all who are similarly armed. This wider segregation it is, however, which we observe taking place with the advance of military organization. The supremacy acquired by the Spartans was largely due to the fact that Lykurgus "established military divisions quite distinct from the civil divisions, whereas in the other states of Greece, until a period much later . . . the two were confounded—the hoplites or horsemen of the same tribe or ward being marshalled together on the field of battle." With the progress of the Roman arms there occurred kindred changes. The divisions came to be related less to rank as dependent on tribal organization, and more to social position as determined by property; so that the kinds of arms to be borne and the services to be rendered, were regulated by the sizes of estates, with the result of "merging all distinctions of a gentile and local nature in the one common levy of the community." In the field, divisions so established stood thus:—
"The four first ranks of each phalanx were formed of the full-armed hoplites of the first class, the holders of an entire hide [?]; in the fifth and sixth were placed the less completely equipped farmers of the second and third class; the two last classes were annexed as rear ranks to the phalanx.

And though political distinctions of clan-origin were not thus directly disregarded in the cavalry, yet they were indirectly interfered with by the addition of a larger troop of non-burgess cavalry. That a system of divisions which tends to obliterate those of rank and locality, has been reproduced during the re-development of military organization in modern times, is a familiar fact.

A concomitant cause of this change has all along been that interfusion of the gentile and tribal groups entailed by aggregation of large numbers. As before pointed out, the Kleisthenian re-organization in Attica, and the Servian re-organization in Rome, were largely determined by the impracticability of maintaining the correspondence between tribal divisions and military obligations; and a redistribution of military obligations naturally proceeded on a numerical basis. By various peoples, we find this step in organization taken for civil purposes or military purposes, or both. To cases named in § 512, may be added that of the Hebrews, who were grouped into tens, fifties, hundreds and thousands. Even the barbarous Araucanians divided themselves into regiments of a thousand, sub-divided into companies of a hundred. Evidently numerical grouping conspires with classing by arms to obliterate the primitive divisions.

This transition from the state of incoherent clusters, each having its own rude organization, to the state of a coherent whole, held together by an elaborate organization running throughout it, of course implies a concomitant progress in the centralization of command. As the primitive horde becomes more efficient for war in proportion as its members grow obedient to the orders of its chief; so, the army formed of aggregated hordes becomes more efficient in proportion as the chiefs of the hordes fall under the power of one supreme chief. And the above-described transition from aggregated tribal and local groups to an army formed of regular divisions and sub-divisions, goes along with the development of grades of commanders, successively subordinated one to another. A controlling system of this kind is developed by the uncivilized, where considerable military efficiency has been reached; as at present among the Araucanians, the Zulus, the Uganda people, who have severally three grades of officers; as in the past among the ancient Peruvians and ancient Mexicans, who had respectively several grades; and as also among the ancient Hebrews.

§ 520. One further general change has to be noticed—the change from a state in which the army now assembles and now disperses, as required, to a state in which it becomes permanently established.

While, as among savages, the male adults are all warriors, the fighting body, existing in its combined form only during war, becomes during peace a dispersed body carrying on in parties or separately, hunting and other occupations; and similarly, as we have seen, during early stages of settled life the armed freemen, owning land jointly or separately, all having to serve as soldiers when called on, return to their farming when war is over: there is no standing army. But though after the compounding of small societies into larger ones by war, and the rise of a central power, a kindred system long continues, there come the beginnings of another system. Of course, irrespective of form of goverment, frequent wars generate permanent military forces; as they did in early times among the Spartans; as later among the Athenians; and as among the Romans, when extension of territory brought frequent needs for repressing rebellions. Recognizing these cases, we may pass to the more usual cases, in which a permanent military force originates from the body of armed attendants surrounding the ruler. Early stages show us this nucleus. In Tahiti the king or chief had warriors among his attendants; and the king of Ashantee has a body-guard clad in skins of wild beasts—leopards, panthers, &c. As was pointed out when tracing the process of political differentiation, there tend everywhere to gather round a predominant chieftain, refugees and others who exchange armed service for support and protection; and so enable the predominant chieftain to become more predominant. Hence the *comites* attached to the *princeps* in the early German community, the húscarlas or housecarls surrounding old English kings, and the antrustions of the Merovingian rulers. These armed followers displayed in little, the characters of a standing army; not simply as being permanently united, but also as being severally bound to their prince or lord by relations of personal fealty, and as being subject to internal government under a code of martial law, apart from the government of the freemen; as was especially shown in the large assemblage of them, amounting to 6,000, which was formed by Cnut.

In this last case we see how small body-guards, growing as the conquering chief or king draws to his standard adventurers, fugitive criminals, men who have fled from injustice, &c., pass unobtrusively into troops of soldiers who fight for pay. The employment of mercenaries goes back to the earliest times—being traceable in the records of the Egyptains at all periods; and it continues to re-appear under certain conditions: a primary condition being that the ruler shall have acquired a considerable revenue. Whether of home origin or foreign origin, these large bodies of professional soldiers can be maintained only by large pecuniary means; and, ordinarily, possession of these means goes along with such power as enables the king to exact dues and fines. In early stages the members of the fighting body, when summoned for service, have severally to provide themselves not only with their appropriate arms, but also with the needful supplies of all kinds: there being, while political organization is little developed, neither the resources nor the administrative machinery required for another system. But the economic resistance to militant action, which, as we have seen, increases as agricultural life spreads, leading to occasional non-attendance, to confiscations, to heavy fines in place of confiscations, then to fixed money-payments in place of personal services, results in the growth of a revenue which serves to pay professional soldiers in place of the vassals who have compounded. And it then becomes possible, instead of hiring many such substitutes for short times, to hire a smaller number continuously—so adding to the original nucleus of a permanent armed force. Every further increase of royal power, increasing the ability to raise money, furthers this differentiation. As Ranke remarks of France, "standing armies, imposts, and loans, all originated together."

Of course the primitive military obligation falling on all freemen, long continues to be shown in modified ways. Among ourselves, for instance, there were the various laws under which men were bound, according to their incomes, to have in readiness specified supplies of horses, weapons, and accoutrements, for themselves and others when demanded. Afterwards came the militia-laws, under which there fell on men in proportion to their means, the obligations to provide duly armed horse-soldiers or foot-soldiers, personally or by substitute, to be called out for exercise at specified intervals for specified numbers of days, and to be provided with subsistence. There may be instanced, again, such laws as those under which in France, in the 15th century, a corps of horsemen was formed by requiring all the parishes to furnish one each. And there are the various more modern forms of conscription, used, now to raise temporary forces, and now to maintain a permanent army. Everywhere, indeed, freemen remain potential soldiers when not actual soldiers.

§ 521. Setting out with that undifferentiated state of the body politic in which the army is co-extensive with the adult male population, we thus observe several ways in which there goes on the evolution which makes it a specialized part.

There is the restriction in relative mass, which, first seen in the growth of a slave-population, engaged in work instead of war, becomes more decided as a settled agricultural life occupies freemen, and increases the obstacles to military service. There is, again, the restriction caused by that growing costliness of the individual soldier accompanying the development of arms, accoutrements, and ancillary appliances of warfare. And there is the yet additional restriction caused by the intenser strain which military action puts on the resources of a nation, in proportion as it is carried on at a greater distance.

With separation of the fighting body from the body-politic at large, there very generally goes acquirement of a separate head. Active militancy ever tends to maintain union of civil rule with military rule, and often causes re-union of them where they have become separate; but with the primary differentiation of civil from military structures, is commonly associated a tendency to the rise of distinct controlling centres for them. This tendency, often defeated by usurpation where wars are frequent, takes effect under opposite conditions; and then produces a military head subordinate to the civil head.

While the whole society is being developed by differentiation of the army from the rest, there goes on a development within the army itself. As in the primitive horde the progress is from the uncombined fighting of individuals to combined fighting under direction of a chief; so, on a larger scale, when small societies are united into great ones, the progress is from the independent fighting of tribal and local groups, to fighting under direction of a general commander. And to effect a centralized control, there arises a graduated system of officers, replacing the set of primitive heads of groups, and a system of divisions which, traversing the original divisions of groups, establish regularly-organized masses having different functions.

With developed structure of the fighting body comes permanence of it. While, as in early times, men are gathered together for small wars and then again dispersed, efficient organization of them is impracticable. It becomes practicable only among men who are constantly kept together by wars or preparations for wars; and bodies of such men growing up, replace the temporarily-summoned bodies.

Lastly, we must not omit to note that while the army becomes otherwise distinguished, it becomes distinguished by retaining and elaborating the system of status; though in the rest of the community, as it advances, the system of contract is spreading and growing definite. Compulsory cooperation continues to be the principle of the military part, however widely the principle of voluntary cooperation comes into play throughout the civil part.

CHAPTER XIII.

JUDICIAL AND EXECUTIVE SYSTEMS.

§ 522. That we may be prepared for recognizing the primitive identity of military institutions with institutions for administering justice, let us observe how close is the kinship between the modes of dealing with external aggression and internal aggression, respectively.

We have the facts, already more than once emphasized, that at first the responsibilities of communities to one another are paralleled by the responsibilities to one another of family-groups within each community; and that the kindred claims are enforced in kindred ways. Various savage tribes show us that, originally, external war has to effect an equalization of injuries, either directly in kind or indirectly by compensations. Among the Chinooks, " has the one party a larger number of dead than the other, indemnification must be made by the latter, or the war is continued;" and among the Arabs " when peace is to be made, both parties count up their dead, and the usual blood-money is paid for excess on either side." By which instances we are shown that in the wars between tribes, as in the family-feuds of early times, a death must be balanced by a death, or else must be compounded for; as it once was in Germany and in England, by specified numbers of sheep and cattle, or by money.

Not only are the wars which societies carry on to effect the righting of alleged wrongs, thus paralleled by family-feuds in the respect that for retaliation in kind there may be substituted a penalty adjudged by usage or authority; but they are paralleled by feuds between individuals in the like respect. From the first stage in which each man avenges himself by force on a transgressing neighbour, as the whole community does on a transgressing community, the transition is to a stage in which he has the alternative of demanding justice at the hands of the ruler. We see this beginning in such places as the Sandwich Islands, where an injured person who is too weak to retaliate, appeals to the king or principal chief; and in quite advanced stages, option between the two methods of obtaining redress survives. The feeling shown down to the 13th century by Italian nobles, who "regarded it as disgraceful to submit to laws rather than do themselves justice by force of arms," is traceable throughout the history of Europe in the slow yielding of private rectification of wrongs to public arbitration. "A capitulary of Charles the Bald bids them [the freemen] go to court armed as for war, for they might have to fight for their jurisdiction;" and our own history furnishes an interesting example in the early form of an action for recovering land: the "grand assize" which tried the cause, originally consisted of knights armed with swords. Again we have evidence in such facts as that in the 12th century in France, legal decisions were so little regarded that trials often issued in duels. Further proof is yielded by such facts as that judicial duels (which were the authorized substitutes for private wars between families) continued in France down to the close of the 14th century; that in England, in 1768, a legislative proposal to abolish trial by battle, was so strongly opposed that the measure was dropped; and that the option of such trial was not disallowed till 1819.

We may observe, also, that this self-protection gradually gives place to protection by the State, only under stress of public needs—especially need for military efficiency. Edicts of Charlemagne and of Charles the Bald, seeking to stop the disorders consequent on private wars, by insisting on appeals to the ordained authorities, and threatening punishment of those who disobeyed, sufficiently imply the motive; and this motive was definitely shown in the feudal period in France, by an ordinance of 1296, which "prohibits private wars and judicial duels so long as the king is engaged in war."

Once more the militant nature of legal protection is seen in the fact that, as at first, so now, it is a replacing of individual armed force by the armed force of the State—always in reserve if not exercised. "The sword of justice" is a phrase sufficiently indicating the truth that action against the public enemy and action against the private enemy are in the last resort the same.

Thus recognizing the original identity of the functions, we shall be prepared for recognizing the original identity of the structures by which they are carried on.

§ 523. For that primitive gathering of armed men which, as we have seen, is at once the council of war and the political assembly, is at the same time the judicial body.

Of existing savages the Hottentots show this. The court of justice "consists of the captain and all the men of the kraal. . . . 'Tis held in the open fields, the men squatting in a circle. . . . All matters are determined by a majority." . . . If the prisoner is "convicted, and the court adjudges him worthy of death, sentence is executed upon the spot." The captain is chief executioner, striking the first blow; and is followed up by the others. The records of various historic peoples yield evidence of kindred meaning. Taking first the Greeks in Homeric days, we read that "sometimes the king separately, sometimes the kings or chiefs or Gerontes, in the plural number, are named as deciding disputes and awarding satisfaction to complainants; always however in public, in the midst of the assembled agora," in which the popular sympathies were expressed: the meeting thus described, being the same with that in which questions of war and peace were debated. That in its early form the Roman gathering of "spearmen," asked by the king to say "yes" or "no" to a proposed military expedition or to some State-measure, also expressed its opinion concerning criminal charges publicly judged, is implied by the fact that "the king could not grant a pardon, for that privilege was vested in the community alone." Describing the gatherings of the primitive Germans, Tacitus says:—"The multitude sits armed in such order as it thinks good . . . It is lawful also in the Assembly to bring matters for trial and to bring charges of capital crimes . . . In the same assembly chiefs are chosen to administer justice throughout the districts and villages. Each chief in so doing has a hundred companions of the commons assigned to him, to strengthen at once his judgment and his dignity." A kindred arrangement is ascribed by Lelevel to the Poles in early times, and to the Slavs at large. Among the Danes, too, "in all secular affairs, justice was administered by the popular tribunal of the Lands-Ting for each province, and by the Herreds-Ting for the smaller districts or sub-divisions." Concerning the Irish in past times, Prof. Leslie quotes Spenser to the effect that it was their usage "to make great assemblies together upon a rath or hill, there to parley about matters and wrongs between township and township, or one private person and another." And then there comes the illustration furnished by old English times The local moots of various kinds had judicial functions; and the witenagemót sometimes acted as a high court of justice.

Interesting evidence that the original military assembly was at the same time the original judicial assembly, is supplied by the early practice of punishing freemen for non-attendance. Discharge of military obligation being imperative the fining of those who did not come to the armed gathering naturally followed; and fining for absence having become the usage, survived when, as for judicial purposes, the need for the presence of all was not imperative. Thence the interpretation of the fact that non-attendance at the hundred-court was thus punishable.

In this connexion it may be added that, in some cases

where the primitive form continued, there was manifested an incipient differentiation between the military assembly and the judicial assembly. In the Carolingian period, judicial assemblies began to be held under cover; and freemen were forbidden to bring their arms. As was pointed out in § 491, among the Scandinavians no one was allowed to come armed when the meeting was for judicial purposes. And since we also read that in Iceland it was disreputable (not punishable) for a freeman to be absent from the annual gathering, the implication is that the imperativeness of attendance diminished with the growing predominance of civil functions.

§ 524. The judicial body being at first identical with the politico-military body, has necessarily the same triune structure; and we have now to observe the different forms it assumes according to the respective developments of its three components. We may expect to find kinship between these forms and the concomitant political forms.

Where, with development of militant organization, the power of the king has become greatly predominant over that of the chiefs and over that of the people, his supremacy is shown by his judicial absoluteness, as well as by his absoluteness in political and military affairs. Such shares as the elders and the multitude originally had in trying causes, almost or quite disappear. But though in these cases the authority of the king as judge, is unqualified by that of his head men and his other subjects, there habitually survive traces of the primitive arrangement. For habitually his decisions are given in public and in the open air. Petitioners for justice bring their cases before him when he makes his appearance out of doors, surrounded by his attendants and by a crowd of spectators; as we have seen in § 372 that they do down to the present day in Kashmere. By the Hebrew rulers, judicial sittings were held "in the gates"—the usual meeting-places of Eastern peoples. Among the early Romans the king administered justice "in the place of public assembly, sitting on a 'chariot-seat.'" Mr. Gomme's *Primitive Folk-Moots* contains sundry illustrations showing that among the Germans in old times, the Königs-stuhl, or king's judgment-seat, was on the green sward; that in other cases the stone steps at the town-gates constituted the seat before which causes were heard by him; and that again, in early French usage, trials often took place under trees. According to Joinville this practice long continued in France.

"Many a time did it happen that, in summer, he [Lewis IX] would go and sit in the forest of Vincennes after mass, and would rest against an oak, and make us sit round him . . . he asked them with his own mouth, 'Is there any one who has a suit?' . . . I have seen him sometimes in summer come to hear his people's suits in the garden of Paris."

And something similar occurred in Scotland under David I. All which customs among various peoples, imply survival of the primitive judicial assembly, changed only by concentration in its head of power originally shared by the leading men and the undistinguished mass.

Where the second component of the triune political structure becomes supreme, this in its turn monopolizes judicial functions. Among the Spartans the oligarchic senate, and in a measure the smaller and chance-selected oligarchy constituted by the ephors, joined judicial functions with their political functions. Similarly in Athens under the aristocratic rule of the Eupatridæ, we find the Areopagus formed of its members, discharging, either itself or through its nine chosen Archons, the duties of deciding causes and executing decisions. In later days, again, we have the case of the Venetian council of ten. And then, certain incidents of the middle ages instructively show us one of the processes by which judicial power, as well as political power, passes from the hands of the freemen at large into the hands of a smaller and wealthier class. In the Carolingian period, besides the bi-annual meetings of the hundred-court, it was—

"convoked at the *Graf's* will and pleasure, to try particular cases . . . in the one case, as in the other, non-attendance was punished . . . it was found that the *Grafs* used their right to summon these extraordinary

Courts in excess, with a view, by repeated fines and amercements, to ruin the small freeholders, and thus to get their abodes into their own hands. Charlemagne introduced a radical law-reform . . . the great body of the freemen were released from attendance at the *Gebotene Dinge*, at which, from thenceforth, justice was to be administered under the presidency, *ex officio*, of the *Centenar*, by . . . permanent jurymen . . . chosen *de melioribus*—*i.e.*, from the more well-to-do freemen."

But in other cases, and especially where concentration in a town renders performance of judicial functions less burdensome, we see that along with retention or acquirement of predominant power by the third element in the triune political structure, there goes exercise of judicial functions by it. The case of Athens, after the replacing of oligarchic rule by democratic rule, is, of course, the most familiar example of this. The Kleisthenian revolution made the annually-appointed magistrates personally responsible to the people judicially assembled; and when, under Perikles, there were established the dikasteries, or courts of paid jurors chosen by lot, the administration of justice was transferred almost wholly to the body of freemen, divided for convenience into committees. Among the Frieslanders, who in early times were enabled by the nature of their habitat to maintain a free form of political organization, there continued the popular judicial assembly:—" When the commons were summoned for any particular purpose, the assembly took the name of the Bodthing. The bodthing was called for the purpose of passing judgment in cases of urgent necessity." And M. de Laveleye, describing the Teutonic mark as still existing in Holland, "especially in Drenthe," a tract "surrounded on all sides by a marsh and bog" (again illustrating the physical conditions favourable to maintenance of primitive free institutions), goes on to say of the inhabitants as periodically assembled:—

"They appeared in arms; and no one could absent himself, under pain of a fine. This assembly directed all the details as to the enjoyment of the common property; appointed the works to be executed; imposed pecuniary penalties for the violation of rules, and nominated the officers charged with the executive power."

The likeness between the judicial form and the political

form is further shown where the government is neither despotic nor oligarchic, nor democratic, but mixed. For in our own case we see a system of administering justice which, like the political system, unites authority that is in a considerable degree irresponsible, with popular authority. In old English times a certain power of making and enforcing local or "bye-laws" was possessed by the township; and in more important and definite ways the hundred-moot and the shire-moot discharged judicial and executive functions: their respective officers being at the same time elected. But the subsequent growth of feudal institutions, followed by the development of royal power, was accompanied by diminution of the popular share in judicial business, and an increasing assignment of it to members of the ruling classes and to agents of the crown. And at present we see that the system, as including the power of juries (which arose by selection of representative men, though not in the interest of the people), is in part popular; that in the summary jurisdiction of unpaid magistrates who, though centrally appointed, mostly belong to the wealthy classes, and especially the landowners, it is in part aristocratic; that in the regal commissioning of judges it continues monarchic; and that yet, as the selection of magistrates and judges is practically in the hands of a ministry executing, on the average, the public will, royal power and class-power in the administration of justice are exercised under popular control.

§ 525. A truth above implied and now to be definitely observed, is that along with the consolidation of small societies into large ones effected by war, there necessarily goes an increasing discharge of judicial functions by deputy.

As the primitive king is very generally himself both commander-in-chief and high priest, it is not unnatural that his delegated judicial functions should be fulfilled both by priests and soldiers. Moreover, since the consultative body, where it becomes established and separated from the multitude, habitually includes members of both these classes, such judicial powers as it exercises cannot at the outset be monopolized by members of either. And this participation is further seen to arise naturally on remembering how, as before shown, priests have in so many societies united military functions with clerical functions; and how, in other cases, becoming local rulers, having the same tenures and obligations with purely military local rulers, they acquire, in common with them, local powers of judgment and execution; as did mediæval prelates. Whether the ecclesiastical class or the class of warrior-chiefs acquires judicial predominance, probably depends mainly on the proportion between men's fealty to the successful soldier, and their awe of the priest as a recipient of divine communications.

Among the Zulus, who, with an undeveloped mythology, have no great deities and resulting organized priesthood, the king " shares his power with two soldiers of his choice. These two form the supreme judges of the country." Similarly with the Eggarahs (Inland Negroes), whose fetish-men do not form an influential order, the first and second judges are " also commanders of the forces in time of war." Passing to historic peoples, we have in Attica, in Solon's time, the nine archons, who, while possessing a certain sacredness as belonging to the Eupatridæ, united judicial with military functions—more especially the polemarch. In ancient Rome, that kindred union of the two functions in the consuls, who called themselves indiscriminately, *prætores* or *judices*, naturally resulted from their inheritance of both functions from the king they replaced; but beyond this there is the fact that though the pontiffs had previously been judges in secular matters as well as in sacred matters, yet, after the establishment of the republic, the several orders of magistrates were selected from the non-clerical patricians,—the original soldier-class. And then throughout the middle ages in Europe, we have the local military chiefs, whether holding positions like those of old English thanes or like those of feudal barons, acting as judges in their respective localities. Perhaps the clearest illustration is that furnished by Japan, where a long-continued and highly-developed military *régime*, has been throughout associated with the monopoly of judicial functions by the military class : the apparent reason being that in presence of the god-descended Mikado, supreme in heaven as on earth, the indigenous Shinto religion never developed a divine ruler whose priests acquired, as his agents, an authority competing with terrestrial authority.

But mostly there is extensive delegation of judicial powers to the sacerdotal class, in early stages. We find it among existing uncivilized peoples, as the Kalmucks, whose priests, besides playing a predominant part in the greatest judicial council, exercise local jurisdiction: in the court of each subordinate chief, one of the high priests is head judge. Of extinct uncivilized or semi-civilized peoples, may be named the Indians of Yucatan, by whom priests were appointed as judges in certain cases—judges who took part in the execution of their own sentences. Originally, if not afterwards, the giving of legal decisions was a priestly function in ancient Egypt; and that the priests were supreme judges among the Hebrews is a familiar fact: the Deuteronomic law condemning to death any one who disregarded their verdicts. In that general assembly of the ancient Germans which, as we have seen, exercised judicial powers, the priests were prominent; and, according to Tacitus, in war "none but the priests are permitted to judge offenders, to inflict bonds or stripes; so that chastisement appears not as an act of military discipline, but as the instigation of the god whom they suppose present with warriors." In ancient Britain, too, according to Cæsar, the druids alone had authority to decide in both civil and criminal cases, and executed their own sentences: the penalty for disobedience to them being excommunication. Grimm tells us that the like held among the Scandinavians. "In their judicial character the priests seem to have exercised a good deal of control over the people . . . In Iceland, even under Christianity, the judges retained the name and several of the functions of heathen goðar." And then we have the illustration furnished by that rise of ecclesiastics to the positions of judges throughout mediæval Europe, which accompanied belief in their divine authority. When, as during the Merovingian period and after, "the fear of hell, the desire of winning heaven," and other motives, prompted donations and bequests to the Church, till a large part of the landed property fell into its hands—when there came increasing numbers of clerical and semi-clerical dependents of the Church, over whom bishops exercised judgment and discipline—when ecclesiastical influence so extended itself that, while priests became exempt from the control of laymen, lay authorities became subject to priests; there was established a judicial power of this divinely-commissioned class to which even kings succumbed. So was it in England too. Before the Conquest, bishops had become the assessors of ealdormen in the scire-gemót, and gave judgments on various civil matters. With that recrudescence of military organization which followed the Conquest, came a limitation of their jurisdiction to spiritual offences and causes concerning clerics. But in subsequent periods ecclesiastical tribunals, bringing under canon law numerous ordinary transgressions, usurped more and more the duties of secular judges: their excommunications being enforced by the temporal magistrates. Moreover, since prelates as feudal nobles were judges in their respective domains; and since many major and minor judicial offices in the central government were filled by prelates; it resulted that the administration of justice was largely, if not mainly, in the hands of priests.

This sharing of delegated judicial functions between the military class and the priestly class, with predominance here of the one and there of the other, naturally continued while there was no other class having wealth and influence. But with the increase of towns and the multiplication of traders, who accumulated riches and acquired education, previously possessed only by ecclesiastics, judicial functions fell more and more into their hands. Sundry causes conspired to produce this transfer. One was lack of culture among the nobles, and their decreasing ability to administer laws, ever increasing in number and in complexity. Another was the political unfitness of ecclesiastics, who grew distasteful to rulers in proportion as they pushed further the powers and privileges which their supposed divine commission gave them. Details need not detain us. The only general fact needing to be emphasized, is that this transfer ended in a differentiation of structures. For whereas in earlier stages, judicial functions were discharged by men who were at the same time either soldiers or priests, they came now to be discharged by men exclusively devoted to them.

§ 526. Simultaneously, the evolution of judicial systems is displayed in several other ways. One of them is the addition of judicial agents who are locomotive to the pre-existing stationary judicial agents.

During the early stages in which the ruler administers justice in person, he does this now in one place and now in another; according as affairs, military or judicial, carry him to this or that place in his kingdom. Societies of various types in various times yield evidence. Historians of ancient Peru tell us that " the Ynca gave sentence according to the crime, for he alone was judge wheresoever he resided, and all persons wronged had recourse to him." Of the German emperor in the 12th century we read that " not only did he receive appeals, but his presence in any duchy or county suspended the functions of the local judges." France in the 15th century supplies an instance. King Charles " spent two or three years in travelling up and down the kingdom . . maintaining justice to the satisfaction of his subjects." In Scotland something similar was done by David I., who " settled marches, forest rights, and rights of pasture:" himself making the marks which recorded his decisions, or seeing them made. In England, "Edgar and Canute had themselves made judicial circuits;" and there is good evidence of such judicial travels in England up to the time of the Great Charter. Sir Henry Maine has quoted documents showing that King John, in common with earlier kings, moved about the country with great activity, and held his court wherever he might happen to be.

Of course with the progress of political integration and consequent growing power of the central ruler, there come more numerous cases in which appeal is made to him to rectify the wrongs committed by local rulers; and as State-business at large augments and complicates, his inability to do this personally leads to doing it by deputy. In France, in Charlemagne's time, there were the "Missi Regii, who held assizes from place to place;" and then, not forgetting that during a subsequent period the chief heralds in royal state, as the king's representatives, made circuits to judge and punish transgressing nobles, we may pass to the fact that in the later feudal period, when the business of the king's court became too great, commissioners were sent into the provinces to judge particular cases in the king's name: a method which does not appear to have been there developed further. But in England, in Henry II.'s time, kindred causes prompted kindred steps which initiated a permanent system. Instead of listening to the increasing number of appeals made to his court, personally or through his lieutenant the justiciar, the king commissioned his constable, chancellor, and co-justiciar to hear pleas in the different counties. Later, there came a larger number of these members of the central judicial court who made these judicial journeys: part of them being clerical and part military. And hence eventually arose the established circuits of judges who, like their prototypes, had to represent the king and exercise supreme authority.

It should be added that here again we meet with proofs that in the evolution of arrangements conducing to the maintenance of individual rights, the obligations are primary and the claims derived. For the business of these travelling judges, like the business of the king's court by which they were commissioned, was primarily fiscal and secondarily judicial. They were members of a central body that was at once Exchequer and *Curia Regis*, in which financial functions at first predominated; and they were sent into the provinces largely, if not primarily, for purposes of assessment: as instance the statement that in 1168, "the four Exchequer officers who assessed the aid *pur fille marier*, acted not only as taxers but as judges." In which facts we see harmony with those before given, showing that support of the ruling agency precedes obtainment of protection from it.

§ 527. With that development of a central government which accompanies consolidation of small societies into a large one, and with the consequent increase of its business, entailing delegation of functions, there goes, in the judicial organization as in the other organizations, a progressive differentiation. The evidence of this is extremely involved; both for the reason that in most cases indigenous judicial agencies have been subordinated but not destroyed by those which conquest has originated, and for the reason that kinds of power, as well as degrees of power, have become distinguished. A few leading traits only of the process can here be indicated.

The most marked differentiation, already partially implied, is that between the lay, the ecclesiastical, and the military tribunals. From those early stages in which the popular assembly, with its elders and chief, condemned military defaulters, decided on ecclesiastical questions, and gave judgments about offences, there has gone on a divergence which, accompanied by disputes and struggles concerning jurisdiction, has parted ecclesiastical courts and courts martial from the courts administering justice in ordinary civil and criminal cases. Just recognizing these cardinal specializations, we may limit our attention to the further specializations which have taken place within the last of the three structures.

Originally the ruler, with or without the assent of the assembled people, not only decides: he executes his decisions, or sees them executed. For example, in Dahomey the king stands by, and if the deputed officer does not please him, takes the sword out of his hand and shows him how to cut off a head. An account of death-punishment among the Bedouins ends with the words—"the executioner being the sheikh himself." Our own early history affords traces of personal executive action by the king; for there came a time when he was interdicted from arresting any one himself, and had thereafter to do it in all cases by deputy. And this interprets for us the familiar truth that, through his deputies the sheriffs, who are bound to act personally if they cannot themselves find deputies, the monarch continues to be theoretically the agent who carries the law into execution: a truth further implied by the fact that execution in criminal cases, nominally authorized by him though actually by his minister, is arrested if his assent is withheld by his minister. And these facts imply that a final power of judgment remains with the monarch, notwithstanding delegation of his judicial functions. How this happens we shall see on tracing the differentiation.

Naturally, when a ruler employs assistants to hear complaints and redress grievances, he does not give them absolute authority; but reserves the power of revising their decisions. We see this even in such rude societies as that of the Sandwich Islands, where one who is dissatisfied with the decision of his chief may appeal to the governor, and from the governor to the king; or as in ancient Mexico, where "none of the judges were allowed to condemn to death without communicating with the king, who had to pass the sentence." And the principle holds where the political headship is compound instead of simple. "When the hegemony of Athens became, in fact, more and more a dominion, the civic body of Attica claimed supreme judicial authority over all the allies. The federal towns only retained their lower courts." Obviously by such changes are produced unlikenesses of degree and differences of kind in the capacities of judicial agencies. As political subordination spreads, the local assemblies which originally judged and executed in cases of all kinds, lose part of their functions; now by restriction in range of jurisdiction, now by subjection of their decisions to supervision, now by denial of executive power. To trace up the process from early stages, as for instance from the stage in which the old English tything-moot discharged administrative, judicial, and executive functions, or from the stage in which the courts of feudal nobles did the like, is here alike impracticable and unnecessary. Reference to such remnants of power as vestries and manorial courts possess, will sufficiently indicate the character of the change. But along with degradation of the small and local judicial agencies, goes development of the great and central ones; and about this something must be said.

Returning to the time when the king with his servants and chief men, surrounded by the people, administers justice in the open air, and passing to the time when his court, held more frequently under cover and consequently with less of the popular element, still consists of king as president and his household officers with other appointed magnates as counsellors (who in fact constitute a small and permanent part of that general consultative body occasionally summoned); we have to note two causes which cooperate to produce a division of these remaining parts of the original triune body —one cause being the needs of subjects, and the other the desire of the king. So long as the king's court is held wherever he happens to be, there is an extreme hindrance to the hearing of suits, and much entailed loss of money and time to suitors. To remedy this evil came, in our own case, the provision included in the Great Charter that the common pleas should no longer follow the king's court, but be held in some certain place. This place was fixed in the palace of Westminster. And then as Blackstone points out—

"This precedent was soon after copied by King Philip the Fair in France, who about the year 1302, fixed the parliament of Paris to abide constantly in that metropolis; which before used to follow the person of the king wherever he went . . . And thus also, in 1495, the Emperor Maximilian I. fixed the imperial chamber, which before always travelled with the court and household, to be constantly at Worms."

As a sequence of these changes it of course happens that suits of a certain kind come habitually to be decided without the king's presence: there results a permanent transfer of part of his judicial power. Again, press of business or love of ease prompts the king himself to hand over such legal matters as are of little interest to him. Thus in France, while we read that Charles V., when regent, sat in his council to administer justice twice a week, and Charles VI. once, we also read that in 1370 the king declared he would no longer try the smaller causes personally. Once initiated and growing into a usage, this judging by commission, becoming more frequent as affairs multiply, is presently otherwise furthered: there arises the doctrine that the king ought not, at any rate in certain cases, to join in judgment. Thus "at the trial of the duke of Brittany in 1378, the peers of France protested against the presence of the king." Again "at the trial of the Marquis of Saluces, under Francis I., that monarch was made to see that he could not sit." When Lewis XIII. wished to be judge in the case of the Duke de la Valette, he was resisted by the judges, who said that it was without precedent. And in our own country there came a time when "James I. was informed by the judges that he had the right to preside .in the court, but not to express his opinion:" a step towards that exclusion finally reached.

While the judicial business of the political head thus lapses into the hands of appointed agencies, these agencies themselves, severally parting with certain of their functions one to another, become specialized. Among ourselves, even before there took place the above-named separation of the permanently-localized court of common pleas, from the king's court which moved about with him, there had arisen within the king's court an incipient differentiation. Causes concerning revenue were dealt with in sittings distinguished from the general sittings of the king's court, by being held in another room; and establishment of this custom produced a division. Adaptation of its parts to unlike ends led to divergence of them; until, out of the original *Curia Regis*, had come the court of exchequer and the court of common pleas; leaving behind the court of king's bench as a remnant of the original body. When the office of justiciar (who, representing the king in his absence, presided over these courts) was abolished, the parting of them became decided; and though, for a length of time, competition for fees led to trenching on one another's functions, yet, eventually, their functions became definitely marked off. A further important development, different but allied, took place. We have seen that when appointing others to judge for him, the king reserves the power of deciding in cases which the law has not previously provided for, and also the power of supervising the decisions made by his deputies. Naturally this power comes to be especially used to over-ride decisions which, technically according to law, are practically unjust: the king acquires an equity jurisdiction. At first exercised personally, this jurisdiction is liable to be deputed; and in our own case was so. The chancellor, one of the king's servants, who "as a baron of the exchequer and as a leading member of the curia" had long possessed judicial functions, and who was the officer to present to the king petitions concerning these "matters of grace and favour," became presently himself the authority who gave decisions in equity qualifying the decisions of law; and thus in time resulted the court of chancery. Minor courts with minor functions also budded out from the original *Curia Regis*. This body included the chief officers of the king's household, each of whom had a jurisdiction in matters pertaining to his special business; and hence resulted the court of the chamberlain, the court of the steward, the court of the earl marshal (now at Herald's College), the court of the constable (no longer extant), the court of the admiral, &c.

In brief, then, we find proofs that, little trace as its structure now shows of such an origin, our complex judicial system, alike in its supreme central parts and in its various small local parts, has evolved by successive changes out of the primitive gathering of people, head men, and chief.

§ 528. Were further detail desirable, there might here be given an account of police-systems; showing their evolution from the same primitive triune body whence originate the several organizations delineated in this and preceding chapters. As using force to subdue internal aggressors, police are like soldiers, who use force to subdue external aggressors; and the two functions, originally one, are not even now quite separated either in their natures or their agents. For besides being so armed that they are in some countries scarcely distinguishable from soldiers, and besides being subject to military discipline, the police are, in case of need, seconded by soldiers in the discharging of their duties. To indicate the primitive identity it will suffice to name two facts. During the Merovingian period in France, armed bands of serfs, attached to the king's household and to the households of dukes, were employed both as police and for garrison purposes; and in feudal England, the *posse comitatus*, consisting of all freemen between fifteen and sixty, under command of the sheriff, was the agent for preserving internal peace at the same time that it was available for repelling invasions, though not for foreign service—an incipient differentiation between the internal and external defenders which became in course of time more marked. Letting this brief indication suffice, it remains only to sum up the conclusions above reached.

Evidences of sundry kinds unite in showing that judicial action and military action, ordinarily having for their common end the rectification of real or alleged wrongs, are closely allied at the outset. The sword is the ultimate resort in either case: use of it being in the one case preceded by a war of words carried on before some authority whose aid is invoked, while in the other case it is not so preceded. As is said by Sir Henry Maine, "the fact seems to be that contention in Court takes the place of contention in arms, but only gradually takes its place."

Thus near akin as the judicial and military actions originally are, they are naturally at first discharged by the same agency—the primitive triune body formed of chief, head men, and people. This which decides on affairs of war and settles questions of public policy, also gives judgments concerning alleged wrongs of individuals and enforces its decisions.

According as the social activities develop one or other element of the primitive triune body, there results one or other form of agency for the administration of law. If continued militancy makes the ruling man all-powerful, he becomes absolute judicially as in other ways: the people lose all share in giving decisions, and the judgments of the chief men who surround him are overridden by his. If conditions favour the growth of the chief men into an oligarchy, the body they form becomes the agent for judging and punishing offences as for other purposes: its acts being little or not at all qualified by the opinion of the mass. While if the surrounding circumstances and mode of life are such as to prevent supremacy of one man, or of the leading men, its primitive judicial power is preserved by the aggregate of freemen—or is regained by it where it re-acquires predominance. And where the powers of these three elements are mingled in the political organization, they are also mingled in the judicial organization.

In those cases, forming the great majority, in which habitual militancy entails subjection of the people, partial or complete, and in which, consequently, political power and judicial power come to be exercised exclusively by the several orders of chief men, the judicial organization which arises as the society enlarges and complicates, is officered by the sacerdotal class, or the military class, or partly the one and partly the other: their respective shares being apparently dependent on the ratio between the degree of conscious subordination to the human ruler and the degree of conscious subordination to the divine ruler, whose will the priests are supposed to communicate. But with the progress of industrialism and the rise of a class which, acquiring property and knowledge, gains consequent influence, the judicial system comes to be largely, and at length chiefly, officered by men derived from this class; and these men become distinguished from their predecessors not only as being of other origin, but also as being exclusively devoted to judicial functions.

While there go on changes of this kind, there go on changes by which the originally-simple and comparatively-uniform judicial system, is rendered increasingly complex. Where, as in ordinary cases, there has gone along with achievement of supremacy by the king, a monopolizing of judicial authority by him, press of business presently obliges him to appoint others to try causes and give judgments: subject of course to his approval. Already his court, originally formed of himself, his chief men, and the surrounding people, has become supreme over courts constituted in analogous ways of local magnates and their inferiors—so initiating a differentiation; and now by delegating certain of his servants or assessors, at first with temporary commissions to hear appeals locally, and then as permanent itinerant judges, a further differentiation is produced. And to this are added yet further differentiations, kindred in nature, by which other assessors of his court are changed into the heads of specialized courts, which divide its business among them. Though this particular course has been taken in but a single case, yet it serves to exemplify the general principle under which, in one way or other, there arises out of the primitive simple judicial body, a centralized and heterogeneous judicial organization.

CHAPTER XIV.

LAWS.

§ 529. If, going back once more to the primitive horde, we ask what happens when increase of numbers necessitates migration—if we ask what it is which causes the migrating part to fall into social arrangements like those of the parent part, and to behave in the same way; the obvious reply is that the inherited natures of its members, regulated by the ideas transmitted from the past, cause these results. That guidance by custom which we everywhere find among rude peoples, is the sole conceivable guidance at the outset.

To recall vividly the truth set forth in § 467, that the rudest men conform their lives to ancestral usages, I may name such further illustrations as that the Sandwich Islanders had " a kind of traditional code . . . followed by general consent;" and that by the Bechuanas, government is carried on according to "long-acknowledged customs." A more specific statement is that made by Mason concerning the Karens, among whom "the elders are the depositaries of the laws, both moral and political, both civil and criminal, and they give them as they receive them, and as they have been brought down from past generations" orally. Here, however, we have chiefly to note that this government by custom, persists through long stages of progress, and even still largely influences judicial administration. Instance the fact that as late as the 14th century in France, an ordinance declared that " the whole kingdom is regulated by 'custom,' and it is as 'custom' that some of our subjects make use of the written law." Instance the fact that our own Common Law is mainly an embodiment of the "customs of the realm," which have gradually become established: its older part, nowhere existing in the shape of enactment, is to be learnt only from text-books; and even parts, such as mercantile law, elaborated in modern times, are known only through reported judgments, given in conformity with usages proved to have been previously followed. Instance again the fact, no less significant, that at the present time custom perpetually re-appears as a living supplementary factor; for it is only after judges' decisions have established precedents which pleaders afterwards quote, and subsequent judges follow, that the application of an act of parliament becomes settled. So that while in the course of civilization written law tends to replace traditional usage, the replacement never becomes complete.

And here we are again reminded that law, whether written or unwritten, formulates the rule of the dead over the living. In addition to that power which past generations exercise over present generations by transmitting their natures, bodily and mental; and in addition to the power they exercise over them by bequeathed private habits and modes of life; there is this power they exercise through these regulations for public conduct handed down orally or in writing. Among savages and in barbarous societies, the authority of laws thus derived is unqualified; and even in advanced stages of civilization, characterized by much modifying of old laws and making of new ones, conduct is controlled in a far greater degree by the body of inherited laws than by those laws which the living make.

I emphasize these obvious truths for the purpose of pointing out that they imply a tacit ancestor-worship. I wish to make it clear that when asking in any case—What is the Law? we are asking—What was the dictate of our forefathers? And my object in doing this is to prepare the way for showing that unconscious conformity to the dictates of the dead, thus shown, is, in early stages, joined with conscious conformity to their dictates.

§ 530. For along with development of the ghost-theory, there arises the practice of appealing to ghosts, and to the gods evolved from ghosts, for directions in special cases, in addition to the general directions embodied in customs. There come methods by which the will of the ancestor, or the dead chief, or the derived deity, is sought; and the reply given, usually referring to a particular occasion, originates in some cases a precedent, from which there results a law added to the body of laws the dead have transmitted.

The seeking of information and advice from ghosts, takes here a supplicatory and there a coercive form. The Veddahs, who ask the spirits of their ancestors for aid, believe that in dreams they tell them where to hunt; and then we read of the Scandinavian diviners, that they " dragged the ghosts of the departed from their tombs and forced the dead to tell them what would happen:" cases which remind us that among the Hebrews, too, there were supernatural directions given in dreams as well as information derived from invoked spirits. This tendency to accept special guidance from the dead, in addition to the general guidance of an inherited code, is traceable in a transfigured shape even among ourselves; for besides conforming to the orally-declared wish of a deceased parent, children are often greatly influenced in their conduct by considering what the deceased parent would have desired or advised: his imagined injunction practically becomes a supplementary law.

Here, however, we are chiefly concerned with that more developed form of such guidance which results where the spirits of distinguished men, regarded with special fear and trust, become deities. Ancient Egyptian hieroglyphics reveal two stages of it. The "Instructions" recorded by King Rash'otephet are given by his father in a dream. "Son of the Sun Amenemhat—deceased :—He says in a dream—unto his son the Lord intact,—he says rising up like a god :—' Listen to what I speak unto thee.'" And then another tablet narrates how Thothmes IV, travelling when a prince, and taking his siesta in the shade of the Sphinx, was spoken to in a dream by that god, who said—"Look at me! . . . Answer me that you will do me what is in my heart" &c.; and when he ascended the throne, Thothmes fulfilled the injunction. Analogous stages were well exemplified among the ancient Peruvians. There is a tradition that Huayna Ccapac, wishing to marry his second sister, applied for assent to the dead body of his father; "but the dead body gave no answer, while fearful signs appeared in the heavens, portending blood." Moreover, as before pointed out in § 477, "the Ynca gave them (the vassals) to understand that all he did with regard to them was by an order and revelation of his father, the Sun." Turning to extant races, we see that in the Polynesian Islands, where the genesis of a pantheon by ancestor worship is variously exemplified, divine direction is habitually sought through priests. Among the Tahitians, one "mode by which the god intimated his will," was to enter the priest, who then "spoke as entirely under supernatural influence." Mariner tells us that in Tonga, too, when the natives wished to consult the gods, there was a ceremony of invocation; and the inspired priest then uttered the divine command. Similar beliefs and usages are described by Turner as existing in Samoa. Passing to another region, we find among the Todas of the Indian hills, an appeal for supernatural guidance in judicial matters.

" When any dispute arises respecting their wives or their buffaloes, it has to be decided by the priest, who affects to become possessed by the Bell-god, and . . . pronounces the deity's decision upon the point in dispute."

These instances serve to introduce and interpret for us those which the records of historic peoples yield. Taking first the Hebrews, we have the familiar fact that the laws for general guidance were supposed to be divinely communicated; and we have the further fact that special directions were often sought. Through the priest who accompanied the army, the commander "inquired of the Lord" about any military movement of importance, and sometimes received very definite orders; as when, before a battle with the Philistines, David is told to "fetch a compass behind them, and come upon them over against the mulberry trees." Sundry Aryan peoples furnish evidence. In common with other Indian codes, the code of Manu, " according to Hindoo mythology, is an emanation from the supreme God." So, too, was it with the Greeks. Not forgetting the tradition that by an ancient Cretan king, a body of laws was brought down from the mountain where Jupiter was said to be buried, we may pass to the genesis of laws from special divine commands, as implied in the Homeric poems. Speaking of these Grote says :—

"The appropriate Greek word for human laws never occurs : amidst a very wavering phraseology, we can detect a gradual transition from the primitive idea of a personal goddess, Themis, attached to Zeus, first to his sentences or orders called Themistes, and next by a still farther remove to various established customs which those sentences were believed to sanctify—the authority of religion and that of custom coalescing into one indivisible obligation."

Congruous in nature was the belief that " Lycurgus obtained not only his own consecration to the office of legislator, but his laws themselves from the mouth of the Delphic God." To which add that we have throughout later Greek times, the obtainment of special information and direction through oracles. Evidence that among the Romans there had occurred a kindred process, is supplied by the story that the ancient laws were received by Numa from the goddess Egeria ; and that Numa appointed augurs by whose interpretation of signs the will of the gods was to be ascertained. Even in the 9th century, under the Carolingians, there were brought before the nobles "articles of law named *capitula*, which the king himself had drawn up by the inspiration of God."

Without following out the influence of like beliefs in later times, as seen in trial by ordeal and trial by judicial combat, in both of which God was supposed indirectly to give judgment, the above evidence makes it amply manifest that, in addition to those injunctions definitely expressed, or embodied in usages tacitly accepted from seniors and through them from remote ancestors, there are further injunctions more consciously attributed to supernatural beings —either the ghosts of parents and chiefs who were personally known, or the ghosts of more ancient traditionally-known chiefs which have been magnified into gods. Whence it follows that originally, under both of its forms, law embodies the dictates of the dead to the living.

§ 531. And here we are at once shown how it happens that throughout early stages of social evolution, no distinction is made between sacred law and secular law. Obedience to established injunctions of whatever kind, originating in reverence for supposed supernatural beings of one or other order, it results that at first all these injunctions have the same species of authority.

The Egyptian wall-sculptures, inscriptions, and papyri, everywhere expressing subordination of the present to the past, show us the universality of the religious sanction for rules of conduct. Of the Assyrians Layard says:—

"The intimate connection between the public and private life of the Assyrians and their religion, is abundantly proved by the sculptures. ... As among most ancient Eastern nations, not only all public and social duties, but even the commonest forms and customs, appear to have been more or less influenced by religion. ... All his [the king's] acts, whether in war or peace, appear to have been connected with the national religion, and were believed to be under the special protection and superintendence of the deity."

That among the Hebrews there existed a like connexion, is conspicuously shown us in the Pentateuch; where, besides the commandments specially so-called, and besides religious ordinances regulating feasts and sacrifices, the doings of the priests, the purification by scapegoat, &c., there are numerous directions for daily conduct—directions concerning kinds of food and modes of cooking; directions for proper farming in respect of periodic fallows, not sowing mingled grain, &c.; directions for the management of those in bondage, male and female, and the payment of hired labourers; directions about trade-transactions and the sales of lands and houses; along with sumptuary laws extending to the quality and fringes of garments and the shaping of beards: instances sufficiently showing that the rules of living, down even to small details, had a divine origin equally with the supreme laws of conduct. The like was true of the Ayrans in early stages. The code of Manu was a kindred mixture of sacred and secular regulations—of moral dictates and rules for carrying on ordinary affairs. Says Tiele of the Greeks after the Doric migration:—" No new political institutions, no fresh culture, no additional games, were established without the sanction of the Pythian oracle." And again we read—

" Chez les Grecs et chez les Romains, comme chez les Hindous, la loi fut d'abord une partie de la religion. Les anciens codes des cités étaient un ensemble de rites de prescriptions liturgiques de prières, en même temps que de dispositions législatives. Les règles du droit de propriété et du droit de succession y étaient éparses au milieu des règles des sacrifices, de la sépulture et du culte des morts."

Originating in this manner, law acquires stability. Possessing a supposed supernatural sanction, its rules have a rigidity enabling them to restrain men's actions in greater degrees than could any rules having an origin recognized as natural. They tend thus to produce settled social arrangements; both directly, by their high authority, and indirectly by limiting the actions of the living ruler. As was pointed out in § 468, early governing agents, not daring to transgress inherited usages and regulations, are practically limited to interpreting and enforcing them: their legislative power being exercised only in respect of matters not already prescribed for. Thus of the ancient Egyptians we read:—" It was not on his [the king's] own will that his occupations depended, but on those rules of duty and propriety which the wisdom of his ancestors had framed, with a just regard for the welfare of the king and of his people." And how persistent is this authority of the sanctified past over the not-yet-sanctified present, we see among ourselves, in the fact that every legislator has to bind himself by oath to maintain certain political arrangements which our ancestors thought good for us.

While the unchangeableness of law, due to its supposed sacred origin, greatly conduces to social order during those early stages in which strong restraints are most needed, there of course results an unadaptiveness which impedes progress when there arise new conditions to be met. Hence come into use those "legal fictions," by the aid of which nominal obedience is reconciled with actual disobedience. Alike in Roman law and in English law, as pointed out by Sir Henry Maine, legal fictions have been the means of modifying statutes which were transmitted as immutable; and so fitting them to new requirements: thus uniting stability with that plasticity which allows of gradual transformation.

§ 532. Such being the origin and nature of laws, it becomes manifest that the cardinal injunction must be obedience. Conformity to each particular direction pre-supposes allegiance to the authority giving it; and therefore the imperativeness of subordination to this authority is primary.

That direct acts of insubordination, shown in treason and rebellion, stand first in degree of criminality, evidently follows. This truth is seen at the present time in South Africa. "According to a horrible law of the Zulu despots, when a chief is put to death they exterminate also his subjects." It was illustrated by the ancient Peruvians, among whom "a rebellious city or province was laid waste, and its inhabitants exterminated;" and again by the ancient Mexicans, by whom one guilty of treachery to the king "was put to death, with all his relations to the fourth degree." A like extension of punishment occurred in past times in Japan, where, when "the offence is committed against the state, punishment is inflicted upon the whole race of the offender." Of efforts thus wholly to extinguish families guilty of disloyalty, the Merovingians yielded an instance: king Guntchram swore that the children of a certain rebel should be destroyed up to the ninth generation. And these examples naturally recall those furnished by Hebrew traditions. When Abraham, treating Jahveh as a terrestrial superior (just as existing Bedouins regard as god the most powerful living ruler known to them) entered into a covenant under which, for territory given, he, Abraham, became a vassal, circumcision was the prescribed badge of subordination; and the sole capital offence named was neglect of circumcision, implying insubordination: Jahveh elsewhere announcing himself as "a jealous god," and threatening punishment "upon the children unto the third and fourth generation of them that hate me." And the truth thus variously illustrated, that during stages in which maintenance of authority is most imperative, direct disloyalty is considered the blackest of crimes, we trace down through later stages in such facts as that, in feudal days, so long as the fealty of a vassal was duly manifested, crimes, often grave and numerous, were overlooked.

Less extreme in its flagitiousness than the direct disobedience implied by treason and rebellion, is, of course, the indirect disobedience implied by breach of commands. This, however, where strong rule has been established, is regarded as a serious offence, quite apart from, and much exceeding, that which the forbidden act intrinsically involves. Its greater gravity was distinctly enunciated by the Peruvians, among whom, says Garcilasso, " the most common punishment was death, for they said that a culprit was not punished for the delinquencies he had committed, but for having broken the commandment of the Ynca, who was respected as God." The like conception meets us in another country where the absolute ruler is regarded as divine. Sir R. Alcock quotes Thunberg to the effect that in Japan, " most crimes are punished with death, a sentence which is inflicted with less regard to the magnitude of the crime than to the audacity of the attempt to transgress the hallowed laws of the empire." And then, beyond the criminality which disobeying the ruler involves, there is the criminality involved by damaging the ruler's property, where his subjects and their services belong wholly or partly to him. In the same way that maltreating a slave, and thereby making him less valuable, comes to be considered as an aggression on his owner—in the same way that even now among ourselves a father's ground for proceeding against a seducer is loss of his daughter's services; so, where the relation of people to monarch is servile, there arises the view that injury done by one person to another, is injury done to the monarch's property. An extreme form of this view is alleged of Japan, where cutting and maiming of the king's dependents "becomes wounding the king, or regicide." And hence the general principle, traceable in European jurisprudence from early days, that a transgression of man against man is punishable mainly, or in large measure, as a transgression against the State. It was thus in ancient Rome: "every one convicted of having broken the public peace, expiated his offence with his life." An early embodiment of the principle occurs in the Salic law, under which "to the *wehrgeld* is added, in a great number of cases, ... the *fred*, a sum paid to the king or magistrate, in reparation for the violation of public peace;" and in later days, the fine paid to the State absorbed the wehrgeld. Our own history similarly shows us that, as authority extends and strengthens, the guilt of disregarding it takes precedence of intrinsic guilt. "'The king's peace' was a privilege which attached to the sovereign's court and castle, but which he could confer on other places and persons, and which at once raised greatly the penalty of misdeeds committed in regard to them." Along with the growing check on the right of private revenge for wrongs—along with the increasing subordination of minor and local jurisdictions—along with that strengthening of a central authority which these changes imply, "offences against the law become offences against the king, and the crime of disobedience a crime of contempt to be expiated by a special sort of fine." And we may easily see how, where a ruler gains absolute power, and especially where he has the *prestige* of divine origin, the guilt of contempt comes to exceed the intrinsic guilt of the forbidden act.

A significant truth may be added. On remembering that Peru, and Japan till lately, above named as countries in which the crime of disobedience to the ruler was considered so great as practically to equalize the flagitiousness of all forbidden acts, had societies in which militant organization, carried to its extreme, assimilated the social government at large to the government of an army; we are reminded that even in societies like our own, there is maintained in the army the doctrine that insubordination is the cardinal offence. Disobedience to orders is penal irrespective of the nature of the orders or the motive for the disobedience; and an act which, considered in itself, is quite innocent, may be visited with death if done in opposition to commands.

While, then, in that enforced conformity to inherited customs which plays the part of law in the earliest stages, we see insisted upon the duty of obedience to ancestors at large, irrespective of the injunctions to be obeyed, which are often trivial or absurd—while in the enforced conformity to special directions given in oracular utterances by priests, or in " themistes," &c., which form a supplementary source of law, we see insisted upon the duty of obedience, in small things as in great, to certain recognized spirits of the dead, or deities derived from them; we also see that obedience to the edicts of the terrestrial ruler, whatever they may be, becomes, as his power grows, a primary duty.

§ 533. What has been said in the foregoing sections brings out with clearness the truth that rules for the regulation of conduct have four sources. Even in early stages we see that beyond the inherited usages which have a quasi-religious sanction; and beyond the special injunctions of deceased leaders, which have a more distinct religious sanction; there is some, though a slight, amount of regulation derived from the will of the predominant man; and there is also the effect, vague but influential, of the aggregate opinion. Not dwelling on the first of these, which is slowly modified by accretions derived from the others, it is observable that in the second we have the germ of the law afterwards distinguished as divine; that in the third we have the germ of the law which gets its sanction from allegiance to the living governor; and that in the fourth we have the germ of the law which eventually becomes recognized as expressing the public will.

Already I have sufficiently illustrated those kinds of laws which originate personally, as commands of a feared invisible ruler and a feared visible ruler. But before going further, it will be well to indicate more distinctly the kind of law which originates impersonally, from the prevailing sentiments and ideas, and which we find clearly shown in rude stages before the other two have become dominant. A few extracts will exhibit it. Schoolcraft says of the Chippewayans—

"Thus, though they have no regular government, as every man is lord in his own family, they are influenced more or less by certain principles which conduce to their general benefit."

Of the unorganized Shoshones Bancroft writes—

"Every man does as he likes. Private revenge, of course, occasionally overtakes the murderer, or, if the sympathies of the tribe be with the murdered man, he may possibly be publicly executed, but there are no fixed laws for such cases."

In like manner the same writer tells us of the Haidahs that—

"Crimes have no punishment by law; murder is settled for with relatives of the victim, by death or by the payment of a large sum; and sometimes general or notorious offenders, especially medicine-men, are put to death by an agreement among leading men."

Even where government is considerably developed, public opinion continues to be an independent source of law. Ellis says that—

"In cases of theft in the Sandwich Islands, those who had been robbed retaliated upon the guilty party, by seizing whatever they could find;

and this mode of obtaining redress was so supported by public opinion, that the latter, though it might be the stronger party, dare not offer resistance."

By which facts we are reminded that where central authority and administrative machinery are feeble, the laws thus informally established by aggregate feeling are enforced by making revenge for wrongs a socially-imposed duty; while failure to revenge is made a disgrace, and a consequent danger. In ancient Scandinavia, "a man's relations and friends who had not revenged his death, would instantly have lost that reputation which constituted their principal security." So that, obscured as this source of law becomes when the popular element in the triune political structure is entirely subordinated, yet it was originally conspicuous, and never ceases to exist. And now having noted the presence of this, along with the other mingled sources of law, let us observe how the several sources, along with their derived laws, gradually become distinguished.

Recalling the proofs above given that where there has been established a definite political authority, inherited from apotheosized chiefs and made strong by divine sanction, laws of all kinds have a religious character; we have first to note that a differentiation takes place between those regarded as sacred and those recognized as secular. An illustration of this advance is furnished us by the Greeks. Describing the state of things exhibited in the Homeric poems, Grote remarks that "there is no sense of obligation then existing, between man and man as such—and very little between each man and the entire community of which he is a member;" while, at the same time, "the tie which binds a man to his father, his kinsman, his guest, or any special promisee towards whom he has taken the engagement of an oath, is conceived in conjunction with the idea of Zeus, as witness and guarantee:" allegiance to a divinity is the source of obligation. But in historical Athens, "the great impersonal authority called 'The Laws' stood out separately, both as guide and sanction, distinct from religious duty or private sympathies." And at the same time there arose the distinction between breach of the sacred law and breach of the secular law: "the murderer came to be considered, first as having sinned against the gods, next as having deeply injured the society, and thus at once as requiring absolution and deserving punishment." A kindred differentiation early occurred in Rome. Though, during the primitive period, the head of the State, at once king and high priest, and in his latter capacity dressed as a god, was thus the mouth-piece of both sacred law and secular law; yet, afterwards, with the separation of the ecclesiastical and political authorities, came a distinction between breaches of divine ordinances and breaches of human ordinances. In the words of Sir Henry Maine, there were "laws punishing *sins*. There were also laws punishing *torts*. The conception of offence against God produced the first class of ordinances; the conception of offence against one's neighbour produced the second; but the idea of offence against the State or aggregate community did not at first produce a true criminal jurisprudence." In explanation of the last statement it should, however, be added that since, during the regal period, according to Mommsen, "judicial procedure took the form of a public or a private process, according as the king interposed of his own motion, or only when appealed to by the injured party;" and since "the former course was taken only in cases which involved a breach of the public peace;" it must be inferred that when kingship ceased, there survived the distinction between transgression against the individual and transgression against the State, though the mode of dealing with this last had not, for a time, a definite form. Again, even among the Hebrews, more persistently theocratic as their social system was, we see a considerable amount of this change, at the same time that we are shown one of its causes. The Mishna contains many detailed civil laws; and these manifestly resulted from the growing complication of affairs. The instance is one showing us that primitive sacred commands, originating as they do in a comparatively undeveloped state of society, fail to cover the cases which arise as institutions become involved. In respect of these there consequently grow up rules having a known human authority only. By accumulation of such rules, is produced a body of human laws distinct from the divine laws; and the offence of disobeying the one becomes unlike the offence of disobeying the other. Though in Christianized Europe, throughout which the indigenous religions were superseded by an introduced religion, the differentiating process was interfered with; yet, on setting out from the stage at which this introduced religion had acquired that supreme authority proper to indigenous religions, we see that the subsequent changes were of like nature with those above described. Along with that mingling of structures shown in the ecclesiasticism of kings and the secularity of prelates, there went a mingling of political and religious legislation. Gaining supreme power, the Church interpreted sundry civil offences as offences against God; and even those which were left to be dealt with by the magistrate were considered as thus left by divine ordinance. But subsequent evolution brought about stages in which various transgressions, held to be committed against both sacred and secular law, were simultaneously expiated by religious penance and civil punishment; and there followed a separation which, leaving but a small remnant of ecclesiastical offences, brought the rest into the category of offences against the State and against individuals.

And this brings us to the differentiation of equal, if not greater, significance, between those laws which derive their obligation from the will of the governing agency, and those laws which derive their obligation from the *consensus* of individual interests—between those laws which, having as their direct end the maintenance of authority, only indirectly thereby conduce to social welfare, and those which, directly and irrespective of authority, conduce to social welfare: of which last, law, in its modern form, is substantially an elaboration. Already I have pointed out that the kind of law initiated by the *consensus* of individual interests, precedes the kind of law initiated by political authority. Already I have said that though, as political authority develops, laws acquire the shape of commands, even to the extent that those original principles of social order tacitly recognized at the outset, come to be regarded as obligatory only because personally enacted, yet that the obligation derived from the *consensus* of individual interests survives, if obscured. And here it remains to show that as the power of the political head declines—as industrialism fosters an increasingly free population—as the third element in the triune political structure, long subordinated, grows again predominant; there again grows predominant this primitive source of law — the *consensus* of individual interests. We have further to note that in its re-developed form, as in its original form, the kind of law hence arising has a character radically distinguishing it from the kinds of law thus far considered. Both the divine laws and the human laws which originate from personal authority, have inequality as their common essential principle; while the laws which originate impersonally, in the *consensus* of individual interests, have equality as their essential principle. Evidence is furnished at the very outset. For what is this *lex talionis* which, in the rudest hordes of men, is not only recognized but enforced by general opinion? Obviously, as enjoining an equalization of injuries or losses, it tacitly assumes equality of claims among the individuals concerned. The principle of requiring "an eye for an eye and a tooth for a tooth," embodies the primitive idea of justice everywhere: the endeavour to effect an exact balance being sometimes quite curious. Thus we read in Arbousset and Daumas:—

"A Basuto whose son had been wounded on the head with a staff, came to entreat me to deliver up the offender,—' with the same staff and on the same spot where my son was beaten, will I give a blow on the head of the man who did it.'"

A kindred effort to equalize in this literal way, the offence and the expiation, occurs in Abyssinia; where, when the murderer is given over to his victim's family, "the nearest of kin puts him to death with the same kind of weapon as that with which he had slain their relative." As the last case shows, this primitive procedure, when it does not assume the form of inflicting injury for injury between individuals, assumes the form of inflicting injury for injury between families or tribes, by taking life for life. With the instances given in § 522 may be joined one from Sumatra.

"When in an affray [between families], there happen to be several persons killed on both sides, the business of justice is only to state the reciprocal losses, in the form of an account current, and order the balance to be discharged if the numbers be unequal."

And then, from this rude justice which insists on a balancing of losses between families or tribes, it results that so long as their mutual injuries are equalized, it matters not whether the blameable persons are or are not those who suffer; and hence the system of vicarious punishment—hence the fact that vengeance is wreaked on any member of the transgressing family or tribe. Moreover, ramifying in these various ways, the principle applies where not life but property is concerned. Schoolcraft tells us that among the Dakotas, "injury to property is sometimes privately revenged by destroying other property in place thereof;" and among the Araucanians, families pillage one another for the purpose of making their losses alike. The idea survives, though changed in form, when crimes come to be compounded for by gifts or payments. Very early we see arising the alternative between submitting to vengeance or making compensation. Kane says of certain North American races, that "horses or other Indian valuables" were accepted in compensation for murder. With the Dakotas "a present of white wampum," if accepted, condones the offence. Among the Araucanians, homicides "can screen themselves from punishment by a composition with the relations of the murdered." Recalling, as these few instances do, the kindred alternatives recognized throughout primitive Europe, they also make us aware of a significant difference. For with the rise of class-distinctions in primitive Europe, the rates of compensation, equal among members of each class, had ceased to be equal between members of different classes. Along with the growth of personally-derived law, there had been a departure from the impersonally-derived law as it originally existed.

But now the truth to be noted is that, with the relative weakening of kingly or aristocratic authority and relative strengthening of popular authority, there revives the partially-suppressed kind of law derived from the *consensus* of individual interests; and the kind of law thus originating tends continually to replace all other law. For the chief business of courts of justice at present, is to enforce, without respect of persons, the principle, recognized before governments arose, that all members of the community, however otherwise distinguished, shall be similarly dealt with when they aggress one upon another. Though the equalization of injuries by retaliation is no longer permitted; and though the Government, reserving to itself the punishment of transgressors, does little to enforce restitution or compensation; yet, in pursuance of the doctrine that all men are equal before the law, it has the same punishment for transgressors of every class. And then in respect of unfulfilled contracts or disputed debts, from the important ones tried at Assizes to the trivial ones settled in County Courts, its aim is to maintain the rights and obligations of citizens without regard for wealth or rank. Of course in our transition state the change is incomplete. But the sympathy with individual claims, and the *consensus* of individual interests accompanying it, lead to an increasing predominance of that kind of law which provides directly for social order; as distinguished from that kind of law which indirectly provides for social order by insisting on obedience to authority, divine or human. With decline of the *régime* of status and growth of the *régime* of contract, personally-derived law more and more gives place to impersonally-derived law; and this of necessity, since a formulated inequality is implied by the compulsory cooperation of the one, while, by the voluntary cooperation of the other, there is implied a formulated equality.

So that, having first differentiated from the laws of supposed divine origin, the laws of recognized human origin subsequently re-differentiate into those which ostensibly have the will of the ruling agency as their predominant sanction, and those which ostensibly have the aggregate of private interests as their predominant sanction; of which two the last tends, in the course of social evolution, more and more to absorb the first. Necessarily, however, while militancy continues, the absorption remains incomplete; since obedience to a ruling will continues to be in some cases necessary.

§ 534. A right understanding of this matter is so important, that I must be excused for briefly presenting two further aspects of the changes described: one concerning the accompanying sentiments, and the other concerning the accompanying theories.

As laws originate partly in the customs inherited from the undistinguished dead, partly in the special injunctions of the distinguished dead, partly in the average will of the undistinguished living, and partly in the will of the distinguished living, the feelings responding to them, allied though different, are mingled in proportions that vary under diverse circumstances.

According to the nature of the society, one or other sanction predominates; and the sentiment appropriate to it obscures the sentiments appropriate to the others, without, however, obliterating them. Thus in a theocratic society, the crime of murder is punished primarily as a sin against God; but not without there being some consciousness of its criminality as a disobedience to the human ruler who enforces the divine command, as well as an injury to a family, and, by implication, to the community. Where, as among the Bedouins or in

Sumatra, there is no such supernaturally-derived injunction, and no consequent reprobation of disobedience to it, the loss entailed on the family of the victim is the injury recognized; and, consequently, murder is not distinguished from manslaughter. Again, in Japan and in Peru, unqualified absoluteness of the living ruler is, or was, accompanied by the belief that the criminality of murder consisted primarily in transgression of his commands; though doubtless the establishment of such commands implied, both in ruler and people, some recognition of evil, individual or general, caused by breach of them. In ancient Rome, the consciousness of injury done to the community by murder was decided; and the feeling enlisted on behalf of public order was that which mainly enforced the punishment. And then among ourselves when a murder is committed, the listener to an account of it shudders not mainly because the alleged command of God has been broken, nor mainly because there has been a breach of "the Queen's peace;" but his strongest feeling of reprobation is that excited by the thought of a life taken away, with which is joined a secondary feeling due to the diminution of social safety which every such act implies. In these different emotions which give to these several sanctions their respective powers, we see the normal concomitants of the social states to which such sanctions are appropriate. More especially we see how that weakening of the sentiments offended by breaches of authority, divine or human, which accompanies growth of the sentiments offended by injuries to individuals and the community, is naturally joined with revival of that kind of law which originates in the *consensus* of individual interests—the law which was dominant before personal authority grew up, and which again becomes dominant as personal authority declines.

At the same time there goes on a parallel change of theory. Along with a rule predominantly theocratic, there is current a tacit or avowed doctrine, that the acts prescribed or forbidden are made right or wrong solely by divine command:

and though this doctrine survives through subsequent stages (as it does still in our own religious world), yet belief in it becomes nominal rather than real. Where there has been established an absolute human authority, embodied in a single individual, or, as occasionally, in a few, there comes the theory that law has no other source than the will of this authority: acts are conceived as proper or improper according as they do or do not conform to its dictates. With progress towards a popular form of government, this theory becomes modified to the extent that though the obligation to do this and refrain from that is held to arise from State-enactment; yet the authority which gives this enactment its force is the public desire. Still it is observable that along with a tacit implication that the *consensus* of individual interests affords the warrant for law, there goes the overt assertion that this warrant is derived from the formulated will of the majority: no question being raised whether this formulated will is or is not congruous with the *consensus* of individual interests. In this current theory there obviously survives the old idea that there is no other sanction for law than the command of embodied authority; though the authority is now a widely different one.

But this theory, much in favour with " philosophical politicians," is a transitional theory. The ultimate theory, which it foreshadows, is that the source of legal obligation is the *consensus* of individual interests itself, and not the will of a majority determined by their opinion concerning it; which may or may not be right. Already, even in legal theory, especially as expounded by French jurists, natural law or law of nature, is recognized as a source of formulated law: the admission being thereby made that, primarily certain individual claims, and secondarily the social welfare furthered by enforcing such claims, furnish a warrant for law, anteceding political authority and its enactments. Already in the qualification of Common Law by Equity, which avowedly proceeds upon the law of " *honesty* and *reason* and of *nations*," there is involved the pre-supposition that, as similarly-constituted beings, men have certain rights in common, maintenance of which, while directly advantageous to them individually, indirectly benefits the community; and that thus the decisions of equity have a sanction independent alike of customary law and parliamentary votes. Already in respect of religious opinions there is practically conceded the right of the individual to disobey the law, even though it expresses the will of a majority. Whatever disapproval there may be of him as a law-breaker, is over-ridden by sympathy with his assertion of freedom of judgment. There is a tacit recognition of a warrant higher than that of State-enactments, whether regal or popular in origin. These ideas and feelings are all significant of progress towards the view, proper to the developed industrial state, that the justification for a law is that it enforces one or other of the conditions to harmonious social cooperation; and that it is unjustified (enacted by no matter how high an authority or how general an opinion) if it traverses these conditions.

And this is tantamount to saying that the impersonally-derived law which revives as personally-derived law declines, and which gives expression to the *consensus* of individual interests, becomes, in its final form, simply an applied system of ethics—or rather, of that part of ethics which concerns men's just relations with one another and with the community.

§ 535. Returning from this somewhat parenthetical discussion, we might here enter on the development of laws, not generally but specially; exhibiting them as accumulating in mass, as dividing and sub-dividing in their kinds, as becoming increasingly definite, as growing into coherent and complex systems, as undergoing adaptations to new conditions. But besides occupying too much space, such an exposition would fall outside the lines of our subject. Present requirements are satisfied by the results above set forth, which may be summarized as follows.

Setting out with the truth, illustrated even in the very rudest tribes, that the ideas conveyed, sentiments inculcated, and usages taught, to children by parents who themselves were similarly taught, eventuate in a rigid set of customs; we recognize the fact that at first, as to the last, law is mainly an embodiment of ancestral injunctions.

To the injunctions of the undistinguished dead, which, qualified by the public opinion of the living in cases not prescribed for, constitute the code of conduct before any political organization has arisen, there come to be added the injunctions of the distinguished dead, when there have arisen chiefs who, in some measure feared and obeyed during life, after death give origin to ghosts still more feared and obeyed. And when, during that compounding of societies effected by war, such chiefs develop into kings, their remembered commands and the commands supposed to be given by their ghosts, become a sacred code of conduct, partly embodying and partly adding to the code pre-established by custom. The living ruler, able to legislate only in respect of matters unprovided for, is bound by these transmitted commands of the unknown and the known who have passed away; save only in cases where the living ruler is himself regarded as divine, in which cases his injunctions become laws having a like sacredness. Hence the trait common to societies in early stages, that the prescribed rules of conduct of whatever kind have a religious sanction. Sacrificial observances, public duties, moral injunctions, social ceremonies, habits of life, industrial regulations, and even modes of dressing, stand on the same footing.

Maintenance of the unchangeable rules of conduct thus originating, which is requisite for social stability during those stages in which the type of nature is yet but little fitted for harmonious social cooperation, pre-supposes implicit obedience; and hence disobedience becomes the blackest crime. Treason and rebellion, whether against the divine or the human ruler bring penalties exceeding all others in severity. The breaking of a law is punished not because of the intrinsic criminality of the act committed, but because of the implied insubordination. And the disregard of governmental authority continues, through subsequent stages, to constitute, in legal theory, the primary element in a transgression.

In societies that become large and complex, there arise forms of activity and intercourse not provided for in the sacred code; and in respect of these the ruler is free to make regulations. As such regulations accumulate there comes into existence a body of laws of known human origin; and though this acquires an authority due to reverence for the men who made it and the generations which approved it, yet it has not the sacredness of the god-descended body of laws: human law differentiates from divine law. But in societies which remain predominantly militant, these two bodies of laws continue similar in the respect that they have a personally-derived authority. The avowed reason for obeying them is that they express the will of a divine ruler, or the will of a human ruler, or, occasionally, the will of an irresponsible oligarchy.

But with the progress of industrialism and growth of a free population which gradually acquires political power, the humanly-derived law begins to sub-divide; and that part which originates in the *consensus* of individual interests, begins to dominate over the part which originates in the authority of the ruler. So long as the social type is one organized on the principle of compulsory cooperation, law, having to maintain this compulsory cooperation, must be primarily concerned in regulating *status*, maintaining inequality, enforcing authority; and can but secondarily consider the individual interests of those forming the mass. But in proportion as the principle of voluntary cooperation more and more characterizes the social type, fulfilment of contracts and implied assertion of equality in men's rights, become the fundamental requirements, and the *consensus* of individual interests the chief source of law: such authority as law otherwise derived coutinues to have, being recognized as secondary, and insisted upon only because maintenance of law for its own sake indirectly furthers the general welfare.

Finally, we see that the systems of laws belonging to these successive stages, are severally accompanied by the sentiments and theories appropriate to them; and that the theories at present current, adapted to the existing compromise between militancy and industrialism, are steps towards the ultimate theory, in conformity with which law will have no other justification than that gained by it as maintainer of the conditions to complete life in the associated state.

CHAPTER XV.

PROPERTY.

§ 536. The fact referred to in § 292, that even intelligent animals display a sense of proprietorship, negatives the belief propounded by some, that individual property was not recognized by primitive men. When we see the claim to exclusive possession understood by a dog, so that he fights in defence of his master's clothes if left in charge of them, it becomes impossible to suppose that even in their lowest state men were devoid of those ideas and emotions which initiate private ownership. All that may be fairly assumed is that these ideas and sentiments were at first less developed than they have since become.

It is true that in some extremely rude hordes, rights of property are but little respected. Lichtenstein tells us that among the Bushmen, "the weaker, if he would preserve his own life, is obliged to resign to the stronger, his weapons, his wife, and even his children;" and there are some degraded North American tribes in which there is no check on the more powerful who choose to take from the less powerful: their acts are held to be legitimized by success. But absence of the idea of property, and the accompanying sentiment, is no more implied by these forcible appropriations than it is implied by the forcible appropriation which a bigger schoolboy makes of the toy belonging to a less. It is also true that even where force is not used, individual claims are in considerable degrees over-ridden or imperfectly maintained. We read of the Chippewayans that "Indian law requires the successful hunter to share the spoils of the chase with all present;" and Hillhouse says of the Arawaks that though individual property is "distinctly marked amongst them," "yet they are perpetually borrowing and lending, without the least care about payment." But such instances merely imply that private ownership is at first ill-defined, as we might expect, *à priori*, that it would be.

Evidently the thoughts and feelings which accompany the act of taking possession, as when an animal clutches its prey, and which at a higher stage of intelligence go along with the grasping of any article indirectly conducing to gratification, are the thoughts and feelings to which the theory of property does but give a precise shape. Evidently the use in legal documents of such expressions as "to have and to hold," and to be "seized" of a thing, as well as the survival up to comparatively late times of ceremonies in which a portion (rock or soil) of an estate bought, representing the whole, actually passed from hand to hand, point back to this primitive physical basis of ownership. Evidently the developed doctrine of property, accompanying a social state in which men's acts have to be mutually restrained, is a doctrine which on the one hand asserts the freedom to take and to keep within specified limits, and denies it beyond those limits—gives positiveness to the claim while restricting it. And evidently the increasing definiteness thus given to rights of individual possession, may be expected to show itself first where definition is relatively easy and afterwards where it is less easy. This we shall find that it does.

§ 537. While in early stages it is difficult, not to say impossible, to establish and mark off individual claims to parts of the area wandered over in search of food, it is not difficult to mark off the claims to movable things and to habitations; and these claims we find habitually recognized. The following passage from Bancroft concerning certain North American savages, well illustrates the distinction :—

"Captain Cook found among the Ahts very 'strict notions of their having a right to the exclusive property of everything that their country produces,' so that they claimed pay for even wood, water, and grass. The limits of tribal property are very clearly defined, but individuals rarely claim any property in land. Houses belong to the men who combine to build them. Private wealth consists of boats and implements for obtaining food, domestic utensils, slaves, and blankets."

A like condition is shown us by the Comanches :—

"They recognize no distinct rights of *meum* and *tuum*, except to personal property; holding the territory they occupy, and the game that depastures upon it, as common to all the tribe: the latter is appropriated only by capture."

And the fact that among these Comanches, as among other peoples, "prisoners of war belong to the captors, and may be sold or released at their will," further shows that the right of property is asserted where it is easily defined. Of the Brazilian Indians, again, Von Martius tells us that,—

"Huts and utensils are considered as private property; but even with regard to them certain ideas of common possession prevail. The same hut is often occupied by more families than one; and many utensils are the joint property of all the occupants. Scarcely anything is considered strictly as the property of an individual except his arms, accoutrements, pipe, and hammock."

Dr. Rink's account of the Esquimaux shows that among them, too, while there is joint ownership of houses made jointly by the families inhabiting them, there is separate ownership of weapons, fishing boats, tools, etc. Thus it is made manifest that private right, completely recognized where recognition of it is easy, is partially recognized where partial recognition only is possible—where the private rights of companions are entangled with it. Instances of other kinds equally prove that among savages claims to possession are habitually marked off when practicable: if not fully, yet partially. Of the Chippewayans "who have no regular government" to make laws or arbitrate, we yet read that,—

"In the former instance [when game is taken in inclosures by a hunting party], the game is divided among those who have been engaged in the pursuit of it. In the latter [when taken in private traps] it is considered as private property; nevertheless, any unsuccessful hunter passing by, may take a deer so caught, leaving the head, skin, and saddle, for the owner."

In cases, still more unlike, but similar in the respect that there exists an obvious connexion between labour expended and benefit achieved, rude peoples re-illustrate this same individualization of property. Burckhardt tells us of the Bedouins that wells "are exclusive property, either of a whole tribe, or of individuals whose ancestors dug the wells."

Taken together such facts make it indisputable that in early stages, private appropriation, carried to a considerable extent, is not carried further because circumstances render extension of it impracticable.

§ 538. Recognition of this truth at once opens the way to explanation of primitive land-ownership; and elucidates the genesis of those communal and family tenures which have prevailed so widely.

While subsistence on wild food continues, the wandering horde inhabiting a given area, must continue to make joint use of the area; both because no claim can be shown by any member to any portion, and because the marking out of small divisions, if sharing were agreed upon, would be impracticable. Where pastoral life has arisen, ability to drive herds hither and thither within the occupied region is necessary. In the absence of cultivation, cattle and their owners could not survive were each owner restricted to one spot: there is nothing feasible but united possession of a wide tract. And when there comes a transition to the agricultural stage, either directly from the hunting stage or indirectly through the pastoral stage, several causes conspire to prevent, or to check, the growth of private land-ownership.

There is first the traditional usage. Joint ownership continues after circumstances no longer render it imperative, because departure from the sacred example of forefathers is resisted. Sometimes the resistance is insuperable; as with the Rechabites and the people of Petra, who by their vow "were not allowed to possess either vineyards or cornfields or houses" but were bound "to continue the nomadic life." And obviously, where the transition to a settled state is effected, the survival of habits and sentiments established during the nomadic state, must long prevent possession of land by individuals. Moreover, apart from opposing ideas and customs, there are physical difficulties in the way. Even did any member of a pastoral horde which had become partially settled, establish a claim to exclusive possession of one part of the occupied area, little advantage could be gained before there existed the means of keeping out the animals belonging to others. Common use of the greater part of the surface must long continue from mere inability to set up effectual divisions. Only small portions can at first be fenced off. Yet a further reason why land-owning by individuals, and land-owning by families, establish themselves very slowly, is that at first each particular plot has but a temporary value. The soil is soon exhausted; and in the absence of advanced arts of culture becomes useless. Such tribes as those of the Indian hills show us that primitive cultivators uniformly follow the practice of clearing a tract of ground, raising from it two or three crops, and then abandoning it: the implication being that whatever private claim had arisen, lapses, and the surface, again becoming wild, reverts to the community.

Thus throughout long stages of incipient civilization, the impediments in the way of private land-ownership are great and the incentives to it small. Besides the fact that primitive men, respecting the connexion between effort expended and benefit gained, and therefore respecting the right of property in things made by labour, recognize no claim thus established by an individual to a portion of land; and besides the fact that in the adhesion to inherited usage and the inability effectually to make bounds, there are both moral and physical obstacles to the establishment of any such individual monopoly; there is the fact that throughout early stages of settled life, no motive to maintain permanent private possession of land comes into play. Manifestly, therefore, it is not from conscious assertion of any theory, or in pursuance of any deliberate policy, that tribal and communal proprietorship of the areas occupied originate; but simply from the necessities of the case.

Hence the prevalence among unrelated peoples of this public ownership of land, here and there partially qualified by temporary private ownership. Some hunting tribes of North America show us a stage in which even the communal possession is still vague. Concerning the Dakotas Schoolcraft says—

"Each village has a certain district of country they hunt in, but do not object to families of other villages hunting with them. Among the Dacotas, I never knew an instance of blood being shed in any disputes or difficulties on the hunting grounds."

Similarly of the Comanches, he remarks that "no dispute ever arises between tribes with regard to their hunting grounds, the whole being held in common." Of the semi-settled and more advanced Iroquois, Morgan tells us that—

"No individual could obtain the absolute title to land, as that was vested by the laws of the Iroquois in all the people; but he could reduce unoccupied lands to cultivation to any extent he pleased; and so long as he continued to use them, his right to their enjoyment was protected and secured."

Sundry pastoral peoples of South Africa show us the survival of such arrangements under different conditions.

"The land which they [the Bechuanas] inhabit is the common property of the whole tribe, as a pasture for their herds."

"Being entirely a pastoral people, the Damaras have no notion of permanent habitations. The whole country is considered public property. . . . There is an understanding that he who arrives first at any given locality, is the master of it as long as he chooses to remain there."

Kaffir custom "does not recognize private property in the soil beyond that of actual possession."

"No one possesses landed property" [among the Koosas]; "he sows his corn wherever he can find a convenient spot."

And various of the uncivilized, who are mainly or wholly

agricultural, exhibit but slight modifications of this usage. Though by the New Zealanders some extra claim of the chief is recognized, yet " all free persons, male and female, constituting the nation, were proprietors of the soil:" there is a qualified proprietorship of land, obtained by cultivation, which does not destroy the proprietorship of the nation or tribe. In Sumatra, cultivation gives temporary ownership but nothing more. We read that the ground "on which a man plants or builds, with the consent of his neighbours, becomes a species of nominal property"; but when the trees which he has planted disappear in the course of nature, "the land reverts to the public." From a distant region may be cited an instance where the usages, though different in form, involve the same principle. Among the modern Indians of Mexico—

"Only a house-place and a garden are hereditary; the fields belong to the village, and are cultivated every year without anything being paid for rent. A portion of the land is cultivated in common, and the proceeds are devoted to the communal expenses."

This joint ownership of land, qualified by individual ownership only so far as circumstances and habits make it easy to mark off individual claims, leads to different modes of using the products of the soil, according as convenience dictates. Anderson tells us that in "Damara-land, the carcases of all animals—whether wild or domesticated—are considered public property." Among the Todas—

"Whilst the land is in each case the property of the village itself, ... the cattle which graze on it are the private property of individuals, being males. ... The milk of the entire herd is lodged in the pálthchi, village dairy, from which each person, male and female, receives for his or her daily consumption; the unconsumed balance being divided, as personal and saleable property, amongst the male members of all ages, in proportion to the number of cattle which each possesses in the herd."

And then in some cases joint cultivation leads to a kindred system of division.

"When harvest is over," the Congo people "put all the kidney-beans into one heap, the Indian wheat into another, and so of other grain: then giving the Macolonte [chief] enough for his maintenance, and laying aside what they design for sowing, the rest is divided at so much to every cottage, according to the number of people each contains. Then all the women together till and sow the land for a new harvest."

In Europe an allied arrangement is exhibited by the southern Slavs. "The fruits of agricultural labour are consumed in common, or divided equally among the married couples; but the produce of each man's industrial labour belongs to him individually." Further, some of the Swiss allmends show us a partial survival of this system; for besides lands which have become in large measure private, there are "communal vineyards cultivated in common," and "there are also cornlands cultivated in the same manner," and "the fruit of their joint labour forms the basis of the banquets, at which all the members of the commune take part."

Thus we see that communal ownership and family ownership at first arose and long continued because, in respect of land, no other could well be established. Records of the civilized show that with them in the far past, as at present with the uncivilized, private possession, beginning with movables, extends itself to immovables only under certain conditions. We have evidence of this in the fact named by Mayer, that "the Hebrew language has no expression for 'landed property;'" and again in the fact alleged by Mommsen of the Romans, that "the idea of property was primarily associated not with immovable estate, but with 'estate in slaves and cattle.'" And if, recalling the circumstances of pastoral life, as carried on alike by Semites and Ayrans, we remember that, as before shown, the patriarchal group is a result of it; we may understand how, in passing into the settled state, there would be produced such forms of land-tenure by the clan and the family as, with minor variations, characterized primitive European societies. It becomes comprehensible why among the Romans "in the earliest times, the arable land was cultivated in common, probably by the several clans; each of these tilled its own land, and thereafter distributed the produce among the several households belonging to it." We are shown that there naturally arose such arrangements as those of the ancient Teutonic mark—a territory held "by a primitive settlement of a family or kindred," each free male member of which had "a right to the enjoyment of the woods, the pastures, the meadow, and the arable land of the mark;" but whose right was "of the nature of usufruct or possession only," and whose allotted private division became each season common grazing land after the crop had been taken off, while his more permanent holding was limited to his homestead and its immediate surroundings. And we may perceive how the community's ownership might readily, as circumstances and sentiments determined, result here in an annual use of apportioned tracts, here in a periodic re-partitioning, and here in tenures of more permanent kinds,—still subject to the supreme right of the whole public.

§ 539. Induction and deduction uniting to show, as they do, that at first land is common property, there presents itself the question—How did possession of it become individualized? There can be little doubt as to the general nature of the answer. Force, in one form or other, is the sole cause adequate to make the members of a society yield up their joint claim to the area they inhabit. Such force may be that of an external aggressor or that of an internal aggressor; but in either case it implies militant activity.

The first evidence of this which meets us is that the primitive system of land-ownership has lingered longest where circumstances have been such as either to exclude war or to minimize it. Already I have referred to a still-extant Teutonic mark existing in Drenthe, "surrounded on all sides by marsh and bog," forming "a kind of island of sand and heath;" and this example, before named as showing the survival of free judicial institutions where free institutions at large survive, simultaneously shows the communal land-ownership which continues while men are unsubordinated. After this typical case may be named one not far distant, and somewhat akin—that, namely, which occurs "in the sandy district of the Campine and beyond the Meuse, in the Ardennes region," where there is great " want of communication:" the implied difficulty of access and the poverty of surface making relatively small the temptation to invade. So that while, says Laveleye, " except in the Ardennes, the lord had succeeded in usurping the eminent domain, without however destroying the inhabitants' rights of user," in the Ardennes itself, the primitive communal possession survived. Other cases show that the mountainous character of a locality, rendering subjugation by external or internal force impracticable, furthers maintenance of this primitive institution, as of other primitive institutions. In Switzerland, and especially in its Alpine parts, the allmends above mentioned, which are of the same essential nature as the Teutonic marks, have continued down to the present day. Sundry kindred regions present kindred facts. Ownership of land by family-communities is still to be found " in the hill-districts of Lombardy." In the poverty-stricken and mountainous portion of Auvergne, as also in the hilly and infertile department of Nièvre, there are still, or recently have been, these original joint-ownerships of land. And the general remark concerning the physical circumstances in which they occur, is that "it is to the wildest and most remote spots that we must go in search of them"—a truth again illustrated " in the small islands of Hœdic and Honat, situated not far from Belle Isle " on the French coast, and also in our own islands of Orkney and Shetland.

Contrariwise, we find that directly by invasion, and indirectly by the chronic resistance to invasion which generates those class-inequalities distinguishing the militant type, there is produced individualization of land-ownership, in one or other form. All the world over, conquest gives a possession that is unlimited because there is no power to dispute it. Along with other spoils of war, the land becomes a spoil; and, according to the nature of the conquering society, is owned wholly by the despotic conqueror, or, partially and in dependent ways, by his followers. Of the first result there are many instances. "The kings of Abyssinia are above all laws ... the land and persons of their subjects are equally their property." "In Kongo the king hath the sole property of goods and lands, which he can grant away at pleasure." And § 479 contains sundry other examples of militant societies in which the monarch, otherwise absolute, is absolute possessor of the soil. Of the second result instances were given in § 458; and I may here add some others. Ancient Mexico supplies one.

"Montezuma possessed in most of the villages . . . and especially in those he had conquered, fiefs which he distributed among those called 'the gallant fellows of Mexico.' These were men who had distinguished themselves in war."

Under a more primitive form the like was done in Iceland by the invading Norsemen.

"When a chieftain had taken possession of a district, he allotted to each of the freemen who accompanied him a certain portion of land, erected a temple (hof), and became, as he had been in Norway, the chief, the pontiff, and the judge of the herad."

But, as was shown when treating of political differentiation, it is not only by external aggressors that the joint possession by all freemen of the area they inhabit is over-ridden. It is over-ridden, also, by those internal aggressors whose power becomes great in proportion as the militancy of the society becomes chronic. With the personal subordination generated by warfare, there goes such subordination of ownership, that lands previously held absolutely by the community, come to be held subject to the claims of the local magnate; until, in course of time, the greater part of the occupied area falls into his exclusive possession, and only a small part continues to be common property.

To complete the statement it must be added that occasionally, though rarely, the passing of land into private hands takes place neither by forcible appropriation, nor by the gradual encroachment of a superior, but by general agreement. Where there exists that form of communal ownership under which joint cultivation is replaced by separate cultivation of parts portioned out—where there results from this a system of periodic redistribution, as of old in certain Greek states, as among the ancient Suevi, and as even down to our own times in some of the Swiss allmends; ownership of land by individuals may and does arise from cessation of the redistribution. Says M. de Laveleye concerning the Swiss allmends—"in the work of M. Rowalewsky, we see how the communal lands became private property by the periodic partitioning becoming more and more rare, and finally falling into desuetude." When not otherwise destroyed, land-owning by the commune tends naturally to end in this way. For besides the inconveniences attendant on re-localization of the members of the commune, positive losses must be entailed by it on many. Out of the whole number, the less skilful and less diligent will have reduced their plots to lower degrees of fertility; and the rest will have a motive for opposing a redistribution which, depriving them of the benefits of past labours, makes over these or parts of them to the relatively unworthy. Evidently this motive is likely, in course of time, to cause refusal to re-divide; and permanent private possession will result.

§ 540. An important factor not yet noticed has cooperated in individualizing property, both movable and fixed; namely, the establishment of measures of quantity and value. Only the rudest balancing of claims can be made before there come into use appliances for estimating amounts. At the outset, ownership exists only in respect of things actually made or obtained by the labour of the owner; and is therefore narrowly limited in range. But when exchange arises and spreads, first under the indefinite form of barter and then under the definite form of sale and purchase by means of a circulating medium, it becomes easy for ownership to extend itself to other things. Observe how clearly this extension depends on the implied progress of industrialism.

It was pointed out in § 319 that during the pastoral stage, it is impracticable to assign to each member of the family-community, or to each of its dependents, such part of the produce or other property as is proportionate to the value of his labour. Though in the case of Jacob and Laban the bargain made for services was one into which some idea of equivalence entered, yet it was an extremely rude idea; and by no such bargains could numerous transactions, or transactions of smaller kinds, be effected. On asking what must happen when the patriarchal group, becoming settled, assumes one or other enlarged form, we see that reverence for traditional usages, and the necessity of union for mutual defence, conspire to maintain the system of joint production and joint consumption: individualization of property is still hindered. Though under such conditions each person establishes private ownership in respect of things on which he has expended separate labour, or things received in exchange for such products of his separate labour; yet only a small amount of property thus distinguished as private, can be acquired. The greater part of his labour, mixed with that of others, brings returns inseparable from the returns of their labours; and the united returns must therefore be enjoyed in common. But as fast as it becomes safer to dispense with the protection of the family-group; and as fast as increasing commercial intercourse opens careers for those who leave their groups; and as fast as the use of money and measures gives definiteness to exchanges; there come opportunities for accumulating individual possessions, as distinguished from joint possessions. And since among those who labour together and live together, there will inevitably be some who feel restive under the imposed restraints, and also some (usually the same) who feel dissatisfied with the equal sharing among those whose labours are not of equal values; it is inferable that these opportunities will be seized: private ownership will spread at the expense of public ownership. Some illustrations may be given. Speaking of the family-communities of the

Southern Slavs, mostly in course of dissolution, M. de Laveleye says—

"The family-group was far more capable of defending itself against the severity of Turkish rule than were isolated individuals. Accordingly, it is in this part of the southern Slav district that family-communities are best preserved, and still form the basis of social order."

The influence of commercial activity as conducing to disintegration, is shown by the fact that these family-communities ordinarily hold together only in rural districts.

"In the neighbourhood of the towns the more varied life has weakened the ancient family-sentiment. Many communities have been dissolved, their property divided and sold, and their members have degenerated into mere tenants and proletarians."

And then the effect of a desire, alike for personal independence and for the exclusive enjoyment of benefits consequent on superiority, is recognized in the remark that these family-communities—

"cannot easily withstand the conditions of a society in which men are striving to improve their own lot, as well as the political and social organization under which they live. . . . Once the desire of self-aggrandisement awakened, man can no longer support the yoke of the *zadruga*. . . . To live according to his own will, to work for himself alone, to drink from his own cup, is now the end preeminently sought."

That this cause of disintegration is general, is implied by passages concerning similar communities still existing in the hill-districts of Lombardy—that is, away from the centres of mercantile activity. Growing averse to the control of the house-fathers, the members of these communities say—

"'Why should we and all our belongings remain in subjection to a master? It were far the best for each to work and think for himself. As the profits derived from any handicraft form a sort of private *peculium*, the associates are tempted to enlarge this at the expense of the common revenue." And then "the craving to live independently carries him away, and he quits the community."

All which evidence shows that the progress of industrialism is the general cause of this growing individualization of property; for such progress is pre-supposed alike by the greater security which makes it safe to live separately, by the increased opportunity for those sales which further the accumulation of a *peculium*, and by the use of measures of quantity and value: these being implied primarily by such sales, and secondarily by the sale and division of all that has been held in common.

Spread of private ownership, which thus goes along with decay of the system of status and growth of the system of contract, naturally passes on from movable property to fixed property. For when the multiplication of trading transactions has made it possible for each member of a family-community to accumulate a *peculium*; and when the strengthening desire for individual domestic life has impelled the majority of the community to sell the land which they have jointly inherited; the several portions of it, whether sold to separate members of the body or to strangers, are thus reduced by definite agreement to the form of individual properties; and private ownership of land thereby acquires a character apparently like that of other private ownership. In other ways, too, this result is furthered by developing industrialism. If, omitting as not relevant the cases in which the absolute ruler allows no rights of property, landed or other, to his subjects, we pass to the cases in which a conqueror recognizes a partial ownership of land by those to whom he has parcelled it out on condition of rendering services and paying dues, we see that the private land-ownership established by militancy is an incomplete one. It has various incompletenesses. The ownership by the suzerain is qualified by the rights he has made over to his vassals; the rights of the vassals are qualified by the conditions of their tenure; and they are further qualified by the claims of serfs and other dependents, who, while bound to specified services, have specified shares of produce. But with the decline of militancy and concomitant disappearance of vassalage, the obligations of the tenure diminish and finally almost lapse out of recognition; while, simultaneously, abolition of serfdom destroys or obscures the other claims which qualified private land-ownership.* As both changes are accompaniments of a developing industrialism, it follows that in these ways also, the individualization of property in land is furthered by it.

At first sight it seems fairly inferable that the absolute ownership of land by private persons, must be the ultimate state which industrialism brings about. But though industrialism has thus far tended to individualize possession of land, while individualizing all other possession, it may be doubted whether the final stage is at present reached. Ownership established by force does not stand on the same footing as ownership established by contract; and though multiplied sales and purchases, treating the two ownerships in the same way, have tacitly assimilated them, the assimilation may eventually be denied. The analogy furnished by assumed rights of possession over human beings, helps us to recognize this possibility. For while prisoners of war, taken by force and held as property in a vague way (being at first much on a footing with other members of a household), were reduced more definitely to the form of property when the buying and selling of slaves became general; and while it might, centuries ago, have been thence inferred that the ownership of man by man was an ownership in course of being permanently established; yet we see that a later stage of civilization, reversing this process, has destroyed ownership of man by man. Similarly, at a stage still more advanced it may be that private ownership of land will disappear. As that primitive freedom of the individual which existed before war established coercive institutions and personal slavery, comes to be re-established as militancy declines; so it seems possible that the primitive ownership of land by the community, which, with the development of coercive institutions, lapsed in large measure or wholly into private ownership, will

* In our own case the definite ending of these tenures took place in 1660; when, for feudal obligations (a burden on landowners) was substituted a beer-excise (a burden on the community).

be revived as industrialism further develops. The *régime* of contract, at present so far extended that the right of property in movables is recognized only as having arisen by exchange of services or products under agreements, or by gift from those who had acquired it under such agreements, may be further extended so far that the products of the soil will be recognized as property only by virtue of agreements between individuals as tenants and the community as landowner. Even now, among ourselves, private ownership of land is not absolute. In legal theory landowners are directly or indirectly tenants of the Crown (which in our day is equivalent to the State, or, in other words, the Community); and the Community from time to time resumes possession after making due compensation. Perhaps the right of the Community to the land, thus tacitly asserted, will in time to come be overtly asserted; and acted upon after making full allowance for the accumulated value artificially given.

§ 541. The rise and development of arrangements which fix and regulate private possession, thus admit of tolerably clear delineation.

The desire to appropriate, and to keep that which has been appropriated, lies deep, not in human nature only, but in animal nature: being, indeed, a condition to survival. The consciousness that conflict, and consequent injury, may probably result from the endeavour to take that which is held by another, ever tends to establish and strengthen the custom of leaving each in possession of whatever he has obtained by labour; and this custom takes among primitive men the shape of an overtly-admitted claim.

This claim to private ownership, fully recognized in respect of movables made by the possessor, and fully or partially recognized in respect of game killed on the territory over which members of the community wander, is not recognized in respect of this territory itself, or tracts of it. Property is individualized as far as circumstances allow individual claims to be marked off with some definiteness; but it is not individualized in respect of land, because, under the conditions, no individual claims can be shown, or could be effectually marked off were they shown.

With the passage from a nomadic to a settled state, ownership of land by the community becomes qualified by individual ownership; but only to the extent that those who clear and cultivate portions of the surface have undisturbed enjoyment of its produce. Habitually the public claim survives; and either when, after a few crops, the cleared tract is abandoned, or when, after transmission to descendants, it has ceased to be used by them, it reverts to the community. And this system of temporary ownership, congruous with the sentiments and usages inherited from ancestral nomads, is associated also with an undeveloped agriculture: land becoming exhausted after a few years.

Where the patriarchal form of organization has been carried from the pastoral state into the settled state, and, sanctified by tradition, is also maintained for purposes of mutual protection, possession of land partly by the clan and partly by the family, long continues; at the same time that there is separate possession of things produced by separate labour. And while in some cases the communal land-ownership, or family land-ownership, survives, it in other cases yields in various modes and degrees to qualified forms of private ownership, mostly temporary, and subject to supreme ownership by the public.

But war, both by producing class-differentiations within each society, and by effecting the subjugation of one society by another, undermines or destroys communal proprietorship of land; and partly or wholly substitutes for it, either the unqualified proprietorship of an absolute conqueror, or proprietorship by a conqueror qualified by the claims of vassals holding it under certain conditions, while their claims are in turn qualified by those of dependents attached to the soil. That is to say, the system of status which militancy develops, involves a graduated ownership of land as it does a graduated ownership of persons.

Complete individualization of ownership is an accompaniment of industrial progress. From the beginning, things identified as products of a man's own labour are recognized as his; and throughout the course of civilization, communal possession and joint household living, have not excluded the recognition of a *peculium* obtained by individual effort. Accumulation of movables privately possessed, arising in this way, increases as militancy is restrained by growing industrialism; because this pre-supposes greater facility for disposing of industrial products; because there come along with it measures of quantity and value, furthering exchange; and because the more pacific relations implied, render it safer for men to detach themselves from the groups in which they previously kept together for mutual protection. The individualization of ownership, extended and made more definite by trading transactions under contract, eventually affects the ownership of land. Bought and sold by measure and for money, land is assimilated in this respect to the personal property produced by labour; and thus becomes, in the general apprehension, confounded with it. But there is reason to suspect that while private possession of things produced by labour, will grow even more definite and sacred than at present; the inhabited area, which cannot be produced by labour, will eventually be distinguished as something which may not be privately possessed. As the individual, primitively owner of himself, partially or wholly loses ownership of himself during the militant *régime*, but gradually resumes it as the industrial *régime* develops; so, possibly, the communal proprietorship of land, partially or wholly merged in the ownership of dominant men during evolution of the militant type, will be resumed as the industrial type becomes fully evolved.

CHAPTER XVI.

REVENUE.

§ 542. Broadly dividing the products of men's labours into the part which remains with them for private purposes and the part taken from them for public purposes; and recognizing the truism that the revenue constituted by this last part must increase with the development of the public organization supported by it; we may be prepared for the fact that in early stages of social evolution, nothing answering to revenue exists.

The political head being at first distinguished from other members of the community merely by some personal superiority, his power, often recognized only during war, is, if recognized at other times, so slight as to bring him no material advantage. Habitually in rude tribes he provides for himself as a private man. Sometimes, indeed, instead of gaining by his distinction he loses by it. Among the Dakotas " the civil-chiefs and war-chiefs are distinguished from the rest by their poverty. They generally are poorer clad than any of the rest." A statement concerning the Abipones shows us why this occasionally happens.

"The cacique has nothing, either in his arms or his clothes, to distinguish him from a common man, except the peculiar oldness and shabbiness of them; for if he appears in the streets with new and handsome apparel, . . . the first person he meets will boldly cry, Give me that dress . . . and unless he immediately parts with it, he becomes the scoff and the scorn of all, and hears himself called covetous."

Among the Patagonians the burdens entailed by relieving and protecting inferiors, lead to abdication. Many "born

Caciques refuse to have any vassals; as they cost them dear, and yield but little profit."

Generally, however, and always where war increases his predominance, the leading warrior begins to be distinguished by wealth accruing to him in sundry ways. The superiority which gains him supremacy, implying as it mostly does greater skill and energy, conduces to accumulation: not uncommonly, as we have seen, (§ 472) the primitive chief is also the rich man. And this possession of much private property grows into a conspicuous attribute when, in the settled state, land held by the community begins to be appropriated by its more powerful members. Rulers habitually become large landowners. In ancient Egypt there were royal lands. Of the primitive Greek king we read that "an ample domain is assigned to him [? taken by him] as an appurtenance of his lofty position." And among other peoples in later times, we find the monarch owning great estates. The income hence derived, continues to the last to represent that revenue which the political head originally had, when he began to be marked off from the rest only by some personal merit.

Such larger amount of private means as thus usually distinguishes the head man at the outset, augments as successful war, increasing his predominance, brings him an increasing portion of the spoils of conquered peoples. In early stages it is the custom for each warrior to keep whatever he personally takes in battle; while that which is taken jointly is in some cases equally divided. But of course the chief is apt to get an extra share; either by actual capture, or by the willing award of his comrades, or, it may be, by forcible appropriation. And as his power grows, this forcible appropriation is yielded to, sometimes tacitly, sometimes under protest; as we are shown by the central incident in the *Iliad*. Through later stages his portion of plunder, reserved before division of the remainder among followers, continues to be a source of revenue. And where he becomes absolute, the property taken from the vanquished, lessened only by such portions as he gives in reward for services, augments his means of supporting his dependents and maintaining his supremacy.

To these sources of income which may be classed as incidental, is simultaneously added a source which is constant. When predominance of the chief has become so decided that he is feared, he begins to receive propitiatory presents; at first occasionally and afterwards periodically. Already in §§ 369-71, when treating of presents under their ceremonial aspects, I have given illustrations; and many more may be added. Describing the king among the Homeric Greeks, Grote writes—"Moreover he receives frequent presents, to avert his enmity, to conciliate his favour, or to buy off his exactions." So, too, of the primitive Germans, we are told by Tacitus that "it is the custom of the states to bestow by voluntary and individual contribution on the chiefs, a present of cattle or of grain, which, while accepted as a compliment, supplies their wants." And gifts to the ruler voluntarily made to obtain good will, or prevent ill will, continue to be a source of revenue until quite late stages. Among ourselves "during the reign of Elizabeth, the custom of presenting New Year's gifts to the sovereign was carried to an extravagant height;" and even "in the reign of James I. the money gifts seem to have been continued for some time."

Along with offerings of money and goods there go offerings of labour. Not unfrequently in primitive communities, it is the custom for all to join in building a new house or clearing a plot of ground for one of their number: such benefits being reciprocated. Of course the growing predominance of a political head, results in a more extensive yielding of gratuitous labour for his benefit, in these and other ways. The same motives which prompt gifts to the ruler prompt offers of help to him more than to other persons; and thus the custom of working for him grows into a usage. We read of the village chief among the Guaranis that "his subjects cultivated for him his plantation, and he enjoyed certain privileges on division of the spoils of the chase. Otherwise he possessed no marks of distinction." And the like practice was followed by some historic races during early stages. In ancient Rome it was "the privilege of the king to have his fields tilled by taskwork of the burgesses."

§ 543. Growth of the regular and definite out of the irregular and indefinite, variously exemplified in the foregoing chapters, is here again exemplified very clearly. For, as already said, it is from propitiatory presents and services, at first spontaneous and incidental, that there eventually come taxes specified in their amounts and times of payment.

It needs but to observe how such a custom as that of making wedding-presents has acquired a partially coercive character, to understand how, when once there begins the practice of seeking the good will of the headman by a gift, this practice is apt to be established. One having gained by it, another follows his example. The more generally the example is followed the greater becomes the disadvantage to those who do not follow it. Until at length all give because none dare stand conspicuous as exceptions. Of course if some repeat the presents upon such occasions as first prompted them, others have to do the like; and at length the periodic obligation becomes so peremptory, that the gift is demanded when it is not offered. In Loango, where presents are expected from all free subjects, "if the king thinks they do not give enough, he sends slaves to their places to take what they have." Among the Tongans, who from time to time give their king or chief "yams, mats, gnatoo, dried fish, live birds, &c.," the quantity is determined "generally by the will of each individual, who will always take care to send as much as he can well afford, lest the superior chief should be offended with him, and deprive him of all that he has." At the present time in Cashmere, at the spring festival, "it is the custom . . . for the Maharajah's servants to bring him a nazar, a present. . . . This has now become so regulated that every one is on these days [festivals] obliged to give from a 10th to a 12th of his monthly pay. . . . The name of each is read from a list, and the amount of his nazar is marked down: those that are absent will have the sum deducted from their pay." Traces of a like transition are seen in the fact that in ancient times crowns of gold, beginning as gifts made by dependent states to Eastern rulers, and by Roman provinces to generals or pro-consuls, became sums of money demanded as of right; and again in the fact that in our own early history, we read of "exactions called benevolences."

Similarly with the labour which, at first voluntarily given to the chief, comes, as his power grows, to be compulsory. Here are some illustrations showing stages in the transition.

A Kafir chief "summons the people to cultivate his gardens, reap his crops, and make his fences; but in this, as in other respects, he has to consult the popular will, and hence the manual labour required by the chiefs has always been of very limited duration."

In the Sandwich Islands, "when a chief wants a house, he requires the labour of all who hold lands under him. . . . Each division of the people has a part of the house allotted by the chief in proportion to its number."

In ancient Mexico "the personal and common service which furnished the water and wood required every day in the houses of the chiefs, was distributed from day to day among the villages and quarters."

It was the same in Yucatan: "the whole community did the sowing for the lord, looked after the seed, and harvested what was required for him and his house."

So in the adjacent regions of Guatemala and San Salvador, " the tribute was paid by means of the cultivation of estates." And in Madagascar " the whole population is liable to be employed on government work, without remuneration, and for any length of time."

Occurring among peoples unallied in blood and unlike in their stages of civilization, these facts show the natural growing up of a forced labour system such as that which existed during feudal times throughout Europe, when labour was exacted from dependents by local rulers, and became also a form of tribute to the central ruler; as instance the specified numbers of days' work which, before the Revolution, had to be given by French peasants to the State under the name of *corvée*.

After presents freely given have passed into presents expected and finally demanded, and volunteered help has passed into exacted service, the way is open for a further step. Change from the voluntary to the compulsory, accompanied as it necessarily is by specification of the amounts of commodities and work required, is apt to be followed eventually by substitution of money payments. During stages in which there has not arisen a circulating medium, the ruler, local or general, is paid his revenue in kind. In Fiji a chief's house is supplied with daily food by his dependents; and tribute is paid by the chiefs to the king "in yams, taro, pigs, fowls, native cloth, &c." In Tahiti, where besides supplies derived from "the hereditary districts of the reigning family," there were "requisitions made upon the people;" the food was generally brought cooked. In early European societies, too, the expected donations to the ruler continued to be made partly in goods, animals, clothes, and valuables of all kinds, long after money was in use. But the convenience both of giver and receiver prompts commutation, when the values of the presents looked for have become settled. And from kindred causes there also comes, as we have seen in a previous chapter, commutation of military services and commutation of labour services. No matter what its nature, that which was at first spontaneously offered, eventually becomes a definite sum taken, if need be, by force—a tax.

§ 544. At the same time his growing power enables the political head to enforce demands of many other kinds. European histories furnish ample proofs.

Besides more settled sources of revenue, there had, in the early feudal period, been established such others as are typically illustrated by a statement concerning the Dukes of Normandy in the 12th century. They profited by escheats (lands reverting to the monarch in default of posterity of the first baron); by guardianships and reliefs; by seizure of the property of deceased prelates, usurers, excommunicated persons, suicides, and certain criminals; and by treasure-trove. They were paid for conceded privileges; and for confirmations of previous concessions. They received bribes when desired to do justice; and were paid fines by those who wished to be maintained in possession of property, or to get liberty to exercise certain rights. In England, under the Norman kings, there were such other sources of revenue as compositions paid by heirs before taking possession; sales of wardships; sales to male heirs of rights to choose their wives; sales of charters to towns, and subsequent re-sales of such charters; sales of permissions to trade; and there was also what was called "moneyage"—a shilling paid every three years by each hearth to induce the king not to debase the coinage. Advantage was taken of every favourable opportunity for making and enforcing a demand; as we see in such facts as that it was customary to mulct a discharged official, and that Richard I. " compelled his father's servants to repurchase their offices."

Showing us, as such illustrations do, that these arbitrary seizures and exactions are numerous and heavy in proportion as the power of the ruler is little restrained, the implication is that they reach their extreme where the social organization is typically militant. Evidence that this is so, was given in § 443; and in the next chapter, under another head, we shall meet with more of it.

§ 545. While in the ways named in the foregoing sections, there arise direct taxes, there simultaneously arise, and insensibly diverge, the taxes eventually distinguished as indirect. These begin as demands made on those who have got considerable quantities of commodities exposed in transit, or on sale; and of which parts, originally offered as presents, are subsequently seized as dues.

Under other heads I have referred to the familiar fact that travellers among rude peoples make propitiatory gifts; and by frequent recurrence the reception of these generates a claim. Narratives of recent African explorers confirm the statements of Livingstone, who describes the Portuguese traders among the Quanga people as giving largely, because "if they did not secure the friendship of these petty chiefs, many slaves might be stolen with their loads while passing through the forests;" and who says of a Balonda chief that "he seemed to regard these presents as his proper dues, and as a cargo of goods had come by Senhor Pascoal, he entered the house for the purpose of receiving his share." Various cases show that instead of attempting to take all at the risk of a fight, the head man enters into a compromise under which part is given without a fight; as instance the habitual arrangement with Bedouin tribes, which compound for robbery of travellers by amounts agreed upon; or as instance the mountain Bhils of India, whose chiefs have "seldom much revenue except plunder," who have officers "to obtain information of unprotected villagers and travellers," and who claim "a duty on goods passing their hills:" apparently a composition accepted when those who carry the goods are too strong to be robbed without danger. Where the protection of individuals depends mainly on family-organizations and clan-organizations, the subject as well as the stranger, undefended when away from his home, similarly becomes liable to this qualified black mail. Now to the local ruler, now to the central ruler, according to their respective powers, he yields up part of his goods, that possession of the rest may be guaranteed him, and his claims on buyers enforced. This state of things was illustrated in ancient Mexico, where —

"Of all the goods which were brought into the market, a certain portion was paid in tribute to the king, who was on his part obliged to do justice to the merchants, and to protect their property and their persons."

We trace the like in the records of early European peoples. Part of the revenue of the primitive Greek king, consisted of "the presents paid for licences to trade"—presents which in all probability were at first portions of the commodities to be sold. At a later period in Greece there obtained a practice that had doubtless descended from this. "To these men [magistrates of markets] a certain toll or tribute was paid by all those who brought anything to sell in the market." In western Europe indirect taxation had a kindred origin. The trader, at the mercy of the ruler whose territory he entered, had to surrender part of his merchandise in consideration of being allowed to pass. As feudal lords, swooping down from their castles on merchants passing along neighbouring roads or navigable rivers, took by force portions of what they had, when they did not take all; so their suzerains laid hands on what they pleased of cargoes entering their ports or passing their frontiers: their shares gradually becoming defined by precedent. In England, though there is no clear proof that the two tuns which the king took from wine-laden ships (wine being then the chief import) was originally an unqualified seizure; yet, since this quantity was called "the king's prisage" we have good reason for suspecting that it was so; and that though, afterwards, the king's officer gave something in return, this, being at his option, was but nominal. The very name "customs," eventually applied to commuted payments on imports, points back to a preceding time when this yielding up of portions of cargoes had become established by usage. Confirmation of this inference is furnished by the fact that internal traders were thus dealt with. So late as 1309 it was complained "that the officers appointed to take articles for the king's use in fairs and markets, took more than they ought, and made a profit of the surplus."

Speaking generally of indirect taxes, we may say that arising when the power of the ruler becomes sufficient to change gifts into exactions, they at first differ from other exactions simply in this, that they are enforced on occasions when the subject is more than usually at the ruler's mercy; either because he is exposing commodities for sale where they can be easily found and a share taken; or because he is transferring them from one part of the territory to another, and can be readily stopped and a portion demanded; or because he is bringing commodities into the territory, and can have them laid hands on at one of the few places of convenient entrance. The shares appropriated by the ruler, originally in kind, are early commuted into money where the commodities are such as, by reason of quantity or distance, he cannot consume: instance the load-penny payable at the pit's mouth on each waggon-load to the old-English kings. And the claim comes to be similarly commuted in other cases, as fast as increasing trade brings a more abundant circulating medium, and a greater quantity of produced and imported commodities; the demanded portions of which it becomes more difficult to transport and to utilize.

§ 546. No great advantage would be gained by here going into details. The foregoing general facts appear to be all that it is needful for us to note.

From the outset the growth of revenue has, like that growth of the political headship which it accompanies, been directly or indirectly a result of war. The property of conquered enemies, at first goods, cattle, prisoners, and at a later stage, land, coming in larger share to the leading warrior, increases his predominance. To secure his good will, which it is now important to do, propitiatory presents and help in labour are given; and these, as his power further grows, become periodic and compulsory. Making him more despotic at the same time that it augments his kingdom, continuance of this process increases his ability to enforce contributions, alike from his original subjects and from tributaries; while the necessity for supplies, now to defend his kingdom, now to invade adjacent kingdoms, is ever made the plea for increasing his demands of established kinds and for making new ones. Under stress of the alleged needs, portions of their goods are taken from subjects whenever they are exposed to view for purposes of exchange. And as the primitive presents of property and labour, once voluntary and variable, but becoming compulsory and periodic, are eventually commuted into direct taxes; so these portions of the trader's goods which were originally given for permission to trade and then seized as of right, come eventually to be transformed into percentages of value paid as tolls and duties.

But to the last as at first, and under free governments as under despotic ones, war continues to be the usual reason for imposing new taxes or increasing old ones; at the same time that the coercive organization in past times developed by war, continues to be the means of exacting them.

CHAPTER XVII.

THE MILITANT TYPE OF SOCIETY.

§ 547. Preceding chapters have prepared the way for framing conceptions of the two fundamentally-unlike kinds of political organization, proper to the militant life and the industrial life, respectively. It will be instructive here to arrange in coherent order, those traits of the militant type already incidentally marked, and to join with them various dependent traits; and in the next chapter to deal in like manner with the traits of the industrial type.

During social evolution there has habitually been a mingling of the two. But we shall find that, alike in theory and in fact, it is possible to trace with due clearness those opposite characters which distinguish them in their respective complete developments. Especially is the nature of the organization which accompanies chronic militancy, capable of being inferred *à priori* and proved *à posteriori* to exist in numerous cases. While the nature of the organization accompanying pure industrialism, of which at present we have little experience, will be made clear by contrast; and such illustrations as exist of progress towards it will become recognizable.

Two liabilities to error must be guarded against. We have to deal with societies compounded and re-compounded in various degrees; and we have to deal with societies which, differing in their stages of culture, have their structures elaborated to different extents. We shall be misled, therefore, unless our comparisons are such as take account of unlikenesses in size and in civilization. Clearly, characteristics of the militant type which admit of being displayed by a vast nation, may not admit of being displayed by a horde of savages, though this is equally militant. Moreover, as institutions take long to acquire their finished forms, it is not to be expected that all militant societies will display the organization appropriate to them in its completeness. Rather may we expect that in most cases it will be incompletely displayed.

In face of these difficulties the best course will be to consider, first, what are the several traits which of necessity militancy tends to produce; and then to observe how far these traits are conjointly shown in past and present nations distinguished by militancy. Having contemplated the society ideally organized for war, we shall be prepared to recognize in real societies the characters which war has brought about.

§ 548. For preserving its corporate life, a society is impelled to corporate action; and the preservation of its corporate life is the more probable in proportion as its corporate action is the more complete. For purposes of offence and defence, the forces of individuals have to be combined; and where every individual contributes his force, the probability of success is greatest. Numbers, natures, and circumstances being equal, it is clear that of two tribes or two larger societies, one of which unites the actions of all its capable members while the other does not, the first will ordinarily be the victor. There must be an habitual survival of communities in which militant cooperation is universal.

This proposition is almost a truism. But it is needful here, as a preliminary, consciously to recognize the truth that the social structure evolved by chronic militancy, is one in which all men fit for fighting act in concert against other societies. Such further actions as they carry on they can carry on separately; but this action they must carry on jointly.

§ 549. A society's power of self-preservation will be great in proportion as, besides the direct aid of all who can fight, there is given the indirect aid of all who cannot fight. Supposing them otherwise similar, those communities will survive in which the efforts of combatants are in the greatest degree seconded by those of non-combatants. In a purely militant society, therefore, individuals who do not bear arms have to spend their lives in furthering the maintenance of those who do. Whether, as happens at first, the non-combatants are exclusively the women; or whether, as happens later, the class includes enslaved captives; or whether, as happens later still, it includes serfs; the implication is the same. For if, of two societies equal in other respects, the first wholly subordinates its workers in this way, while the workers in the second are allowed to retain for themselves the produce of their labour, or more of it than is needful for maintaining them; then, in the second, the warriors, not otherwise supported, or supported less fully than they might else be, will have partially to support themselves, and will be so much the less available for war purposes. Hence in the struggle for existence between such societies, it must usually happen that the first will vanquish the second. The social type produced by survival of the fittest, will be one in which the fighting part includes all who can bear arms and be trusted with arms, while the remaining part serves simply as a permanent commissariat.

An obvious implication, of a significance to be hereafter pointed out, is that the non-combatant part, occupied in supporting the combatant part, cannot with advantage to the self-preserving power of the society increase beyond the limit at which it efficiently fulfils its purpose. For, otherwise, some who might be fighters are superfluous workers; and the fighting power of the society is made less than it might be. Hence, in the militant type, the tendency is for the body of warriors to bear the largest practicable ratio to the body of workers.

§ 550. Given two societies of which the members are all

either warriors or those who supply the needs of warriors, and, other things equal, supremacy will be gained by that in which the efforts of all are most effectually combined. In open warfare joint action triumphs over individual action. Military history is a history of the successes of men trained to move and fight in concert.

Not only must there be in the fighting part a combination such that the powers of its units may be concentrated, but there must be a combination of the subservient part with it. If the two are so separated that they can act independently, the needs of the fighting part will not be adequately met. If to be cut off from a temporary base of operations is dangerous, still more dangerous is it to be cut off from the permanent base of operations; namely, that constituted by the body of non-combatants. This has to be so connected with the body of combatants that its services may be fully available. Evidently, therefore, development of the militant type involves a close binding of the society into a whole. As the loose group of savages yields to the solid phalanx, so, other things equal, must the society of which the parts are but feebly held together, yield to one in which they are held together by strong bonds.

§ 551. But in proportion as men are compelled to co-operate, their self-prompted actions are restrained. By as much as the unit becomes merged in the mass, by so much does he lose his individuality as a unit. And this leads us to note the several ways in which evolution of the militant type entails subordination of the citizen.

His life is not his own, but is at the disposal of his society. So long as he remains capable of bearing arms he has no alternative but to fight when called on; and, where militancy is extreme, he cannot return as a vanquished man under penalty of death.

Of course, with this there goes possession of such liberty only as military obligations allow. He is free to pursue his private ends only when the tribe or nation has no need of him; and when it has need of him, his actions from hour to hour must conform, not to his own will but to the public will.

So, too, with his property. Whether, as in many cases, what he holds as private he so holds by permission only, or whether private ownership is recognized, it remains true that in the last resort he is obliged to surrender whatever is demanded for the community's use.

Briefly, then, under the militant type the individual is owned by the State. While preservation of the society is the primary end, preservation of each member is a secondary end—an end cared for chiefly as subserving the primary end.

§ 552. Fulfilment of these requirements, that there shall be complete corporate action, that to this end the non-combatant part shall be occupied in providing for the combatant part, that the entire aggregate shall be strongly bound together, and that the units composing it must have their individualities in life, liberty, and property, thereby subordinated, presupposes a coercive instrumentality. No such union for corporate action can be achieved without a powerful controlling agency. On remembering the fatal results caused by division of counsels in war, or by separation into factions in face of an enemy, we see that chronic militancy tends to develop a despotism; since, other things equal, those societies will habitually survive in which, by its aid, the corporate action is made complete.

And this involves a system of centralization. The trait made familiar to us by an army, in which, under a commander-in-chief there are secondary commanders over large masses, and under these tertiary ones over smaller masses, and so on down to the ultimate divisions, must characterize the social organization at large. A militant society requires a regulative structure of this kind, since, otherwise, its corporate action cannot be made most effectual. Without such grades of governing centres diffused throughout the non-combatant part as well as the combatant part, the entire forces of the aggregate cannot be promptly put forth. Unless the workers are under a control akin to that which the fighters are under, their indirect aid cannot be insured in full amount and with due quickness.

And this is the form of a society characterized by *status*— a society, the members of which stand one towards another in successive grades of subordination. From the despot down to the slave, all are masters of those below and subjects of those above. The relation of the child to the father, of the father to some superior, and so on up to the absolute head, is one in which the individual of lower status is at the mercy of one of higher status.

§ 553. Otherwise described, the process of militant organization is a process of regimentation, which, primarily taking place in the army, secondarily affects the whole community.

The first indication of this we trace in the fact everywhere visible, that the military head grows into a civil head— usually at once, and, in exceptional cases, at last, if militancy continues. Beginning as leader in war he becomes ruler in peace; and such regulative policy as he pursues in the one sphere, he pursues, so far as conditions permit, in the other. Being, as the non-combatant part is, a permanent commissariat, the principle of graduated subordination is extended to it. Its members come to be directed in a way like that in which the warriors are directed—not literally, since by dispersion of the one and concentration of the other exact parallelism is prevented; but, nevertheless, similarly in principle. Labour is carried on under coercion; and supervision spreads everywhere.

To suppose that a despotic military head, daily maintaining regimental control in conformity with inherited traditions, will not impose on the producing classes a kindred control, is to suppose in him sentiments and ideas entirely foreign to his circumstances.

§ 554. The nature of the militant form of government will be further elucidated on observing that it is both positively regulative and negatively regulative. It does not simply restrain; it also enforces. Besides telling the individual what he shall not do, it tells him what he shall do.

That the government of an army is thus characterised needs no showing. Indeed, commands of the positive kind given to the soldier are more important than those of the negative kind: fighting is done under the one, while order is maintained under the other. But here it chiefly concerns us to note that not only the control of military life but also the control of civil life, is, under the militant type of government, thus characterized. There are two ways in which the ruling power may deal with the private individual. It may simply limit his activities to those which he can carry on without aggression, direct or indirect, upon others; in which case its action is negatively regulative. Or, besides doing this, it may prescribe the how, and the where, and the when, of his activities—may force him to do things which he would not spontaneously do—may direct in greater or less detail his mode of living; in which case its action is positively regulative. Under the militant type this positively regulative action is widespread and peremptory. The civilian is in a condition as much like that of the soldier as difference of occupation permits.

And this is another way of expressing the truth that the fundamental principle of the militant type is compulsory co-operation. While this is obviously the principle on which the members of the combatant body act, it no less certainly must be the principle acted on throughout the non-combatant body, if military efficiency is to be great; since, otherwise, the aid which the non-combatant body has to furnish cannot be insured.

§ 555. That binding together by which the units of a militant society are made into an efficient fighting structure, tends to fix the position of each in rank, in occupation, and in locality.

In a graduated regulative organization there is resistance to change from a lower to a higher grade. Such change is made difficult by lack of the possessions needed for filling superior positions; and it is made difficult by the opposition of those who already fill them, and can hold inferiors down. Preventing intrusion from below, these transmit their respective places and ranks to their descendants; and as the principle of inheritance becomes settled, the rigidity of the social structure becomes decided. Only where an "egalitarian despotism" reduces all subjects to the same political status—a condition of decay rather than of development—does the converse state arise.

The principle of inheritance, becoming established in respect of the classes which militancy originates, and fixing the general functions of their members from generation to generation, tends eventually to fix also their special functions. Not only do men of the slave-classes and the artizan-classes succeed to their respective ranks, but they succeed to the particular occupations carried on in them. This, which is a result of the tendency towards regimentation, is ascribable primarily to the fact that a superior, requiring from each kind of worker his particular product, has an interest in replacing him at death by a capable successor; while the worker, prompted to get aid in executing his tasks, has an interest in bringing up a son to his own occupation: the will of the son being powerless against these conspiring interests. Under the system of compulsory cooperation, therefore, the principle of inheritance, spreading through the producing organization, causes a relative rigidity in this also.

A kindred effect is shown in the entailed restraints on movement from place to place. In proportion as the individual is subordinated in life, liberty, and property, to his society, it is needful that his whereabouts shall be constantly known. Obviously the relation of the soldier to his officer, and of this officer to his superior, is such that each must be ever at hand; and where the militant type is fully developed the like holds throughout the society. The slave cannot leave his appointed abode; the serf is tied to his allotment; the master is not allowed to absent himself from his locality without leave.

So that the corporate action, the combination, the cohesion, the regimentation, which efficient militancy necessitates, imply a structure which strongly resists change.

§ 556. A further trait of the militant type, naturally accompanying the last, is that organizations other than those forming parts of the State-organization, are wholly or partially repressed. The public combination occupying all fields, excludes private combinations.

For the achievement of complete corporate action there must, as we have seen, be a centralized administration, not only throughout the combatant part but throughout the non-combatant part; and if there exist unions of citizens which act independently, they in so far diminish the range of this centralized administration. Any structures which are not portions of the State-structure, serve more or less as limitations to it, and stand in the way of the required unlimited subordination. If private combinations are allowed to exist, it will be on condition of submitting to an official regulation such as greatly restrains independent action; and since private combinations officially regulated are inevitably hindered from doing things not conforming to established routine, and are thus debarred from improvement, they cannot habitually thrive and grow. Obviously, indeed, such combinations, based on the principle of voluntary cooperation, are incongruous with social arrangements based on the principle of compulsory cooperation. Hence the militant type is characterized by the absence, or comparative rarity, of bodies of citizens associated for commercial purposes, for propagating special religious views, for achieving philanthropic ends, &c.

Private combinations of one kind, however, are congruous with the militant type—the combinations, namely, which are formed for minor defensive or offensive purposes. We have, as examples, those which constitute factions, very general in militant societies; those which belong to the same class as primitive guilds, serving for mutual protection; and those which take the shape of secret societies. Of such bodies it may be noted that they fulfil on a small scale ends like those which the whole society fulfils on a large scale—the ends of self-preservation, or aggression, or both. And it may be further noted that these small included societies are organized on the same principle as the large including society—the principle of compulsory cooperation. Their governments are coercive: in some cases even to the extent of killing those of their members who are disobedient.

§ 557. A remaining fact to be set down is that a society of the militant type tends to evolve a self-sufficient sustaining organization. With its political autonomy there goes what we may call an economic autonomy. Evidently if it carries on frequent wars against surrounding societies, its commercial intercourse with them must be hindered or prevented: exchange of commodities can go on to but a small extent between those who are continually fighting. A militant society must, therefore, to the greatest degree practicable, provide internally the supplies of all articles needful for carrying on the lives of its members. Such an economic state as that which existed during early feudal times, when, as in France, "the castles made almost all the articles used in them," is a state evidently entailed on groups, small or large, which are in constant antagonism with surrounding groups. If there does not already exist within any group so circumstanced, an agency for producing some necessary article, inability to obtain it from without will lead to the establishment of an agency for obtaining it within.

Whence it follows that the desire "not to be dependent on foreigners" is one appropriate to the militant type of society. So long as there is constant danger that the supplies of needful things derived from other countries will be cut off by the breaking out of hostilities, it is imperative that there shall be maintained a power of producing these supplies at home, and that to this end the required structures shall be maintained. Hence there is a manifest direct relation between militant activities and a protectionist policy.

§ 558. And now having observed the traits which may be expected to establish themselves by survival of the fittest during the struggle for existence among societies, let us observe how these traits are displayed in actual societies, similar in respect of their militancy but otherwise dissimilar.

Of course in small primitive groups, however warlike they may be, we must not look for more than rude outlines of the structure proper to the militant type. Being loosely aggregated, definite arrangement of their parts can be carried but to a small extent. Still, so far as it goes, the evidence is to the point. The fact that habitually the fighting body is co-extensive with the adult male population, is so familiar that no illustrations are needed. An equally familiar fact is that the women, occupying a servile position, do all the unskilled labour and bear the burdens; with which may be joined the fact that not unfrequently during war they carry the supplies, as in Asia among the Bhils and Khonds, as in Polynesia among the New Caledonians and Sandwich Islanders, as in America among the Comanches, Mundrucus, Patagonians: their office as forming the permanent commissariat being thus clearly shown. We see, too, that where the enslaving of captives has arisen, these also serve to support and aid the combatant class; acting during peace as producers and during war joining the women in attendance on the army, as among the New Zealanders, or, as among the Malagasy, being then exclusively the carriers of provisions, &c. Again, in these first stages, as in later stages, we are shown that private claims are, in the militant type, over-ridden by public claims. The life of each man is held subject to the needs of the group; and, by implication, his freedom of action is similarly held. So, too, with his goods; as instance the remark made of the Brazilian Indians, that personal property, recognized but to a limited extent during peace, is scarcely at all recognized during war; and as instance Hearne's statement concerning certain hyperborean tribes of North America when about to make war, that "property of every kind that could be of general use now ceased to be private." To which add the cardinal truth, once more to be repeated, that where no political subordination exists war initiates it. Tacitly or overtly a chief is temporarily acknowledged; and he gains permanent power if war continues. From these beginnings of the militant type which small groups show us, let us pass to its developed forms as shown in larger groups.

"The army, or what is nearly synonymous, the nation of Dahome," to quote Burton's words, furnishes us with a good example: the excessive militancy being indicated by the fact that the royal bedroom is paved with skulls of enemies. Here the king is absolute, and is regarded as supernatural in character—he is the "spirit;" and of course he is the religious head—he ordains the priests. He absorbs in himself all powers and all rights: "by the state-law of Dahome . . . all men are slaves to the king." He "is heir to all his subjects;" and he takes from living subjects whatever he likes. When we add that there is a frequent killing of victims to carry messages to the other world, as well as occasions on which numbers are sacrificed to supply deceased kings with attendants, we are shown that life, liberty, and property, are at the entire disposal of the State as represented by its head. In both the civil and military organizations, the centres and sub-centres of control are numerous. Names, very generally given by the king and replacing surnames, change "with every rank of the holder;" and so detailed is the regimentation that "the dignities seem interminable." There are numerous sumptuary laws; and, according to Waitz, no one wears any other clothing or weapons than what the king gives him or allows him. Under penalty of slavery or death, "no man must alter the construction of his house, sit upon a chair, or be carried on a hammock, or drink out of a glass," without permission of the king.

The ancient Peruvian empire, gradually established by the conquering Yncas, may next be instanced. Here the ruler, divinely descended, sacred, absolute, was the centre of a system which minutely controlled all life. His headship was at once military, political, ecclesiastical, judicial; and the entire nation was composed of those who, in the capacity of soldiers, labourers, and officials, were slaves to him and his deified ancestors. Military service was obligatory on all taxable Indians who were capable; and those of them who had served their prescribed terms, formed into reserves, had then to work under State-superintendence. The army having heads over groups of ten, fifty, a hundred, five hundred, a thousand, ten thousand, had, besides these, its superior commanders of Ynca blood. The community at large was subject to a parallel regimentation: the inhabitants registered in groups, being under the control of officers over tens, fifties, hundreds, and so on. And through these successive grades of centres, reports ascended to the Ynca-governors of great divisions, passing on from them to the Ynca; while his orders descended "from rank to rank till they reached the lowest." There was an ecclesiastical organization similarly elaborate, having, for example, five classes of diviners; and there was an organization of spies to examine and report upon the doings of the other officers. Everything was under public

Inspection. There were village-officers who overlooked the ploughing, sowing, and harvesting. When there was a deficiency of rain, measured quantities of water were supplied by the State. All who travelled without authority were punished as vagabonds; but for those who were authorized to travel for public purposes, there were establishments supplying lodging and necessaries. "It was the duty of the decurions to see that the people were clothed;" and the kinds of cloth, decorations, badges, &c., to be worn by the different ranks were prescribed. Besides this regulation of external life there was regulation of domestic life. The people were required to "dine and sup with open doors, that the judges might be able to enter freely;" and these judges had to see that the house, clothes, furniture, &c., were kept clean and in order, and the children properly disciplined: those who mismanaged their houses being flogged. Subject to this minute control, the people laboured to support this elaborate State-organization. The political, religious, and military classes were exempt from tribute; while the labouring classes when not serving in the army, had to yield up all produce beyond that required for their bare sustenance. Of the whole empire, one-third was allotted for supporting the State, one-third for supporting the priesthood who ministered to the manes of ancestors, and the remaining third had to support the workers. Besides giving tribute by tilling the lands of the Sun and the King, the workers had to till the lands of the soldiers on duty, as well as those of incapables. And they also had to pay tribute of clothes, shoes, and arms. Of the lands on which the people maintained themselves, a tract was apportioned to each man according to the size of his family. Similarly with the produce of the flocks. Such moiety of this in each district as was not required for supplying public needs, was periodically shorn, and the wool divided by officials. These arrangements were in pursuance of the principle that "the private property of each man was held by favour of the Ynca, and according to their laws he had no other title to it." Thus the people, completely possessed by the State in person, property, and labour, transplanted to this or that locality as the Ynca directed, and, when not serving as soldiers, living under a discipline like that within the army, were units in a centralized regimented machine, moved throughout life to the greatest practicable extent by the Ynca's will, and to the least practicable extent by their own wills. And, naturally, along with militant organization thus carried to its ideal limit, there went an almost entire absence of any other organization. They had no money; "they neither sold clothes, nor houses, nor estates;" and trade was represented among them by scarcely anything more than some bartering of articles of food.

So far as accounts of it show, ancient Egypt presented phenomena allied in their general, if not in their special, characters. Its predominant militancy during remote unrecorded times, is sufficiently implied by the vast population of slaves who toiled to build the pyramids; and its subsequent continued militancy we are shown alike by the boasting records of its kings, and the delineations of their triumphs on its temple-walls. Along with this form of activity we have, as before, the god-descended ruler, limited in his powers only by the usages transmitted from his divine ancestors, who was at once political head, high-priest, and commander-in-chief. Under him was a centralized organization, of which the civil part was arranged in classes and sub-classes as definite as were those of the militant part. Of the four great social divisions—priests, soldiers, traders, and common people, beneath whom came the slaves—the first contained more than a score different orders; the second, some half-dozen beyond those constituted by military grades; the third, nearly a dozen; and the fourth, a still greater number. Though within the ruling classes the castes were not so rigorously defined as to prevent change of function in successive generations, yet Herodotus and Diodorus state that industrial occupations descended from father to son: "every particular trade and manufacture was carried on by its own craftsmen, and none changed from one trade to another." How elaborate was the regimentation may be judged from the detailed account of the staff of officers and workers engaged in one of their vast quarries: the numbers and kinds of functionaries paralleling those of an army. To support this highly-developed regulative organization, civil, military, and sacerdotal (an organization which held exclusive possession of the land) the lower classes laboured. "Overseers were set over the wretched people, who were urged to hard work more by the punishment of the stick than words of warning." And whether or not official oversight included domiciliary visits, it at any rate went to the extent of taking note of each family. "Every man was required under pain of death to give an account to the magistrate of how he earned his livelihood."

Take, now, another ancient society, which, strongly contrasted in sundry respects, shows us, along with habitual militancy, the assumption of structural traits allied in their fundamental characters to those thus far observed. I refer to Sparta. That warfare did not among the Spartans evolve a single despotic head, while in part due to causes which, as before shown, favour the development of compound political heads, was largely due to the accident of their double kingship: the presence of two divinely-descended chiefs prevented the concentration of power. But though from this cause there continued an imperfectly centralized government, the relation of this government to members of the community was substantially like that of militant governments in general. Notwithstanding the serfdom, and in towns the slavery, of the Helots, and notwithstanding the political subordination of the Periœki, they all, in common with the Spartans proper, were under obligation to military service: the working function of the first, and the trading function, so far as it existed, which was carried on by the second, were subordinate to the militant function, with which the third was exclusively occupied. And the civil divisions thus marked re-appeared in the military divisions: "at the battle of Platæa every Spartan hoplite had seven Helots, and every Pericœki hoplite one Helot to attend him." The extent to which, by the daily military discipline, prescribed military mess, and fixed contributions of food, the individual life of the Spartan was subordinated to public demands, from seven years upwards, needs mention only to show the rigidity of the restraints which here, as elsewhere, the militant type imposes—restraints which were further shown in the prescribed age for marriage, the prevention of domestic life, the forbidding of industry or any money-seeking occupation, the interdict on going abroad without leave, and the authorized censorship under which his days and nights were passed. There was fully carried out in Sparta the Greek theory of society, that "the citizen belongs neither to himself nor to his family, but to his city." So that though in this exceptional case, chronic militancy was prevented from developing a supreme head, owning the individual citizen in body and estate, yet it developed an essentially identical relation between the community as a whole and its units. The community, exercising its power through a compound head instead of through a simple head, completely enslaved the individual. While the lives and labours of the Helots were devoted exclusively to the support of those who formed the military organization, the lives and labours of those who formed the military organization were exclusively devoted to the service of the State: they were slaves with a difference.

Of modern illustrations, that furnished by Russia will suffice. Here, again, with the wars which effected conquests and consolidations, came the development of the victorious commander into the absolute ruler, who, if not divine by alleged origin, yet acquired something like divine *prestige*. "All men are equal before God, and the Russians' God is the Emperor," says De Custine: "the supreme governor is so raised above earth, that he sees no difference between the serf and the lord." Under the stress of Peter the Great's wars, which, as the nobles complained, took them away from their homes, "not, as formerly, for a single campaign, but for long years," they became "servants of the State, without privileges, without dignity, subjected to corporal punishment, and burdened with onerous duties from which there was no escape." "Any noble who refused to serve ['the State in the Army, the Fleet, or the Civil Administration, from boyhood to old age,'] was not only deprived of his estate, as in the old times, but was declared to be a traitor, and might be condemned to capital punishment." "Under Peter," says Wallace, "all offices, civil and military," were "arranged in fourteen classes or ranks;" and he "defined the obligations of each with microscopic minuteness. After his death the work was carried on in the same spirit, and the tendency reached its climax in the reign of Nicholas." In the words of De Custine, "the tchinn [the name for this organization] is a nation formed into a regiment; it is the military system applied to all classes of society, even to those who never go to war." With this universal regimentation in structure went a regimental discipline. The conduct of life was dictated to the citizens at large in the same way as to soldiers. In the reign of Peter and his successors, domestic entertainments were appointed and regulated ; the people were compelled to change their costumes ; the clergy to cut off their beards ; and even the harnessing of horses was according to pattern. Occupations were controlled to the extent that "no boyard could enter any profession, or forsake it when embraced, or retire from public to private life, or dispose of his property, or travel into any foreign country, without the permission of the Czar." This omnipresent rule is well expressed in the close of certain rhymes, for which a military officer was sent to Siberia:—

> "Tout se fait par ukase ici ;
> C'est par ukase que l'on voyage,
> C'est par ukase que l'on rit."

Taking thus the existing barbarous society of Dahomey, formed of negroes, the extinct semi-civilized empire of the

Yncas, whose subjects were remote in blood from these, the ancient Egyptian empire peopled by yet other races, the community of the Spartans, again unlike in the type of its men, and the existing Russian nation made up of Slavs and Tatars, we have before us cases in which such similarities of social structure as exist, cannot be ascribed to inheritance of a common character by the social units. The immense contrasts between the populations of these several societies, too, varying from millions at the one extreme to thousands at the other, negative the supposition that their common structural traits are consequent on size. Nor can it be supposed that likenesses of conditions in respect of climate, surface, soil, flora, fauna, or likenesses of habits caused by such conditions, can have had anything to do with the likenesses of organization in these societies; for their respective habitats present numerous marked unlikenesses. Such traits as they one and all exhibit, not ascribable to any other cause, must thus be ascribed to the habitual militancy characteristic of them all. The results of induction alone would go far to warrant this ascription; and it is fully warranted by their correspondence with the results of deduction, as set forth above.

§ 559. Any remaining doubts must disappear on observing how continued militancy is followed by further development of the militant organization. Three illustrations will suffice.

When, during Roman conquests, the tendency for the successful general to become despot, repeatedly displayed, finally took effect—when the title *imperator*, military in its primary meaning, became the title for the civil ruler, showing us on a higher platform that genesis of political headship out of military headship visible from the beginning—when, as usually happens, an increasingly divine character was acquired by the civil ruler, as shown in the assumption of the sacred name Augustus, as well as in the growth of an actual worship of him; there simultaneously became more pronounced those further traits which characterize the militant type in its developed form. Practically, if not nominally, the other powers of the State were absorbed by him. In the words of Duruy, he had—

"The right of proposing, that is, of making laws; of receiving and trying appeals, *i.e.* the supreme jurisdiction; of arresting by the tribunitian veto every measure and every sentence, *i.e.* of putting his will in opposition to the laws and magistrates; of summoning the senate or the people and presiding over it, *i.e.* of directing the electoral assemblies as he thought fit. And these prerogatives he will have not for a single year but for life; not in Rome only . . . but throughout the empire; not shared with ten colleagues, but exercised by himself alone; lastly, without any account to render, since he never resigns his office."

Along with these changes went an increase in the number and definiteness of social divisions. The Emperor—

"Placed between himself and the masses a multitude of people regularly classed by categories, and piled one above the other in such a way that this hierarchy, pressing with all its weight upon the masses underneath, held the people and factious individuals powerless. What remained of the old patrician nobility had the foremost rank in the city; . . . below it came the senatorial nobility, half hereditary; below that the moneyed nobility or equestrian order—three aristocracies superposed. . . . The sons of senators formed a class intermediate between the senatorial and the equestrian order. . . . In the 2nd century the senatorial families formed an hereditary nobility with privileges."

At the same time the administrative organization was greatly extended and complicated.

"Augustus created a large number of new offices, as the superintendence of public works, roads, aqueducts, the Tiber-bed, distribution of corn to the people. . . . He also created numerous offices of procurators for the financial administration of the empire, and in Rome there were 1,060 municipal officers."

The structural character proper to an army spread in a double way: military officers acquired civil functions and functionaries of a civil kind became partially military. The magistrates appointed by the Emperor, tending to replace those appointed by the people, had, along with their civil authority, military authority; and while "under Augustus the prefects of the pretorium were only military chiefs, . . . they gradually possessed themselves of the whole civil authority, and finally became, after the Emperor, the first personages in the empire." Moreover, the governmental structures grew by incorporating bodies of functionaries who were before independent. "In his ardour to organize everything, he aimed at regimenting the law itself, and made an official magistracy of that which had always been a free profession." To enforce the rule of this extended administration, the army was made permanent, and subjected to severe discipline. With the continued growth of the regulating and coercing organization, the drafts on producers increased; and, as shown by extracts in a previous chapter concerning the Roman *régime* in Egypt and in Gaul, the working part of the community was reduced more and more to the form of a permanent commissariat. In Italy the condition eventually arrived at was one in which vast tracts were "intrusted to freedmen, whose only consideration was . . . how to extract from their labourers the greatest amount of work with the smallest quantity of food."

An example under our immediate observation may next be taken—that of the German Empire. Such traits of the militant type in Germany as were before manifest, have, since the late war, become still more manifest. The army, active and passive, including officers and attached functionaries, has been increased by about 100,000 men; and changes in 1875 and 1880, making certain reserves more available, have practically caused a further increase of like amount. Moreover, the smaller German States, having in great part surrendered the administration of their several contingents, the German army has become more consolidated; and even the armies of Saxony, Würtemberg, and Bavaria, being subject to Imperial supervision, have in so far ceased to be independent. Instead of each year granting military supplies, as had been the practice in Prussia before the formation of the North German Confederation, the Parliament of the Empire was, in 1871, induced to vote the required annual sum for three years thereafter; in 1874 it did the like for the succeeding seven years; and again in 1880 the greatly increased amount for the augmented army was authorized for the seven years then following: steps obviously surrendering popular checks on Imperial power. Simultaneously, military officialism has been in two ways replacing civil officialism. Subaltern officers are rewarded for long services by appointments to civil posts—local communes being forced to give them the preference to civilians; and not a few members of the higher civil service, and of the universities, as well as teachers in the public schools, having served as "volunteers of one year," become commissioned officers of the Landwehr. During the struggles of the so-called Kulturkampf, the ecclesiastical organization became more subordinated by the political. Priests suspended by bishops were maintained in their offices; it was made penal for a clergyman publicly to take part against the government; a recalcitrant bishop had his salary stopped; the curriculum for ecclesiastics was prescribed by the State, and examination by State-officials required; church discipline was subjected to State-approval; and a power of expelling rebellious clergy from the country was established. Passing to the industrial activities we may note, first, that through sundry steps, from 1873 onwards, there has been a progressive transfer of railways into the hands of the State; so that, partly by original construction (mainly of lines for military purposes), and partly by purchase, three-fourths of all Prussian railways have been made government property; and the same percentage holds in the other German States: the aim being eventually to make them all Imperial. Trade interferences have been extended in various ways—by protectionist tariffs, by revival of the usury laws, by restrictions on Sunday labour. Through its postal service the State has assumed industrial functions—presents acceptances, receives money on bills of exchange that are due, as also on ordinary bills, which it gets receipted; and until stopped by shopkeepers' protests, undertook to procure books from publishers. Lastly there come the measures for extending, directly and indirectly, the control over popular life. On the one hand there are the laws under which, up to the middle of last year, 224 socialist societies have been closed, 180 periodicals suppressed, 317 books, &c., forbidden; and under which sundry places have been reduced to a partial state of siege. On the other hand may be named Prince Bismarck's scheme for re-establishing guilds (bodies which by their regulations coerce their members), and his scheme of State-insurance, by the help of which the artizan would, in a considerable degree, have his hands tied. Though these measures have not been carried in the forms proposed, yet the proposal of them sufficiently shows the general tendency. In all which changes we see progress towards a more integrated structure, towards increase of the militant part as compared with the industrial part, towards the replacing of civil organization by military organization, towards the strengthening of restraints over the individual and regulation of his life in greater detail.*

The remaining example to be named is that furnished by our own society since the revival of military activity—a revival which has of late been so marked that our illustrated papers are, week after week, occupied with little else than scenes of warfare. Already in the first volume of *The Principles of Sociology,* I have pointed out many ways in which the system of compulsory cooperation characterizing the militant type, has been trenching on the system of voluntary cooperation characterizing the industrial type; and since those passages appeared (July, 1876), other changes in the same direction have taken place. Within the military organization itself, we may note the increasing assimilation of the volunteer forces to the regular army, now going to the extent of proposing to make them available abroad, so that instead of defensive action for which they were created, they

* This chapter was originally published in the *Contemporary Review* for Sept., 1881. Since that date a further movement of German society in the same general direction has been shown by the pronounced absolutism of the imperial rescript of Jan., 1882, endorsing Prince Bismarck's scheme of State-socialism.

can be used for offensive action; and we may also note that the tendency shown in the army during the past generation to sink the military character whenever possible, by putting on civilian dresses, is now checked by an order to officers in garrison towns to wear their uniforms when off duty, as they do in more militant countries. Whether, since the date named, usurpations of civil functions by military men (which had in 1873–4 gone to the extent that there were 97 colonels, majors, captains, and lieutenants employed from time to time as inspectors of science and art classes) have gone further, I cannot say; but there has been a manifest extension of the militant spirit and discipline among the police, who, wearing helmet-shaped hats, beginning to carry revolvers, and looking upon themselves as half soldiers, have come to speak of the people as "civilians." To an increasing extent the executive has been over-riding the other governmental agencies; as in the Cyprus business, and as in the doings of the Indian Viceroy under secret instructions from home. In various minor ways are shown endeavours to free officialism from popular checks; as in the desire expressed in the House of Lords that the hanging of convicts in prisons, entrusted entirely to the authorities, should have no other witnesses; and as in the advice given by the late Home Secretary (on 11th May, 1878) to the Derby Town Council, that it should not interfere with the chief constable (a military man) in his government of the force under him—a step towards centralizing local police control in the Home Office. Simultaneously we see various actual or prospective extensions of public agency, replacing or restraining private agency. There is the "endowment of research," which, already partially carried out by a government fund, many wish to carry further; there is the proposed act for establishing a registration of authorized teachers; there is the bill which provides central inspection for local public libraries; there is the scheme for compulsory insurance—a scheme showing us in an instructive manner the way in which the regulating policy extends itself: compulsory charity having generated improvidence, there comes compulsory insurance as a remedy for the improvidence. Other proclivities towards institutions belonging to the militant type, are seen in the increasing demand for some form of protection, and in the lamentations uttered by the "society papers" that duelling has gone out. Nay, even through the party which by position and function is antagonistic to militancy, we see that militant discipline is spreading; for the caucus-system, established for the better organization of liberalism, is one which necessarily, in a greater or less degree, centralizes authority and controls individual action.

Besides seeing, then, that the traits to be inferred *à priori* as characterizing the militant type, constantly exist in societies which are permanently militant in high degrees, we also see that in other societies increase of militant activity is followed by development of such traits.

§ 560. In some places I have stated, and in other places implied, that a necessary relation exists between the structure of a society and the natures of its citizens. Here it will be well to observe in detail the characters proper to, and habitually exemplified by, the members of a typically militant society.

Other things equal, a society will be successful in war in proportion as its members are endowed with bodily vigour and courage. And, on the average, among conflicting societies there will be a survival and spread of those in which the physical and mental powers called for in battle, are not only most marked but also most honoured. Egyptian and Assyrian sculptures and inscriptions, show us that prowess was the thing above all others thought most worthy of record. Of the words good, just, &c., as used by the ancient Greeks, Grote remarks that they "signify the man of birth, wealth, influence and daring, whose arm is strong to destroy or to protect, whatever may be the turn of his moral sentiments;

while the opposite epithet, bad, designates the poor, lowly, and weak, from whose dispositions, be they ever so virtuous society has little to hope or to fear." In the identification of virtue with bravery among the Romans, we have a like implication. During early turbulent times throughout Europe, the knightly character, which was the honourable character, primarily included fearlessness: lacking this, good qualities were of no account; but with this, sins of many kinds, great though they might be, were condoned.

If, among antagonist groups of primitive men, some tolerated more than others the killing of their members—if, while some always retaliated others did not; those which did not retaliate, continually aggressed on with impunity, would either gradually disappear or have to take refuge in undesirable habitats. Hence there is a survival of the unforgiving. Further, the *lex talionis*, primarily arising between antagonist groups, becomes the law within the group; and chronic feuds between component families and clans, everywhere proceed upon the general principle of life for life. Under the militant *régime* revenge becomes a virtue, and failure to revenge a disgrace. Among the Fijians, who foster anger in their children, it is not infrequent for a man to commit suicide rather than live under an insult; and in other cases the dying Fijian bequeathes the duty of inflicting vengeance to his children. This sentiment and the resulting practices we trace among peoples otherwise wholly alien, who are, or have been, actively militant. In the remote East may be instanced the Japanese. They are taught that "with the slayer of his father a man may not live under the same heaven; against the slayer of his brother a man must never have to go home to fetch a weapon; with the slayer of his friend a man may not live in the same State." And in the West may be instanced France during feudal days, when the relations of one killed or injured were required by custom to retaliate on any relations of the offender—even those living at a distance and knowing nothing of the matter. Down to the time of the Abbé Brantôme, the spirit was such that that ecclesiastic, enjoining on his nephews by his will to avenge any unredressed wrongs done to him in his old age, says of himself—" I may boast, and I thank God for it, that I never received an injury without being revenged on the author of it." That where militancy is active, revenge, private as well as public, becomes a duty, is well shown at the present time among the Montenegrins—a people who have been at war with the Turks for centuries. "Dans le Montenegro," says Boué, "on dira d'un homme d'une natrie [clan] ayant tué un individu d'une autre : Cette natrie nous doit une tête, et il faut que cette dette soit acquittée, car qui ne se venge pas ne se sancitie pas."

Where activity in destroying enemies is chronic, destruction will become a source of pleasure; where success in subduing fellow-men is above all things honoured, there will arise delight in the forcible exercise of mastery; and with pride in spoiling the vanquished, will go disregard for the rights of property at large. As it is incredible that men should be courageous in face of foes and cowardly in face of friends, so it is incredible that the other feelings fostered by perpetual conflicts abroad should not come into play at home. We have just seen that with the pursuit of vengeance outside the society, there goes the pursuit of vengeance inside the society; and whatever other habits of thought and action constant war necessitates, must show their effects on the social life at large. Facts from various places and times prove that in militant communities the claims to life, liberty, and property, are little regarded. The Dahomans, warlike to the extent that both sexes are warriors, and by whom slave-hunting invasions are, or were, annually undertaken "to furnish funds for the royal exchequer," show their bloodthirstiness by their annual "customs," at which multitudinous victims are publicly slaughtered for the popular gratification. The Fijians, again, highly militant in their activities and type of organization, who display their recklessness of life not only by killing their own people for cannibal feasts, but by destroying immense numbers of their infants and by sacrificing victims on such trivial occasions as launching a new canoe, so much applaud ferocity that to commit a murder is a glory. Early records of Asiatics and Europeans show us the like relation. What accounts there are of the primitive Mongols, who, when united, massacred western peoples wholesale, show us a chronic reign of violence, both within and without their tribes; while domestic assassinations, which from the beginning have characterized the militant Turks, continue to characterize them down to our own day. In proof that it was so with the Greek and Latin races it suffices to instance the slaughter of the two thousand helots by the Spartans, whose brutality was habitual, and the murder of large numbers of suspected citizens by jealous Roman emperors, who also, like their subjects, manifested their love of bloodshed in their arenas. That where life is little regarded there can be but little regard for liberty, follows necessarily. Those who do not hesitate to end another's activities by killing him, will still less hesitate to restrain his activities by holding him in bondage. Militant savages, whose captives, when not eaten, are enslaved, habitually show us this absence of regard for fellow-men's freedom, which characterizes the members of militant societies in general. How little, under the *régime* of war, more or less markedly displayed in all early historic societies, there was any sentiment against depriving men of their liberties, is sufficiently shown by the fact that even in the teachings of primitive Christianity there was no express condemnation of slavery. Naturally the like holds with the right of property. Where mastery established by force is honourable, claims to possession by the weaker are likely to be little respected by the stronger. In Fiji it is considered chief-like to seize a subject's goods; and theft is virtuous if undiscovered. Among the Spartans "the ingenious and successful pilferer gained applause with his booty." In mediæval

Europe, with perpetual robberies of one society by another there went perpetual robberies within each society. Under the Merovingians "the murders and crimes it [*The Ecclesiastical History of the Franks*] relates, have almost all for their object the possession of the treasure of the murdered persons." And under Charlemagne plunder by officials was chronic: the moment his back was turned, "the provosts of the king appropriated the funds intended to furnish food and clothing for the artisans."

Where warfare is habitual, and the required qualities most needful and therefore most honoured, those whose lives do not display them are treated with contempt, and their occupations regarded as dishonourable. In early stages labour is the business of women and of slaves—conquered men and the descendants of conquered men; and trade of every kind, carried on by subject classes, long continues to be identified with lowness of origin and nature. In Dahomey, "agriculture is despised because slaves are employed in it." "The Japanese nobles and placemen, even of secondary rank, entertain a sovereign contempt for traffic." Of the ancient Egyptians Wilkinson says, "their prejudices against mechanical employments, as far as regarded the soldier, were equally strong as in the rigid Sparta." "For trade and commerce the [ancient] Persians were wont to express extreme contempt," writes Rawlinson. That progress of class-differentiation which accompanied the conquering wars of the Romans, was furthered by establishment of the rule that it was disgraceful to take money for work, as also by the law forbidding senators and senators' sons from engaging in speculation. And how great has been the scorn expressed by the militant classes for the trading classes throughout Europe, down to quite recent times, needs no showing.

That there may be willingness to risk life for the benefit of the society, there must be much of the feeling called patriotism. Though the belief that it is glorious to die for one's country cannot be regarded as essential, since mercenaries fight without it; yet it is obvious that such a belief conduces greatly to success in war; and that entire absence of it is so unfavourable to offensive and defensive action that failure and subjugation will, other things equal, be likely to result. Hence the sentiment of patriotism is habitually established by the survival of societies the members of which are most characterized by it.

With this has to be united the sentiment of obedience. The possibility of that united action by which, other things equal, war is made successful, depends on the readiness of individuals to subordinate their wills to the will of a commander or ruler. Loyalty is essential. In early stages the manifestation of it is but temporary; as among the Araucanians who, ordinarily showing themselves "repugnant to all subordination, are then [when war is impending] prompt to obey, and submissive to the will of their military sovereign" appointed for the occasion. And with development of the militant type this sentiment becomes permanent. Erskine tells us that the Fijians are intensely loyal: men buried alive in the foundations of a king's house, considered themselves honoured by being so sacrificed; and the people of a slave district " said it was their duty to become food and sacrifice for the chiefs." So in Dahomey, there is felt for the king "a mixture of love and fear, little short of adoration." In ancient Egypt again, where " blind obedience was the oil which caused the harmonious working of the machinery " of social life, the monuments on every side show with wearisome iteration the daily acts of subordination—of slaves and others to the dead man, of captives to the king, of the king to the gods. Though for reasons already pointed out, chronic war did not generate in Sparta a supreme political head, to whom there could be shown implicit obedience, yet the obedience shown to the political agency which grew up was profound: individual wills were in all things subordinate to the public will expressed by the established authorities. Primitive Rome, too, though without a divinely-descended king to whom submission could be shown, displayed great submission to an appointed king, qualified only by expressions of opinion on special occasions; and the principle of absolute obedience, slightly mitigated in the relations of the community as a whole to its ruling agency, was unmitigated within its component groups. That throughout European history, alike on small and on large scales, we see the sentiment of loyalty dominant where the militant type of structure is pronounced, is a truth that will be admitted without detailed proof.

From these conspicuous traits of nature, let us turn to certain consequent traits which are less conspicuous, and which have results of less manifest kinds. Along with loyalty naturally goes faith—the two being, indeed, scarcely separable. Readiness to obey the commander in war, implies belief in his military abilities; and readiness to obey him during peace, implies belief that his abilities extend to civil affairs also. Imposing on men's imaginations, each new conquest augments his authority. There come more frequent and more decided evidences of his regulative action over men's lives; and these generate the idea that his power is boundless. Unlimited confidence in governmental agency is fostered. Generations brought up under a system which controls all affairs, private and public, tacitly assume that affairs can only thus be controlled. Those who have experience of no other *régime* are unable to imagine any other *régime*. In such societies as that of ancient Peru, for example, where, as we have seen, regimental rule was universal, there were no materials for framing the thought of an industrial life spontaneously carried on and spontaneously regulated.

By implication there results repression of individual initiative, and consequent lack of private enterprise. In proportion as an army becomes organized, it is reduced to a state in which the independent action of its members is forbidden. And in proportion as regimentation pervades the society at large, each member of it, directed or restrained at every turn, has little or no power of conducting his business otherwise than by established routine. Slaves can do only what they are told by their masters; their masters cannot do anything that is unusual without official permission; and no permission is to be obtained from the local authority until superior authorities through their ascending grades have been consulted. Hence the mental state generated is that of passive acceptance and expectancy. Where the militant type is fully developed, everything must be done by public agencies; not only for the reason that these occupy all spheres, but for the further reason that did they not occupy them, there would arise no other agencies: the prompting ideas and sentiments having been obliterated.

There must be added a concomitant influence on the intellectual nature, which cooperates with the moral influences just named. Personal causation is alone recognized, and the conception of impersonal causation is prevented from developing. The primitive man has no idea of cause in the modern sense. The only agents included in his theory of things are living persons and the ghosts of dead persons. All unusual occurrences, together with those usual ones liable to variation, he ascribes to supernatural beings. And this system of interpretation survives through early stages of civilization; as we see, for example, among the Homeric Greeks, by whom wounds, deaths, and escapes in battle, were ascribed to the enmity or the aid of the gods, and by whom good and bad acts were held to be divinely prompted. Continuance and development of militant forms and activities maintain this way of thinking. In the first place, it indirectly hinders the discovery of causal relations. The sciences grow out of the arts—begin as generalizations of truths which practice of the arts makes manifest. In proportion as processes of production multiply in their kinds and increase in their complexities, more numerous uniformities come to be recognized; and the ideas of necessary relation and physical cause arise and develop. Consequently, by discouraging industrial progress, militancy checks the replacing of ideas of personal agency by ideas of impersonal agency. In the second place, it does the like by direct repression of intellectual culture. Naturally a life occupied in acquiring knowledge, like a life occupied in industry, is regarded with contempt by a people devoted to arms. The Spartans clearly exemplified this relation in ancient times; and it was again exemplified during feudal ages in Europe, when learning was scorned as proper only for clerks and the children of mean people. And obviously, in proportion as warlike activities are antagonistic to study and the spread of knowledge, they further retard that emancipation from primitive ideas which ends in recognition of natural uniformities. In the third place, and chiefly, the effect in question is produced by the conspicuous and perpetual experience of personal agency which the militant *régime* yields. In the army, from the commander-in-chief down to the private undergoing drill, every movement is directed by a superior; and throughout the society, in proportion as its regimentation is elaborate, things are hourly seen to go thus or thus according to the regulating wills of the ruler and his subordinates. In the interpretation of social affairs, personal causation is consequently alone recognized. History comes to be made up of the doings of remarkable men; and it is tacitly assumed that societies have been formed by them. Wholly foreign to the habit of mind as is the thought of impersonal causation, the course of social evolution is unperceived. The natural genesis of social structures and functions is an utterly alien conception, and appears absurd when alleged. The notion of a self-regulating social process is unintelligible. So that militancy moulds the citizen into a form not only morally adapted but intellectually adapted—a form which cannot think away from the entailed system.

§ 561. In three ways, then, we are shown the character of the militant type of social organization. Observe the congruities which comparison of results discloses.

Certain conditions, manifest *à priori*, have to be fulfilled by a society fitted for preserving itself in presence of antagonist societies. To be in the highest degree efficient, the corporate action needed for preserving the corporate life must be joined in by every one. Other things equal, the fighting power will be greatest where those who cannot fight, labour exclusively to support and help those who can: an evident implication being that the working part shall be no larger than is required for these ends. The efforts of all being utilized directly or indirectly for war, will be most effectual when they are most combined; and, besides union among the combatants, there must be such union of the non-combatants with them as renders the aid of these fully and promptly available. To satisfy these requirements, the life, the actions, and the possessions, of each individual must be held at the service of the society. This universal service, this combination, and this merging of individual claims, pre-suppose a despotic controlling agency. That the will of the soldier-chief may be operative when the aggregate is large, there must be sub-centres and sub-sub-centres in descending grades, through whom orders may be conveyed and enforced, both throughout the combatant part and the non-combatant part. As the commander tells the soldier both what he shall not do and what he shall do; so, throughout the militant community at large, the rule is both negatively regulative and positively regulative: it not only restrains, but it directs: the citizen as well as the soldier lives under a system of compulsory cooperation. Development of the militant type involves increasing rigidity, since the cohesion, the combination, the subordination, and the regulation, to which the units of a society are subjected by it, inevitably decrease their ability to change their social positions, their occupations, their localities.

On inspecting sundry societies, past and present, large and small, which are, or have been, characterized in high degrees by militancy, we are shown, *à posteriori*, that amid the differences due to race, to circumstances, and to degrees of development, there are fundamental similarities of the kinds above inferred *à priori*. Modern Dahomey and Russia, as well as ancient Peru, Egypt, and Sparta, exemplify that owning of the individual by the State in life, liberty, and goods, which is proper to a social system adapted for war. And that with changes further fitting a society for warlike activities, there spread throughout it an officialism, a dictation, and a superintendence, akin to those under which soldiers live, we are shown by imperial Rome, by imperial Germany, and by England since its late aggressive activities.

Lastly comes the evidence furnished by the adapted characters of the men who compose militant societies. Making success in war the highest glory, they are led to identify goodness with bravery and strength. Revenge becomes a sacred duty with them; and acting at home on the law of retaliation which they act on abroad, they similarly, at home as abroad, are ready to sacrifice others to self: their sympathies, continually deadened during war, cannot be active during peace. They must have a patriotism which regards the triumph of their society as the supreme end of action; they must possess the loyalty whence flows obedience to authority; and that they may be obedient they must have abundant faith. With faith in authority and consequent readiness to be directed, naturally goes relatively little power of initiation. The habit of seeing everything officially controlled fosters the belief that official control is everywhere needful; while a course of life which makes personal causation familiar and negatives experience of impersonal causation, produces an inability to conceive of any social processes as carried on under self-regulating arrangements. And these traits of individual nature, needful concomitants as we see of the militant type, are those which we observe in the members of actual militant societies.

CHAPTER XVIII.

THE INDUSTRIAL TYPE OF SOCIETY.

§ 562. Having nearly always to defend themselves against external enemies, while they have to carry on internally the processes of sustentation, societies, as remarked in the last chapter, habitually present us with mixtures of the structures adapted to these diverse ends. Disentanglement is not easy. According as either structure predominates it ramifies through the other: instance the fact that where the militant type is much developed, the worker, ordinarily a slave, is no more free than the soldier; while, where the industrial type is much developed, the soldier, volunteering on specified terms, acquires in so far the position of a free worker. In the one case the system of status, proper to the fighting part, pervades the working part; while in the other the system of contract, proper to the working part, affects the fighting part. Especially does the organization adapted for war obscure that adapted for industry. While, as we have seen, the militant type as theoretically constructed, is so far displayed in many societies as to leave no doubt about its essential nature, the industrial type has its traits so hidden by those of the still-dominant militant type, that its nature is nowhere more than very partially exemplified. Saying thus much to exclude expectations which cannot be fulfilled, it will be well also to exclude certain probable misconceptions.

In the first place, industrialism must not be confounded with industriousness. Though the members of an industrially-organized society are habitually industrious, and are, indeed, when the society is a developed one, obliged to be so; yet it must not be assumed that the industrially-organized society is one in which, of necessity, much work is done. Where the society is small, and its habitat so favourable that life may be comfortably maintained with but little exertion, the social relations which characterize the industrial type may coexist with but very moderate productive activities. It is not the diligence of its members which constitutes the society an industrial one in the sense here intended, but the form of cooperation under which their labours, small or great in amount, are carried on. This distinction will be best understood on observing that, conversely, there may be, and often is, great industry in societies framed on the militant type. In ancient Egypt there was an immense labouring population and a large supply of commodities, numerous in their kinds, produced by it. Still more did ancient Peru exhibit a vast community purely militant in its structure, the members of which worked unceasingly. We are here concerned, then, not with the quantity of labour but with the mode of organization of the labourers. A regiment of soldiers can be set to construct earth-works; another to cut down wood; another to bring in water; but they are not thereby reduced for the time being to an industrial society. The united individuals do these several things under command; and having no private claims to the products, are, though industrially occupied, not industrially organized. And the same holds throughout the militant society as a whole, in proportion as the regimentation of it approaches completeness.

The industrial type of society, properly so called, must also be distinguished from a type very likely to be confounded with it—the type, namely, in which the component individuals, while exclusively occupied in production and distribution, are under a regulation such as that advocated by socialists and communists. For this, too, involves in another form the principle of compulsory cooperation. Directly or indirectly, individuals are to be prevented from severally and independently occupying themselves as they please; are to be prevented from competing with one another in supplying goods for money; are to be prevented from hiring themselves out on such terms as they think fit. There can be no artificial system for regulating labour which does not interfere with the natural system. To such extent as men are debarred from making whatever engagements they like, they are to that extent working under dictation. No matter in what way the controlling agency is constituted, it stands towards those controlled in the same relation as does the controlling agency of a militant society. And how truly the *régime* which those who declaim against competition would establish, is thus characterized, we see both in the fact that communistic forms of organization existed in early societies which were predominantly warlike, and in the fact that at the present time communistic projects chiefly originate among, and are most favoured by, the more warlike societies.

A further preliminary explanation may be needful. The structures proper to the industrial type of society must not be looked for in distinct forms when they first appear. Contrariwise, we must expect them to begin in vague unsettled forms. Arising, as they do, by modification of pre-existing structures, they are necessarily long in losing all trace of these. For example, transition from the state in which the labourer, owned like a beast, is maintained that he may work exclusively for his master's benefit, to the condition in which he is completely detached from master, soil, and locality, and free to work anywhere and for anyone, is through gradations. Again, the change from the arrangement proper to militancy, under which subject-persons receive, in addition to maintenance, occasional presents, to the arrangement under which, in place of both, they received fixed wages, or salaries, or fees, goes on slowly and unobtrusively. Once more it is observable that the process of exchange, originally indefinite, has become definite only where industrialism is considerably developed. Barter began, not with a distinct intention of giving one thing for another thing equivalent in value, but it began by making a present and receiving a present in return; and even now in the East there continue traces of this primitive transaction. In Cairo the purchase of articles from a shopkeeper is preceded by his offer of coffee and cigarettes; and during the negotiation which ends in the engagement of a *dahabeah*, the dragoman brings gifts and expects to receive them. Add to which that there exists under such conditions none of that definite equivalence which characterizes exchange among ourselves: prices are not fixed, but vary widely with every fresh transaction. So that throughout our interpretations we must keep in view the truth, that the structures and functions proper to the industrial type distinguish themselves but gradually from those proper to the militant type.

Having thus prepared the way, let us now consider what are, *à priori*, the traits of that social organization which, entirely unfitted for carrying on defence against external enemies, is exclusively fitted for maintaining the life of the society by subserving the lives of its units. As before in treating of the militant type, so here in treating of the industrial type, we will consider first its ideal form.

§ 563. While corporate action is the primary requirement in a society which has to preserve itself in presence of hostile societies, conversely, in the absence of hostile societies, corporate action is no longer the primary requirement.

The continued existence of a society implies, first, that it shall not be destroyed bodily by foreign foes, and implies, second, that it shall not be destroyed in detail by failure of its members to support and propagate themselves. If danger of destruction from the first cause ceases, there remains only danger of destruction from the second cause. Sustentation of the society will now be achieved by the self-sustentation and multiplication of its units. If his own welfare and the welfare of his offspring is fully achieved by each, the welfare of the society is by implication achieved. Comparatively little corporate activity is now required. Each man may maintain himself by labour, may exchange his products for the products of others, may give aid and receive payment, may enter into this or that combination for carrying on an undertaking, small or great, without the direction of the society as a whole. The remaining end to be achieved by public action is to keep private actions within due bounds; and the amount of public action needed for this becomes small in proportion as private actions become duly self-bounded.

So that whereas in the militant type the demand for corporate action is intrinsic, such demand for corporate action as continues in the industrial type is mainly extrinsic—is called for by those aggressive traits of human nature which chronic warfare has fostered, and may gradually diminish as, under enduring peaceful life, these decrease.

§ 564. In a society organized for militant action, the individuality of each member has to be so subordinated in life, liberty, and property, that he is largely, or completely, *owned* by the State; but in a society industrially organized, no such subordination of the individual is called for. There remain no occasions on which he is required to risk his life while destroying the lives of others; he is not forced to leave his occupation and submit to a commanding officer; and it ceases to be needful that he should surrender for public purposes whatever property is demanded of him.

Under the industrial *régime* the citizen's individuality, instead of being sacrificed by the society, has to be defended by the society. Defence of his individuality becomes the society's essential duty. That after external protection is no longer called for, internal protection must become the cardinal function of the State, and that effectual discharge of this function must be a predominant trait of the industrial type, may be readily shown.

For it is clear that, other things equal, a society in which life, liberty, and property, are secure, and all interests justly regarded, must prosper more than one in which they are not; and, consequently, among competing industrial societies, there must be a gradual replacing of those in which personal rights are imperfectly maintained, by those in which they are perfectly maintained. So that by survival of the fittest must be produced a social type in which individual claims, considered as sacred, are trenched on by the State no further than is requisite to pay the cost of maintaining them, or rather, of arbitrating among them. For the aggressiveness of nature fostered by militancy having died out, the corporate function becomes that of deciding between those conflicting claims, the equitable adjustment of which is not obvious to the persons concerned.

§ 565. With the absence of need for that corporate action by which the efforts of the whole society may be utilized for war, there goes the absence of need for a despotic controlling agency.

Not only is such an agency unnecessary, but it cannot exist. For since, as we see, it is an essential requirement of the industrial type, that the individuality of each man shall have the fullest play compatible with the like play of other men's individualities, despotic control, showing itself as it must by otherwise restricting men's individualities, is necessarily excluded. Indeed, by his mere presence an autocratic ruler is an aggressor on citizens. Actually or potentially exercising power not given by them, he in so far restrains their wills more than they would be restrained by mutual limitation merely.

§ 566. Such control as is required under the industrial type, can be exercised only by an appointed agency for ascertaining and executing the average will; and a representative agency is the one best fitted for doing this.

Unless the activities of all are homogeneous in kind, which they cannot be in a developed society with its elaborate division of labour, there arises a need for conciliation of divergent interests; and to the end of insuring an equitable adjustment, each interest must be enabled duly to express itself. It is, indeed, supposable that the appointed agency should be a single individual. But no such single individual could arbitrate justly among numerous classes variously occupied, without hearing evidence: each would have to send representatives setting forth its claims. Hence the choice would lie between two systems, under one of which the representatives privately and separately stated their cases to an arbitrator on whose single judgment decisions depended; and under the other of which these representatives stated their cases in one another's presence, while judgments were openly determined by the general *consensus*. Without insisting on the fact that a fair balancing of class-interests is more likely to be effected by this last form of representation than by the first, it is sufficient to remark that it is more congruous with the nature of the industrial type; since men's individualities are in the smallest degree trenched upon. Citizens who, appointing a single ruler for a prescribed time, may have a majority of their wills traversed by his during this time, surrender their individualities in a greater degree than do those who, from their local groups, depute a number of rulers; since these, speaking and acting under public inspection and mutually restrained, habitually conform their decisions to the wills of the majority.

§ 567. The corporate life of the society being no longer in danger, and the remaining business of government being that of maintaining the conditions requisite for the highest individual life, there comes the question—What are these conditions?

Already they have been implied as comprehended under the administration of justice; but so vaguely is the meaning of this phrase commonly conceived, that a more specific statement must be made. Justice then, as here to be understood, means preservation of the normal connexions between acts and results—the obtainment by each of as much benefit as his efforts are equivalent to—no more and no less. Living and working within the restraints imposed by one another's presence, justice requires that individuals shall severally take the consequences of their conduct, neither increased nor decreased. The superior shall have the good of his superiority; and the inferior the evil of his inferiority. A veto is therefore put on all public action which abstracts from some men part of the advantages they have earned, and awards to other men advantages they have not earned.

That from the developed industrial type of society there are excluded all forms of communistic distribution, the inevitable trait of which is that they tend to equalize the lives of good and bad, idle and diligent, is readily proved. For when, the struggle for existence between societies by war having ceased, there remains only the industrial struggle for existence, the final survival and spread must be on the part of those societies which produce the largest number of the best individuals—individuals best adapted for life in the industrial state. Suppose two societies, otherwise equal, in one of which the superior are allowed to retain, for their own benefit and the benefit of their offspring, the entire proceeds of their labour; but in the other of which the superior have taken from them part of these proceeds for the benefit of the inferior and their offspring. Evidently the superior will thrive and multiply more in the first than in the second. A greater number of the best children will be reared in the first; and eventually it will outgrow the second. It must not be inferred that private and voluntary aid to the inferior is negatived, but only public and enforced aid. Whatever effects the sympathies of the better for the worse spontaneously produce, cannot, of course, be interfered with; and will, on the whole, be beneficial. For while, on the average, the better will not carry such efforts so far as to impede their own multiplication, they will carry them far enough to mitigate the ill-fortunes of the worse without helping them to multiply.

§ 568. Otherwise regarded, this system under which the efforts of each bring neither more nor less than their natural returns, is the system of contract.

We have seen that the *régime* of status is in all ways proper to the militant type. It is the concomitant of that graduated subordination by which the combined action of a fighting body is achieved, and which must pervade the fighting society at large to insure its corporate action. Under this *régime*, the relation between labour and produce is traversed by authority. As in the army, the food, clothing, &c., received by each soldier are not direct returns for work done, but are arbitrarily apportioned, while duties are arbitrarily enforced; so throughout the rest of the militant society, the superior dictates the labour and assigns such share of the returns as he pleases. But as, with declining militancy and growing industrialism, the power and range of authority decrease while uncontrolled action increases, the relation of contract becomes general; and in the fully-developed industrial type it becomes universal.

Under this universal relation of contract when equitably administered, there arises that adjustment of benefit to effort which the arrangements of the industrial society have to achieve. If each as producer, distributor, manager, adviser, teacher, or aider of other kind, obtains from his fellows such payment for his service as its value, determined by the demand, warrants; then there results that correct apportioning of reward to merit which ensures the prosperity of the superior.

§ 569. Again changing the point of view, we see that whereas public control in the militant type is both positively regulative and negatively regulative, in the industrial type it is negatively regulative only. To the slave, to the soldier, or to other member of a community organized for war, authority says—"Thou shalt do this; thou shalt not do that." But to the member of the industrial community, authority gives only one of these orders—"Thou shalt not do that."

For people who, carrying on their private transactions by voluntary cooperation, also voluntarily cooperate to form and support a governmental agency, are, by implication, people who authorize it to impose on their respective activities, only those restraints which they are all interested in maintaining—the restraints which check aggressions. Omitting criminals (who under the assumed conditions must be very few, if not a vanishing quantity), each citizen will wish to preserve uninvaded his sphere of action, while not invading others' spheres, and to retain whatever benefits are achieved within it. The very motive which prompts all to unite in upholding a public protector of their individualities, will also prompt them to unite in preventing any interference with their individualities beyond that required for this end.

Hence it follows that while, in the militant type, regimentation in the army is paralleled by centralized administration throughout the society at large; in the industrial type, administration, becoming decentralized, is at the same time narrowed in its range. Nearly all public organizations save that for administering justice, necessarily disappear; since they have the common character that they either aggress on the citizen by dictating his actions, or by taking from him more property than is needful for protecting him, or by both. Those who are forced to send their children to this or that school, those who have, directly or indirectly, to help in supporting a State priesthood, those from whom rates are demanded that parish officers may administer public charity, those who are taxed to provide gratis reading for people who will not save money for library subscriptions, those whose businesses are carried on under regulation by inspectors, those who have to pay the costs of State science-and-art-teaching, State emigration, &c., all have their individualities trenched upon, either by compelling them to do what they would not spontaneously do, or by taking away money which else would have furthered their private ends. Coercive arrangements of such kinds, consistent with the militant type, are inconsistent with the industrial type.

§ 570. With the relatively narrow range of public organizations, there goes, in the industrial type, a relatively wide range of private organizations. The spheres left vacant by the one are filled by the other.

Several influences conspire to produce this trait. Those motives which, in the absence of that subordination necessitated by war, make citizens unite in asserting their individualities subject only to mutual limitations, are motives which make them unite in resisting any interference with their freedom to form such private combinations as do not involve aggression. Moreover, beginning with exchanges of goods and services under agreements between individuals, the principle of voluntary cooperation is simply carried out in a larger way by individuals who, incorporating themselves, contract with one another for jointly pursuing this or that business or function. And yet again, there is entire congruity between the representative constitutions of such private combinations, and that representative constitution of the public combination which we see is proper to the industrial type. The same law of organization pervades the society in general and in detail. So that an inevitable trait of the industrial type is the multiplicity and heterogeneity of associations, political, religious, commercial, professional, philanthropic, and social, of all sizes.

§ 571. Two indirectly resulting traits of the industrial type must be added. The first is its relative plasticity.

So long as corporate action is necessitated for national self-preservation—so long as, to effect combined defence or offence, there is maintained that graduated subordination which ties all inferiors to superiors, as the soldier is tied to his officer—so long as there is maintained the relation of status, which tends to fix men in the positions they are severally born to; there is insured a comparative rigidity of social organization. But with the cessation of those needs that initiate and preserve the militant type of structure, and with the establishment of contract as the universal relation under which efforts are combined for mutual advantage, social organization loses its rigidity. No longer determined by the principle of inheritance, places and occupations are now determined by the principle of efficiency; and changes of structure follow when men, not bound to prescribed functions, acquire the functions for which they have proved themselves most fit. Easily modified in its arrangements, the industrial type of society is therefore one which adapts itself with facility to new requirements.

§ 572. The other incidental result to be named is a tendency towards loss of economic autonomy.

While hostile relations with adjacent societies continue, each society has to be productively self-sufficing; but with the establishment of peaceful relations, this need for self-sufficingness ceases. As the local divisions composing one of our great nations, had, while they were at feud, to produce each for itself almost everything it required, but now permanently at peace with one another, have become so far mutually dependent that no one of them can satisfy its wants without aid from the rest; so the great nations themselves, at present forced in large measure to maintain their economic autonomies, will become less forced to do this as war decreases, and will gradually become necessary to one another. While, on the one hand, the facilities possessed by each for certain kinds of production, will render exchange mutually advantageous; on the other hand, the citizens of each will, under the industrial *régime*, tolerate no such restraints on their individualities as are implied by interdicts on exchange or impediments to exchange.

With the spread of industrialism, therefore, the tendency is towards the breaking down of the divisions between nationalities, and the running through them of a common organization: if not under a single government, then under a federation of governments.

§ 573. Such being the constitution of the industrial type of society to be inferred from its requirements, we have now to inquire what evidence is furnished by actual societies that approach towards this constitution accompanies the progress of industrialism.

As, during the peopling of the Earth, the struggle for existence among societies, from small hordes up to great nations, has been nearly everywhere going on; it is, as before said, not to be expected that we should readily find examples of the social type appropriate to an exclusively industrial life. Ancient records join the journals of the day in proving that thus far no civilized or semi-civilized nation has fallen into circumstances making needless all social structures for resisting aggression; and from every region travellers' accounts bring evidence that almost universally among the uncivilized, hostilities between tribes are chronic. Still, a few examples exist which show, with tolerable clearness, the outline of the industrial type in its rudimentary form—the form which it assumes where culture has made but little progress. We will consider these first; and then proceed to disentangle the traits distinctive of the industrial type as exhibited by large nations which have become predominantly industrial in their activities.

Among the Indian hills there are many tribes belonging to different races, but alike in their partially-nomadic habits. Mostly agricultural, their common practice is to cultivate a patch of ground while it yields average crops, and when it is exhausted to go elsewhere and repeat the process. They have fled before invading peoples, and have here and there found localities in which they are able to carry on their peaceful occupations unmolested: the absence of molestation being, in somes cases, due to their ability to live in a malarious atmosphere which is fatal to the Aryan races. Already, under other heads, I have referred to the Bodo and to the Dhimáls as wholly unmilitary, as lacking political organization, as being without slaves or social grades, and as aiding one another in their heavier undertakings; to the Todas, who, leading tranquil lives, are "without any of those bonds of union which man in general is induced to form from a sense of danger," and who settle their disputes by arbitration or by a council of five; to the Mishmies as being unwarlike, as having but nominal chiefs, and as administering justice by an assembly; and I have joined with these the case of a people remote in locality and race—the ancient Pueblos of North America—who, sheltering in their walled vil'ages and fighting only when invaded, similarly united with their habitually industrial life a free form of government: "the governor and his council are [were] annually elected by the people." Here I may add sundry kindred examples. As described in the Indian Government Report for 1869—70, "the 'white Karens' are of a mild and peaceful disposition, . . . their chiefs are regarded as patriarchs, who have little more than a nominal authority;" or, as said of them by Lieut. McMahon, "they possess neither laws nor dominant authority." Instance, again, the "fascinating" Lepchas; not industrious, but yet industrial in the sense that their social relations are of the non-militant type. Though I find nothing specific said about the system under which they live in their temporary villages; yet the facts told us sufficiently imply its uncoercive character. They have no castes; "family and political feuds are alike unheard of amongst them;" "they are averse to soldiering;" they prefer taking refuge in the jungle and living on wild food "to enduring any injustice or harsh treatment"—traits which negative ordinary political control. Take next the "quiet, unoffensive" Santals, who, while they fight if need be with infatuated bravery to resist aggression, are essentially unaggressive. These people "are industrious cultivators, and enjoy their existence unfettered by caste." Though, having become tributaries, there habitually exists in each village a head appointed by the Indian Government to be responsible for the tribute, &c.; yet the nature of their indigenous government remains sufficiently clear. While there is a patriarch who is honoured, but who rarely interferes, "every village has its council place, . . . where the committee assemble and discuss the affairs of the village and its inhabitants. All petty disputes, both of a civil and criminal nature, are settled there." What little is told us of tribes living in the Shervaroy Hills is, so far as it goes, to like effect. Speaking generally of them, Shortt says they "are essentially a timid and harmless people, addicted chiefly to pastoral and agricultural pursuits;" and more specifically describing one division of them, he says "they lead peaceable lives among themselves, and any dispute that may arise is usually settled by arbitration." Then, to show that these social traits are not peculiar to any one variety of man, but are dependent on conditions, I may recall the before-named instance of the Papuan Arafuras, who, without any divisions of rank or hereditary chieftainships, live in harmony, controlled only by the decisions of their assembled elders. In all which cases we may discern the leading traits above indicated as proper to societies not impelled to corporate action by war. Strong centralized control not being required, such government as exists is exercised by a council, informally approved—a rude representative government; class-distinctions do not exist, or are but faintly indicated— the relation of *status* is absent; whatever transactions take place between individuals are by agreement; and the function which the ruling body has to perform, becomes substantially limited to protecting private life by settling such disputes as arise, and inflicting mild punishments for small offences.

Difficulties meet us when, turning to civilized societies, we seek in them for traits of the industrial type. Consolidated and organized as they have all been by wars actively carried on throughout the earlier periods of their existence, and mostly continued down to recent times; and having simultaneously been developing within themselves organizations for producing and distributing commodities, which have little by little become contrasted with those proper to militant activities; the two are everywhere presented so mingled that clear separation of the first from the last is, as said at the outset, scarcely practicable. Radically opposed, however, as is compulsory cooperation, the organizing principle of the militant type, to voluntary cooperation, the organizing principle of the industrial type, we may, by observing the decline of institutions exhibiting the one, recognize, by implication, the growth of institutions exhibiting the other. Hence if, in passing from the first states of civilized nations in which war is the business of life, to states in which hostilities are but occasional, we simultaneously pass to states in which the ownership of the individual by his society is not so constantly and strenuously enforced, in which the subjection of rank to rank is mitigated, in which political rule is no longer autocratic, in which the regulation of citizens' lives is diminished in range and rigour, while the protection of them is increased; we are, by implication, shown the traits of a developing industrial type. Comparisons of several kinds disclose results which unite in verifying this truth.

Take, first, the broad contrast between the early condition of the more civilized European nations at large, and their later condition. Setting out from the dissolution of the Roman empire, we observe that for many centuries during which conflicts were effecting consolidations, and dissolutions, and re-consolidations in endless variety, such energies as were not directly devoted to war were devoted to little else than supporting the organizations which carried on war: the working part of each community did not exist for its own sake, but for the sake of the fighting part. While militancy was thus high and industrialism undeveloped, the reign of superior strength, continually being established by societies one over another, was equally displayed within each society. From slaves and serfs, through vassals of different grades up to dukes and kings, there was an enforced subordination by which the individualities of all were greatly restricted. And at the same time that, to carry on external aggression or resistance, the ruling power in each group sacrificed the personal claims of its members, the function of defending its members from one another was in but small degree discharged by it: they were left to defend themselves. If with these traits of European societies in mediæval times, we compare their traits in modern times, we see the following essential differences. First, with the formation of nations covering large areas, the perpetual wars within each area have ceased; and though the wars between nations which from time to time occur are on larger scales, they are less frequent, and they are no longer the business of all freemen. Second, there has grown up in each country a relatively large population which carries on production and distribution for its own maintenance; so that whereas of old, the working part existed for the benefit of the fighting part, now the fighting part exists mainly for the benefit of the working part—exists ostensibly to protect it in the quiet pursuit of its ends. Third, the system of status, having under some of its forms disappeared and under others become greatly mitigated, has been almost universally replaced by the system of contract. Only among those who, by choice or by conscription, are incorporated in the military organization, does the system of status in its primitive rigour still hold so long as they remain in this organization. Fourth, with this decrease of compulsory coöperation and increase of voluntary coöperation, there have diminished or ceased many minor restraints over individual actions. Men are less tied to their localities than they were; they are not obliged to profess certain religious opinions; they are less debarred from expressing their political views; they no longer have their dresses and modes of living dictated to them; they are comparatively little restrained from forming private combinations and holding meetings for one or other purpose—political, religious, social. Fifth, while the individualities of citizens are less aggressed upon by public agency, they are more protected by public agency against aggression. Instead of a *régime* under which individuals rectified their private wrongs by force as well as they could, or else bribed the ruler, general or local, to use his power in their behalf, there has come a *régime* under which, while much less self-protection is required, a chief function of the ruling power and its agents is to administer justice. In all ways, then, we are shown that with this relative decrease of militancy and relative increase of industrialism, there has been a change from a social order in which individuals exist for the benefit of the State, to a social order in which the State exists for the benefit of individuals.

When, instead of contrasting early European communities at large with European communities at large as they now exist, we contrast the one in which industrial development has been less impeded by militancy with those in which it has been more impeded by militancy, parallel results are apparent. Between our own society and continental societies, as for example, France, the differences which have gradually arisen may be cited in illustration. After the conquering Normans had spread over England, there was established here a much greater subordination of local rulers to the general ruler than existed in France; and, as a result, there was not nearly so much internal dissension. Says Hallam, speaking of this period, "we read very little of private wars in England." Though from time to time, as under Stephen, there were rebellions, and though there were occasional fights between nobles, yet for some hundred and fifty years, up to the time of King John, the subjection maintained secured comparative order. Further, it is to be noted that such general wars as occurred were mostly carried on abroad. Descents on our coasts were few and unimportant, and conflicts with Wales, Scotland, and Ireland, entailed but few intrusions on English soil. Consequently, there was a relatively small hindrance to industrial life and the growth of social forms appropriate to it. Meanwhile, the condition of France was widely different. During this period and long after, besides wars with England (mostly fought out on French soil) and wars with other countries, there were going on everywhere local wars. From the 10th to the 14th century perpetual fights between suzerains and their vassals occurred, as well as fights of vassals with one another. Not until towards the middle of the 14th century did the king begin greatly to predominate over the nobles; and only in the 15th century was there established a supreme ruler strong enough to prevent the quarrels of local rulers. How great was the repression of industrial development caused by internal conflicts, may be inferred from the exaggerated language of an old writer, who says of this period, during which the final struggle of monarchy with feudalism was going on, that "agriculture, traffic, and all the mechanical arts ceased." Such being the contrast between the small degree in which industrial life was impeded by war in England, and the great degree in which it was impeded by war in France, let us ask—what were the political contrasts which arose. The first fact to be noted is that in the middle of the 13th century there began in England a mitigation of villeinage, by limitation of labour-services and commutation of them for money, and that in the 14th century the transformation of a servile into a free population had in great measure taken place; while in France, as in other continental countries, the old condition survived and became worse. As Mr. Freeman says of this period—"in England villeinage was on the whole dying out, while in many other countries it was getting harder and harder." Besides this spreading substitution of contract for status, which, taking place first in the industrial centres, the towns, afterwards went on in the rural districts, there was going on an analogous enfranchisement of the noble class. The enforced military obligations of vassals were more and more replaced by money payments or scutages; so that by King John's time, the fighting services of the upper class had been to a great extent compounded for, like the labour services of the lower class. After diminished restraints over persons, there came diminished invasions of property. By the Charter, arbitrary tallages on towns and non-military king's tenants were checked; and while the aggressive actions of the State were thus decreased, its protective actions were extended: provisions were made that justice should be neither sold, delayed, nor denied. All which changes were towards those social arrangements which we see characterize the industrial type. Then, in the next place, we have the subsequently-occurring rise of a representative government; which, as shown in a preceding chapter by another line of inquiry, is at once the product of industrial growth and the form proper to the industrial type. But in France none of these changes took place. Villeinage remaining unmitigated continued to comparatively late times; compounding for military obligation of vassal to suzerain was less general; and when there arose tendencies towards the establishment of an assembly expressing the popular will, they proved abortive. Detailed comparisons of subsequent periods and their changes would detain us too long: it must suffice to indicate the leading facts. Beginning with the date at which, under the influences just indicated, parliamentary government was finally established in England, we find that for a century and a half, down to the Wars of the Roses, the internal disturbances were few and unimportant compared with those which took place in France; and at the same time (remembering that the wars between England and France, habitually taking place on French soil, affected the state of France more than that of England) we note that

France carried on serious wars with Flanders, Castille and Navarre besides the struggle with Burgundy: the result being that while in England popular power as expressed by the House of Commons became settled and increased, such power as the States General had acquired in France, dwindled away. Not forgetting that by the Wars of the Roses, lasting over thirty years, there was initiated a return towards absolutism; let us contemplate the contrasts which subsequently arose. For a century and a half after these civil conflicts ended, there were but few and trivial breaches of internal peace; while such wars as went on with foreign powers, not numerous, took place as usual out of England. During this period the retrograde movement which the Wars of the Roses set up, was reversed, and popular power greatly increased; so that in the words of Mr. Bagehot, "the slavish parliament of Henry VIII. grew into the murmuring parliament of Queen Elizabeth, the mutinous Parliament of James I., and the rebellious parliament of Charles I." Meanwhile France, during the first third of this period, had been engaged in almost continuous external wars with Italy, Spain, and Austria; while during the remaining two-thirds, it suffered from almost continuous internal wars, religious and political: the accompanying result being that, notwithstanding resistances from time to time made, the monarchy became increasingly despotic. Fully to make manifest the different social types which had been evolved under these different conditions, we have to compare not only the respective political constitutions but also the respective systems of social control. Observe what these were at the time when there commenced that reaction which ended in the French revolution. In harmony with the theory of the militant type, that the individual is in life, liberty, and property, owned by the State, the monarch was by some held to be the universal proprietor. The burdens he imposed upon landowners were so grievous that a part of them preferred abandoning their estates to paying. Then besides the taking of property by the State, there was the taking of labour. One-fourth of the working days in the year went to the *corvées*, due now to the king and now to the feudal lord. Such liberties as were allowed, had to be paid for and again paid for: the municipal privileges of towns being seven times in twenty-eight years withdrawn and re-sold to them. Military services of nobles and people were imperative to whatever extent the king demanded; and conscripts were drilled under the lash. At the same time that the subjection of the individual to the State was pushed to such an extreme by exactions of money and services that the impoverished people cut the grain while it was green, ate grass, and died of starvation in multitudes, the State did little to guard their persons and homes. Contemporary writers enlarge on the immense numbers of highway robberies, burglaries, assassinations, and torturings of people to discover their hoards. Herds of vagabonds, levying blackmail, roamed about; and when, as a remedy, penalties were imposed, innocent persons denounced as vagabonds were sent to prison without evidence. No personal security could be had either against the ruler or against powerful enemies. In Paris there were some thirty prisons where untried and unsentenced people might be incarcerated; and the "brigandage of justice" annually cost suitors forty to sixty millions of francs. While the State, aggressing on citizens to such extremes, thus failed to protect them against one another, it was active in regulating their private lives and labours. Religion was dictated to the extent that Protestants were imprisoned, sent to the galleys, or whipped, and their ministers hanged. The quantity of salt (on which there was a heavy tax) to be consumed by each person was prescribed; as were also the modes of its use. Industry of every kind was supervised. Certain crops were prohibited; and vines destroyed that were on soils considered unfit. The wheat that might be bought at market was limited to two bushels; and sales took place in presence of dragoons. Manufacturers were regulated in their processes and products to the extent that there was destruction of improved appliances and of goods not made according to law, as well as penalties upon inventors. Regulations succeeded one another so rapidly that amid their multiplicity, government agents found it difficult to carry them out; and with increasing official orders there came increasing swarms of public functionaries. Turning now to England at the same period, we see that along with progress towards the industrial type of political structure, carried to the extent that the House of Commons had become the predominant power, there had gone a progress towards the accompanying social system. Though the subjection of the individual to the State was considerably greater than now, it was far less than in France. His private rights were not sacrificed in the same unscrupulous way; and he was not in danger of a *lettre de cachet*. Though justice was very imperfectly administered, still it was not administered so wretchedly: there was a fair amount of personal security, and aggressions on property were kept within bounds. The disabilities of Protestant dissenters were diminished early in the century; and, later on, those of Catholics. Considerable freedom of the press was acquired, showing itself in the discussion of political questions, as well as in the publication of parliamentary debates; and, about the same time, there came free speech in public meetings. While thus the State aggressed on the individual less and protected him more, it interfered to a smaller extent with his daily transactions. Though there was much regulation of commerce and industry, yet it was pushed to no such extreme as that which in France subjected agriculturists, manufacturers, and merchants, to an army of officials who directed their acts at every turn. In brief, the contrast between our state and that of France was such as to excite the surprise and admiration of various French writers of the time; from whom Mr. Buckle quotes numerous passages showing this.

Most significant of all, however, are the changes in England itself, first retrogressive and then progressive, that occurred during the war-period which extended from 1775 to 1815, and during the subsequent period of peace. At the end of the last century and the beginning of this, reversion towards ownership of the individual by the society had gone a long way. "To statesmen, the State, as a unit, was all in all, and it is really difficult to find any evidence that the people were thought of at all, except in the relation of obedience." "The Government regarded the people with little other view than as a taxable and soldier-yielding mass." While the militant part of the community had greatly developed, the industrial part had approached towards the condition of a permanent commissariat. By conscription and by press-gangs, was carried to a relatively vast extent that sacrifice of the citizen in life and liberty which war entails; and the claims to property were trenched on by merciless taxation, weighing down the middle classes so grievously that they had greatly to lower their rate of living, while the people at large were so distressed (partly no doubt by bad harvests) that "hundreds ate nettles and other weeds." With these major aggressions upon the individual by the State, went numerous minor aggressions. Irresponsible agents of the executive were empowered to suppress public meetings and seize their leaders: death being the punishment for those who did not disperse when ordered. Libraries and news-rooms could not be opened without licence; and it was penal to lend books without permission. There were "strenuous attempts made to silence the press;" and booksellers dared not publish works by obnoxious authors. "Spies were paid, witnesses were suborned, juries were packed, and the *habeas corpus* Act being constantly suspended, the Crown had the power of imprisoning without inquiry and without limitation." While the Government taxed and coerced and restrained the citizen to this extent, its protection of him was inefficient. It is true that the penal code was made more extensive and more severe. The definition of treason was enlarged, and numerous offences were made capital which were not capital before; so that there was "a vast and absurd variety of offences for which men and women were sentenced to death by the score:" there was "a devilish levity in dealing with human life." But at the same time there was not an increase, but rather a decrease, of security. As says Mr. Pike in his *History of Crime in England*, "it became apparent that the greater the strain of the conflict the greater is the danger of a reaction towards violence and lawlessness." Turn now to the opposite picture. After recovery from the prostration which prolonged wars had left, and after the dying away of those social perturbations caused by impoverishment, there began a revival of traits proper to the industrial type. Coercion of the citizen by the State decreased in various ways. Voluntary enlistment replaced compulsory military service; and there disappeared some minor restraints over personal freedom, as instance the repeal of laws which forbade artizans to travel where they pleased, and which interdicted trades-unions. With these manifestations of greater respect for personal freedom, may be joined those shown in the amelioration of the penal code: the public whipping of females being first abolished; then the long list of capital offences being reduced until there finally remained but one; and, eventually, the pillory and imprisonment for debt being abolished. Such penalties on religious independence as remained disappeared; first by removal of those directed against Protestant Dissenters, and then of those which weighed on Catholics, and then of some which told specially against Quakers and Jews. By the Parliamentary Reform Bill and the Municipal Reform Bill, vast numbers were removed from the subject classes to the governing classes. Interferences with the business-transactions of citizens were diminished by allowing free trade in bullion, by permitting joint-stock banks, by abolishing multitudinous restrictions on the importation of commodities—leaving eventually but few which pay duty. Moreover while these and kindred changes, such as the removal of restraining burdens on the press, decreased the impediments to free actions of citizens, the protective action of the State was increased. By a greatly-improved police system, by county courts, and so forth, personal safety and claims to property were better secured.

Not to elaborate the argument further by adding the case of the United States, which repeats with minor differences the same relations of phenomena, the evidence given adequately supports the proposition laid down. Amid all the complexities and perturbations, comparisons show us with sufficient clearness that in actually-existing societies those attributes which we inferred must distinguish the industrial type, show themselves clearly in proportion as the social activities are predominantly characterized by exchange of services under agreement.

§ 574. As, in the last chapter, we noted the traits of character proper to the members of a society which is habitually at war; so here, we have to note the traits of character proper to the members of a society occupied exclusively in peaceful pursuits. Already in delineating above, the rudiments of the industrial type of social structure as exhibited in certain small groups of unwarlike peoples, some indications of the accompanying personal qualities have been given; but it will be well now to emphasize these and add to them, before observing the kindred personal qualities in more advanced industrial communities.

Absence of a centralized coercive rule, implying as it does feeble political restraints exercised by the society over its units, is accompanied by a strong sense of individual freedom, and a determination to maintain it. The amiable Bodo and Dhimáls, as we have seen, resist "injunctions injudiciously urged with dogged obstinacy." The peaceful Lepchas "undergo great privations rather than submit to oppression or injustice." The "simple-minded Santál" has a "strong natural sense of justice, and should any attempt be made to coerce him, he flies the country." Similarly of a tribe not before mentioned, the Jakuns of the South Malayan Peninsula, who, described as "entirely inoffensive," personally brave but peaceful, and as under no control but that of popularly-appointed heads who settle their disputes, are also described as "extremely proud:" the so-called pride being exemplified by the statement that their remarkably good qualities "induced several persons to make attempts to domesticate them, but such essays have generally ended in the Jakuns' disappearance on the slightest coercion."

With a strong sense of their own claims, these unwarlike men display unusual respect for the claims of others. This is shown in the first place by the rarity of personal collisions among them. Hodgson says that the Bodo and the Dhimáls "are void of all violence towards their own people or towards their neighbours." Of the peaceful tribes of the Neilgherry Hills, Colonel Ouchterlony writes:—"drunkenness and violence are unknown amongst them." Campbell remarks of the Lepchas, that "they rarely quarrel among themselves." The Jakuns, too, "have very seldom quarrels among themselves;" and such disputes as arise are settled by their popularly-chosen heads "without fighting or malice." In like manner the Arafuras "live in peace and brotherly love with one another." Further, in the accounts of these peoples we read nothing about the *lex talionis*. In the absence of hostilities with adjacent groups there does not exist within each group that "sacred duty of blood-revenge" universally recognized in military tribes and nations. Still more significantly, we find evidence of the opposite doctrine and practice. Says Campbell of the Lepchas—"they are singularly forgiving of injuries . . . making mutual amends and concessions."

Naturally, with respect for others' individualities thus shown, goes respect for their claims to property. Already in the preliminary chapter I have quoted testimonies to the great honesty of the Bodo and the Dhimáls, the Lepchas, the Santáls, the Todas, and other peoples kindred in their form of social life; and here I may add further ones. Of the Lepchas,

Hooker remarks:—"in all my dealings with these people, they proved scrupulously honest." "Among the pure Santáls," writes Hunter, "crime and criminal officers are unknown;" while of the Hos, belonging to the same group as the Santáls, Dalton says, "a reflection on a man's honesty or veracity may be sufficient to send him to self-destruction." Shortt testifies that "the Todas, as a body, have never been convicted of heinous crimes of any kind;" and concerning other tribes of the Shervaroy Hills, he states that "crime of a serious nature is unknown amongst them." Again of the Jakuns we read that "they are never known to steal anything, not even the most insignificant trifle." And so of certain natives of Malacca who "are naturally of a commercial turn," Jukes writes:—"no part of the world is freer from crime than the district of Malacca;" "a few petty cases of assault, or of disputes about property . . . are all that occur."

Thus free from the coercive rule which warlike activities necessitate, and without the sentiment which makes the needful subordination possible—thus maintaining their own claims while respecting the like claims of others—thus devoid of the vengeful feelings which aggressions without and within the tribe generate; these peoples, instead of the bloodthirstiness, the cruelty, the selfish trampling upon inferiors, characterizing militant tribes and societies, display, in unusual degrees, the humane sentiments. Insisting on their amiable qualities, Hodgson describes the Bodo and the Dhimáls as being "almost entirely free from such as are unamiable." Remarking that "while courteous and hospitable he is firm and free from cringing," Hunter tells us of the Santál that he thinks "uncharitable men" will suffer after death. Saying that the Lepchas are "ever foremost in the forest or on the bleak mountain, and ever ready to help, to carry, to encamp, collect, or cook," Hooker adds—"they cheer on the traveller by their unostentatious zeal in his service;" and he also adds that, "a present is divided equally amongst many, without a syllable of discontent or grudging look or word." Of the Jakuns, too, Favre tells us that "they are generally kind, affable, inclined to gratitude and to beneficence:" their tendency being not to ask favours but to confer them. And then of the peaceful Arafuras we learn from Kolff that—

"They have a very excusable ambition to gain the name of rich men, by paying the debts of their poorer villagers. The officer [M. Bik], whom I quoted above, related to me a very striking instance of this. At Affara he was present at the election of the village chiefs, two individuals aspiring to the station of Orang Tua. The people chose the elder of the two, which greatly afflicted the other, but he soon afterwards expressed himself satisfied with the choice the people had made, and said to M. Bik, who had been sent there on a commission, 'What reason have I to grieve; whether I am Orang Tua or not, I still have it in my power to assist my fellow villagers.' Several old men agreed to this, apparently to comfort him. Thus the only use they make of their riches is to employ it in settling differences."

With these superiorities of the social relations in permanently peaceful tribes, go superiorities of the domestic relations. As I have before pointed out (§ 327), while the status of women is habitually very low in tribes given to war and in more advanced militant societies, it is habitually very high in these primitive peaceful societies. The Bodo and the Dhimáls, the Kocch, the Santáls, the Lepchas, are monogamic, as were also the Pueblos; and along with their monogamy habitually goes a superior sexual morality. Of the Lepchas Hooker says—"the females are generally chaste, and the marriage tie is strictly kept." Among the Santáls "unchastity is almost unknown," and "divorce is rare." By the Bodo and the Dhimáls, "polygamy, concubinage and adultery are not tolerated;" "chastity is prized in man and woman, married and unmarried." Further it is to be noted that the behaviour to women is extremely good. "The Santál treats the female members of his family with respect;" the Bodo and the Dhimáls "treat their wives and daughters with confidence and kindness; they are free from all out-door work whatever." And even among the Todas, low as are the forms of their sexual relations, "the wives are treated by their husbands with marked respect and attention." Moreover, we are told concerning sundry of these unwarlike peoples that the status of children is also high; and there is none of that distinction of treatment between boys and girls which characterizes militant peoples.

Of course on turning to the civilized to observe the form of individual character which accompanies the industrial form of society, we encounter the difficulty that the personal traits proper to industrialism, are, like the social traits, mingled with those proper to militancy. It is manifestly thus with ourselves. A nation which, besides its occasional serious wars, is continually carrying on small wars with uncivilized tribes—a nation which is mainly ruled in Parliament and through the press by men whose school-discipline led them during six days in the week to take Achilles for their hero, and on the seventh to admire Christ—a nation which, at its public dinners, habitually toasts its army and navy before toasting its legislative bodies; has not so far emerged out of militancy that we can expect either the institutions or the characteristics proper to industrialism to be shown with clearness. In independence, in honesty, in truthfulness, in humanity, its citizens are not likely to be the equals of the uncultured but peaceful peoples above described. All we may anticipate is an approach to those moral qualities appropriate to a state undisturbed by international hostilities; and this we find.

In the first place, with progress of the *régime* of contract has come growth of independence. Daily exchange of services under agreement, involving at once the maintenance of personal claims and respect for the claims of others, has fostered a normal self-assertion and consequent resistance to unauthorized power. The facts that the word "independence," in its modern sense, was not in use among us before the middle of the last century, and that on the continent independence is less markedly displayed, suggest the connexion between this trait and a developing industrialism.

The trait is shown in the multitudinousness of religious sects, in the divisions of political parties, and, in minor ways, by the absence of those "schools" in art, philosophy, &c., which, among continental peoples, are formed by the submission of disciples to an adopted master. That Englishmen show, more than their neighbours, a jealousy of dictation, and a determination to act as they think fit, will not, I think, be disputed.

The diminished subordination to authority, which is the obverse of this independence, of course implies decrease of loyalty. Worship of the monarch, at no time with us reaching the height it did in France early in the last century, or in Russia down to recent times, has now changed into a respect depending very much on the monarch's personal character. Our days witness no such extreme servilities of expression as were used by ecclesiastics in the dedication of the Bible to King James, nor any such exaggerated adulations as those addressed to George III. by the House of Lords. The doctrine of divine right has long since died away; belief in an indwelling supernatural power (implied by the touching for king's evil, &c.) is named as a curiosity of the past; and the monarchical institution has come to be defended on grounds of expediency. So great has been the decrease of this sentiment which, under the militant *régime*, attaches subject to ruler, that now-a-days the conviction commonly expressed is that, should the throne be occupied by a Charles II. or a George IV., there would probably result a republic. And this change of feeling is shown in the attitude towards the Government as a whole. For not only are there many who dispute the authority of the State in respect of sundry matters besides religious beliefs, but there are some who passively resist what they consider unjust exercises of its authority, and pay fines or go to prison rather than submit.

As this last fact implies, along with decrease of loyalty has gone decrease of faith, not in monarchs only but in governments. Such belief in royal omnipotence as existed in ancient Egypt, where the power of the ruler was supposed to extend to the other world, as it is even now supposed to do in China, has had no parallel in the West; but still, among European peoples in past times, that confidence in the soldier-king essential to the militant type, displayed itself among other ways in exaggerated conceptions of his ability to rectify mischiefs, achieve benefits, and arrange things as he willed. If we compare present opinion among ourselves with opinion in early days, we find a decline in these credulous expectations. Though, during the late retrograde movement towards militancy, State-power has been invoked for various ends, and faith in it has increased; yet, up to the commencement of this reaction, a great change had taken place in the other direction. After the repudiation of a State-enforced creed, there came a denial of the State's capacity for determining religious truth, and a growing movement to relieve it from the function of religious teaching; held to be alike needless and injurious. Long ago it had ceased to be thought that Government could do any good by regulating people's food, clothing, and domestic habits; and over the multitudinous processes carried on by producers and distributors, constituting immensely the larger part of our social activities, we no longer believe that legislative dictation is beneficial. Moreover, every newspaper by its criticisms on the acts of ministers and the conduct of the House of Commons, betrays the diminished faith of citizens in their rulers. Nor is it only by contrasts between past and present among ourselves that we are shown this trait of a more developed industrial state. It is shown by kindred contrasts between opinion here and opinion abroad. The speculations of social reformers in France and in Germany, prove that the hope for benefits to be achieved by State-agency is far higher with them than with us.

Along with decrease of loyalty and concomitant decrease of faith in the powers of governments, has gone decrease of patriotism—patriotism, that is, under its original form. To fight "for king and country" is an ambition which now-a-days occupies but a small space in men's minds; and though there is among us a majority whose sentiment is represented by the exclamation—" Our country, right or wrong!" yet there are large numbers whose desire for human welfare at large, so far overrides their desire for national prestige, that they object to sacrificing the first to the last. The spirit of self-criticism, which in sundry respects leads us to make unfavourable comparisons between ourselves and our continental neighbours, leads us more than heretofore to blame ourselves for wrong conduct to weaker peoples. The many and strong reprobations of our dealings with the Afghans, the Zulus, and the Boers, show that there is a large amount of the feeling reprobated by the "Jingo"-class as unpatriotic.

That adaptation of individual nature to social needs, which, in the militant state, makes men glory in war and despise peaceful pursuits, has partially brought about among us a converse adjustment of the sentiments. The occupation of the soldier has ceased to be so much honoured, and that of the civilian is more honoured. During the forty years' peace, the popular sentiment became such that "soldiering" was spoken of contemptuously; and those who enlisted, habitually the idle and the dissolute, were commonly regarded as having completed their disgrace. Similarly in America before the late civil war, such small military gatherings and exercises as from time to time occurred, excited general ridicule. Meanwhile we see that labours, bodily and mental, useful to self and others, have come to be not only honourable but in a considerable degree imperative. In America the adverse comments on a man who does nothing, almost force him into some active pursuit; and among ourselves the respect for industrial life has become such that men of high rank put their sons into business.

While, as we saw, the compulsory cooperation proper to militancy, forbids, or greatly discourages, individual initiative, the voluntary cooperation which distinguishes industrialism, gives free scope to individual initiative, and develops it by letting enterprise bring its normal advantages. Those who are successfully original in idea and act, prospering and multiplying in a greater degree than others, produce, in course of time, a general type of nature ready to undertake new things. The speculative tendencies of English and American capitalists, and the extent to which large undertakings, both at home and abroad, are carried out by them, sufficiently indicate this trait of character. Though, along with considerable qualifications of militancy by industrialism on the continent, there has occurred there, too, an extension of private enterprise; yet the fact that while many towns in France and Germany have been supplied with gas and water by English companies, there is in England but little of kindred achievement by foreign companies, shows that among the more industrially-modified English, individual initiative is more decided.

There is evidence that the decline of international hostilities, associated as it is with the decline of hostilities between families and between individuals, is followed by a weakening of revengeful sentiments. This is implied by the fact that in our own country the more serious of these private wars early ceased, leaving only the less serious in the form of duels, which also have at length ceased: their cessation coinciding with the recent great development of industrial life—a fact with which may be joined the fact that in the more militant societies, France and Germany, they have not ceased. So much among ourselves has the authority of the *lex talionis* waned, that a man whose actions are known to be prompted by the wish for vengeance on one who has injured him, is reprobated rather than applauded.

With decrease of the aggressiveness shown in acts of violence and consequent acts of retaliation, has gone decrease of the aggressiveness shown in criminal acts at large. That this change has been a concomitant of the change from a more militant to a more industrial state, cannot be doubted by one who studies the history of crime in England. Says

Mr. Pike in his work on that subject, "the close connexion between the military spirit and those actions which are now legally defined to be crimes, has been pointed out, again and again, in the course of this history." If we compare a past age in which the effects of hostile activities had been less qualified by the effects of peaceful activities than they are in our own age, we see a marked contrast in respect of the numbers and kinds of offences against person and property. We have no longer any English buccaneers; wreckers have ceased to be heard of; and travellers do not now prepare themselves to meet highwaymen. Moreover, that flagitiousness of the governing agencies themselves, which was shown by the venality of ministers and members of Parliament, and by the corrupt administration of justice, has disappeared. With decreasing amount of crime has come increasing reprobation of crime. Biographies of pirate captains, suffused with admiration of their courage, no longer find a place in our literature; and the sneaking kindness for "gentlemen of the road," is, in our days, but rarely displayed. Many as are the transgressions which our journals report, they have greatly diminished; and though in trading transactions there is much dishonesty (chiefly of the indirect sort) it needs but to read Defoe's *English Tradesman*, to see how marked has been the improvement since his time. Nor must we forget that the change of character which has brought a decrease of unjust actions, has brought an increase of beneficent actions; as seen in paying for slave-emancipation, in nursing the wounded soldiers of our fighting neighbours, in philanthropic efforts of countless kinds.

§ 575. As with the militant type then, so with the industrial type, three lines of evidence converge to show us its essential nature. Let us set down briefly the several results, that we may observe the correspondences among them.

On considering what must be the traits of a society organized exclusively for carrying on internal activities, so as most efficiently to subserve the lives of citizens, we find them to be these. A corporate action subordinating individual actions by uniting them in joint effort, is no longer requisite. Contrariwise, such corporate action as remains has for its end to guard individual actions against all interferences not necessarily entailed by mutual limitation: the type of society in which this function is best discharged, being that which must survive, since it is that of which the members will most prosper. Excluding, as the requirements of the industrial type do, a despotic controlling agency, they imply, as the only congruous agency for achieving such corporate action as is needed, one formed of representatives who serve to express the aggregate will. The function of this controlling agency, generally defined as that of administering justice, is more specially defined as that of seeing that each citizen gains neither more nor less of benefit than his activities normally bring; and there is thus excluded all public action involving any artificial distribution of benefits. The *régime* of status proper to militancy having disappeared, the *régime* of contract which replaces it has to be universally enforced; and this negatives interferences between efforts and results by arbitrary apportionment. Otherwise regarded, the industrial type is distinguished from the militant type as being not both positively regulative and negatively regulative, but as being negatively regulative only. With this restricted sphere for corporate action comes an increased sphere for individual action; and from that voluntary cooperation which is the fundamental principle of the type, arise multitudinous private combinations, akin in their structures to the public combination of the society which includes them. Indirectly it results that a society of the industrial type is distinguished by plasticity; and also that it tends to lose its economic autonomy, and to coalesce with adjacent societies.

The question next considered was, whether these traits of the industrial type as arrived at by deduction are inductively verified; and we found that in actual societies they are visible more or less clearly in proportion as industrialism is more or less developed. Glancing at those small groups of uncultured people who, wholly unwarlike, display the industrial type in its rudimentary form, we went on to compare the structures of European nations at large in early days of chronic militancy, with their structures in modern days characterized by progressing industrialism; and we saw the differences to be of the kind implied. We next compared two of these societies, France and England, which were once in kindred states, but of which the one has had its industrial life much more repressed by its militant life than the other; and it became manifest that the contrasts which, age after age, arose between their institutions, were such as answer to the hypothesis. Lastly, limiting ourselves to England itself, and first noting how recession from such traits of the industrial type as had shown themselves, occurred during a long war-period, we observed how, during the subsequent long period of peace beginning in 1815, there were numerous and decided approaches to that social structure which we concluded must accompany developed industrialism.

We then inquired what type of individual nature accompanies the industrial type of society; with the view of seeing whether, from the character of the unit as well as from the character of the aggregate, confirmation is to be derived. Certain uncultured peoples whose lives are passed in peaceful occupations, proved to be distinguished by independence, resistance to coercion, honesty, truthfulness, forgivingness, kindness. On contrasting the characters of our ancestors during more warlike periods with our own characters, we see that, with an increasing ratio of industrialism to militancy, have come a growing independence, a less-marked loyalty, a smaller faith in governments, and a more qualified patriotism; and while, by enterprising action, by diminished faith in authority, by resistance to irresponsible power, there has been shown a strengthening assertion of individuality, there has accompanied it a growing respect for the individualities of others, as is implied by the diminution of aggressions upon them and the multiplication of efforts for their welfare.

To prevent misapprehension it seems needful, before closing, to explain that these traits are to be regarded less as the immediate results of industrialism than as the remote results of non-militancy. It is not so much that a social life passed in peaceful occupations is positively moralizing, as that a social life passed in war is positively demoralizing. Sacrifice of others to self is in the one incidental only; while in the other it is necessary. Such aggressive egoism as accompanies the industrial life is extrinsic; whereas the aggressive egoism of the militant life is intrinsic. Though generally unsympathetic, the exchange of services under agreement is now, to a considerable extent, and may be wholly, carried on with a due regard to the claims of others—may be constantly accompanied by a sense of benefit given as well as benefit received; but the slaying of antagonists, the burning of their houses, the appropriation of their territory, cannot but be accompanied by vivid consciousness of injury done them, and a consequent brutalizing effect on the feelings—an effect wrought, not on soldiers only, but on those who employ them and contemplate their deeds with pleasure. The last form of social life, therefore, inevitably deadens the sympathies and generates a state of mind which prompts crimes of trespass; while the first form, allowing the sympathies free play if it does not directly exercise them, favours the growth of altruistic sentiments and the resulting virtues.

NOTE.—This reference to the natural genesis of a higher moral nature, recalls a controversy some time since carried on. In a "Symposium" published in the *Nineteenth Century* for April and May, 1877, was discussed "the influence upon morality of a decline in religious belief:" the question eventually raised being whether morality can exist without religion. Not much difficulty in answering this question will be felt by those who, from the conduct of the rude tribes described in this chapter, turn to that of Europeans during a great part of the Christian era; with its innumerable and immensurable public and private atrocities, its bloody aggressive wars, its ceaseless family-vendettas, its bandit barons and fighting bishops, its massacres, political and religious, its torturings and burnings, its all-pervading crime from the assassinations of and by kings down to the lyings and petty thefts of slaves and serfs. Nor do the contrasts between our own conduct at the present time and the conduct of these so-called savages, leave us in doubt concerning the right answer. When, after reading police reports, criminal assize proceedings, accounts of fraudulent bankruptcies, &c., which in our journals accompany advertisements of sermons and reports of religious meetings, we learn that the "amiable" Bodo and Dhimáls, who are so "honest and truthful," "have no word for God, for soul, for heaven, for hell" (though they have ancestor-worship and some derivative beliefs), we find ourselves unable to recognize the alleged connexion. If, side by side with narratives of bank-frauds, railway-jobbings, turf-chicaneries, &c., among people who are anxious that the House of Commons should preserve its theism untainted, we place descriptions of the "fascinating" Lepchas, who are so "wonderfully honest," but who "profess no religion, though acknowledging the existence of good and bad spirits" (to the last of whom only they pay any attention), we do not see our way to accepting the dogma which our theologians think so obviously true; nor will acceptance of it be made easier when we add the description of the conscientious Santál, who "never thinks of making money by a stranger," and "feels pained if payment is pressed upon him" for food offered; but concerning whom we are told that "of a supreme and beneficent God the Santál has no conception." Admission of the doctrine that right conduct depends on theological conviction, becomes difficult on reading that the Veddahs who are "almost devoid of any sentiment of religion" and have no idea "of a Supreme Being," nevertheless "think it perfectly inconceivable that any person should ever take that which does not belong to him, or strike his fellow, or say anything that is untrue." After finding that among the select of the select who profess our established creed, the standard of truthfulness is such that the statement of a minister concerning cabinet transactions is distinctly falsified by the statement of a seceding minister; and after then recalling the marvellous veracity of these godless Bodo and Dhimáls, Lepchas, and other peaceful tribes having kindred beliefs, going to such extent that an imputation of falsehood is enough to make one of the Hos destroy himself; we fail to see that in the absence of a theistic belief there can be no regard for truth. When, in a weekly journal specially representing the university culture shared in by our priests, we find a lament over the moral degradation shown by our treatment of the Boers—when we are held degraded because we have not slaughtered them for successfully resisting our trespasses—when we see that the "sacred duty of blood revenge," which the cannibal savage insists upon, is insisted upon by those to whom the Christian religion was daily taught throughout their education; and when, from contemplating this fact, we pass to the fact that the unreligious Lepchas "are singularly forgiving of injuries," the assumed relation between humanity and theism appears anything but congruous with the evidence. If, with the ambitions of our church-going citizens, who (not always in very honourable ways) strive to get fortunes that they may make great displays, and gratify themselves by thinking that at death they will "cut up well," we compare the ambitions of the Arafuras, among whom wealth is desired that its possessor may pay the debts of poorer men and settle differences, we are obliged to reject the assumption that "brotherly love" can exist only as a consequence of divine injunctions, with promised rewards and threatened punishments; for of these Arafuras we read that—

"Of the immortality of the soul they have not the least conception. To all my enquiries on the subject they answered, 'No Arafura has ever returned to us after death, therefore we know nothing of a future state, and this is the first time we have heard of it.' Their idea was, when you are dead there is an end of you. Neither have they any notion of the creation of the world. They only answered, 'None of us were aware of this, we have never heard anything about it, and therefore do not know who has done it all.'"

The truth disclosed by the facts is that, so far as men's moral states are concerned, theory is almost nothing and practice is almost everything. No matter how high their nominal creed, nations given to political burglaries to get "scientific frontiers," and the like, will have among their members many who "annex" others' goods for their own convenience; and with the organized crime of aggressive war, will go criminality in the behaviour of one citizen to another. Conversely, as these uncultivated tribes prove, no matter how devoid they are of religious beliefs, those who, generation after generation remaining unmolested, inflict no injuries upon others, have their altruistic sentiments fostered by the sympathetic intercourse of a peaceful daily life, and display the resulting virtues. We need teaching that it is impossible to join injustice and brutality abroad with justice and humanity at home. What a pity these Heathens cannot be induced to send missionaries among the Christians!

CHAPTER XIX.

POLITICAL RETROSPECT AND PROSPECT.

§ 576. In the foregoing chapters little has been said concerning the doctrine of Evolution at large, as re-illustrated by political evolution; though doubtless the observant reader has occasionally noted how the transformations described conform to the general law of transformation. Here, in summing up, it will be convenient briefly to indicate their conformity. Already in Part II, when treating of Social Growth, Social Structures, and Social Functions, the outlines of this correspondence were exhibited; but the materials for exemplifying it in a more special way, which have been brought together in this Part, may fitly be utilized to emphasize afresh a truth not yet commonly admitted.

That under its primary aspect political development is a process of integration, is clear. By it individuals originally separate are united into a whole; and the union of them into a whole is variously shown. In the earliest stages the groups of men are small, they are loose, they are not unified by subordination to a centre. But with political progress comes the compounding, re-compounding, and re-re-compounding of groups until great nations are produced. Moreover, with that settled life and agricultural development accompanying political progress, there is not only a formation of societies covering wider areas, but an increasing density of their populations. Further, the loose aggregation of savages passes into the coherent connexion of citizens; at one stage coercively bound to one another and to their localities by family-ties and class-ties, and at a later stage voluntarily bound together by their mutually-dependent occupations. Once more, there is that merging of individual wills in a governmental will, which reduces a society, as it reduces an army, to a consolidated body.

An increase of heterogeneity at the same time goes on in many ways. Everywhere the horde, when its members co-operate for defence or offence, begins to differentiate into a predominant man, a superior few, and an inferior many. With that massing of groups which war effects, there grow out of these, head chief, subordinate chiefs, and warriors; and at higher stages of integration, kings, nobles, and people: each of the two great social strata presently becoming differentiated within itself. When small societies have been united, the respective triune governing agencies of them grow unlike: the local political assemblies falling into subordination to a central political assembly. Though, for a time, the central one continues to be constituted after the same manner as the local ones, it gradually diverges in character by loss of its popular element. While these local and central bodies are becoming contrasted in their powers and structures, they are severally becoming differentiated in another way. Originally each is at once military, political, and judicial; but by and by the assembly for judicial business, no longer armed, ceases to be like the politico-military assembly; and the politico-military assembly eventually gives origin to a consultative body, the members of which, when meeting for political deliberation, come unarmed. Within each of these divisions, again, kindred changes subsequently occur. While themselves assuming more specialized forms, local judicial agencies fall under the control of a central judicial agency; and the central judicial agency, which has separated from the original consultative body, subdivides into parts or courts which take unlike kinds of business. The central political body, too, where its powers do not disappear by absorption in those of the supreme head, tends to complicate; as in our own case by the differentiation of a privy council from the original consultative body, and again by the differentiation of a cabinet from the privy council: accompanied, in the other direction, by division of the consultative body into elective and non-elective parts. While these metamorphoses are going on, the separation of the three organizations, legislative, judicial, and executive, progresses. Moreover, with progress in these major political changes goes that progress in minor political changes which, out of family-governments and clan-governments, evolves such governments as those of the tything, the gild, and the municipality. Thus in all directions from primitive simplicity there is produced ultimate complexity, through modifications upon modifications.

With this advance from small incoherent social aggregates to great coherent ones, which, while becoming integrated pass from uniformity to multiformity, there goes an advance from indefiniteness of political organization to definiteness of political organization. Save inherited ideas and usages, nothing is fixed in the primitive horde. But the differentiations above described, severally beginning vaguely, grow in their turns gradually more marked. Class-divisions, absent at first and afterwards undecided, eventually acquire great distinctness: slaves, serfs, freemen, nobles, king, become separated, often by impassable barriers, and their positions shown by mutilations, badges, dresses, &c. Powers and obligations which were once diffused are parted off and rigorously maintained. The various parts of the political machinery come to be severally more and more restricted in their ranges of duties; and usage, age by age accumulating precedents, brings every kind of official action within prescribed bounds. This increase of definiteness is everywhere well shown by the development of laws. Beginning as inherited sacred injunctions briefly expressed, these have to be applied after some prescribed method, and their meanings in relation to particular cases made clear. Rules of procedure become step by step detailed and formal, while interpretations change the general command into specialized commands to meet incidental circumstances; and gradually there grows up a legal system everywhere precise and fixed. How pronounced is this tendency is interestingly shown in our system of Equity, which, arising to qualify the unduly defined and rigid applications of Law, itself slowly multiplied its technicalities until it grew equally defined and rigid.

To meet an obvious criticism it must be added that these changes from societies which are small, loose, uniform, and vague in structure, to societies which are large, compact, multiform, and distinct in structure, present varieties of characters under varieties of conditions, and alter as the conditions alter. Different parts of a society display the transformation, according as the society's activities are of one or other kind. Chronic war generates a compulsory cohesion, and produces an ever-greater heterogeneity and definiteness in that controlling organization by which unity of action is secured; while that part of the organization which carries on production and distribution, exhibits these traits of evolution in a relatively small degree. Conversely, when joint action of the society against other societies decreases, the traits of the structure developed for carrying it on begin to fade; while the traits of the structure for carrying on production and distribution become more decided: the increasing cohesion, heterogeneity, and definiteness, begin now to be shown throughout the industrial organization. Hence the phenomena become complicated by a simultaneous evolution of one part of the social organization and dissolution of another part—a mingling of changes well illustrated in our own society.

§ 577. With this general conception before us, which, without more detailed recapitulation of the conclusions reached, will sufficiently recall them, we may turn from retrospect to prospect; and ask through what phases political evolution is likely hereafter to pass.

Such speculations concerning higher political types as we may allow ourselves, must be taken with the understanding that such types are not likely to become universal. As in the past so in the future, local circumstances must be influential in determining governmental arrangements; since these depend in large measure on the modes of life which the climate, soil, flora, and fauna, necessitate. In regions like those of Central Asia, incapable of supporting considerable populations, there are likely to survive wandering hordes under simple forms of control. Large areas such as parts of Africa present, which prove fatal to the higher races of men, and the steaming atmospheres of which cause enervation, may continue to be inhabited by lower races of men, subject to political arrangements adapted to them. And in conditions such as those furnished by small Pacific Islands, mere deficiency of numbers must negative the forms of government which become alike needful and possible in large nations. Recognizing the fact that with social organisms as with individual organisms, the evolution of superior types does not entail the extinction of all inferior ones, but leaves many of these to survive in habitats not available by the superior, we may here restrict ourselves to the inquiry—What are likely to be the forms of political organization and action in societies that are favourably circumstanced for carrying social evolution to its highest stage?

Of course deductions respecting the future must be drawn from inductions furnished by the past. We must assume that hereafter social evolution will conform to the same principles as heretofore. Causes which have everywhere produced certain effects must, if they continue at work, be expected to produce further effects of like kinds. If we see that political transformations which have arisen under certain conditions, admit of being carried further in the same directions, we must conclude that they will be carried further if the conditions are maintained; and that they will go on until they reach limits beyond which there is no scope for them.

Not indeed that any trustworthy forecast can be made concerning proximate changes. All that has gone before unites to prove that political institutions, fundamentally determined in their forms by the predominance of one or other of the antagonist modes of social action, the militant and the industrial, will be moulded in this way or in that way according as there is frequent war or habitual peace. Hence we must infer that throughout approaching periods, everything will depend on the courses which societies happen to take in their behaviour to one another—courses which cannot be predicted. On the one hand, in the present state of armed preparation throughout Europe, an untoward accident may bring about wars which, lasting perhaps for a generation, will re-develop the coercive forms of political control. On the other hand, a long peace is likely to be accompanied by so vast an increase of manufacturing and commercial activity, with accompanying growth of the appropriate political structures within each nation, and strengthening of those ties between nations which mutual dependence generates, that hostilities will be more and more resisted and the organization adapted for the carrying them on will decay.

Leaving, however, the question—What are likely to be the proximate political changes in the most advanced nations? and inferring from the changes which civilization has thus far wrought out, that at some time, more or less distant, the industrial type will become permanently established, let us now ask—What is to be the ultimate political *régime?*

§ 578. Having so recently contemplated at length the political traits of the industrial type as inferable *à priori*, and as partially exemplified *à posteriori* in societies most favourably circumstanced for evolving them, there remains only to present these under a united and more concrete form, with some dependent ones which have not been indicated. We will glance first at the implied political structures, and next at the implied political functions.

What forms of governmental organization must be the outcome of voluntary cooperation carried to its limit? We have already seen that in the absence of those appliances for coercion which accompany the militant type, whatever legislative and administrative structures exist, must be, in general and in detail, of directly or indirectly representative origin. The presence in them of functionaries not deriving their powers from the aggregate will, and not changeable by the aggregate will, would imply partial continuance of that *régime* of status which the *régime* of contract has, by the hypothesis, entirely replaced. But assuming the exclusion of all irresponsible agents, what particular structures will best serve to manifest and execute the aggregate will? This is a question to which only approximate answers can be given. There are various possible organizations through which the general *consensus* of feeling and opinion may display itself and issue in action; and it is very much a question of convenience, rather than of principle, which of these shall be adopted. Let us consider some of their varieties.

The representatives constituting the central legislature may form one body or they may form two. If there is but one, it may consist of men directly elected by all qualified citizens; or its members may be elected by local bodies which have themselves arisen by direct election; or it may include members some of whom are elected in the one way and some in the other. If there are two chambers, the lower one may arise in the first of the three ways named; while the second arises in one of several ways. It may consist of members chosen by local representative bodies; or it may be chosen by the lower chamber out of its own number. Its members may either have no test of eligibility, or they may be required to have special qualifications: experience in administration, for example. Then besides these various forms of the legislature, there are the various modes in which it may be partially or wholly replaced. Entire dissolution and re-election of one body or of both bodies may occur at intervals, either the same for the two or different for the two, and either simultaneously or otherwise; or the higher body, though representative, may be permanent, while the lower is changeable; or the changing of one or both, at given intervals, may be partial instead of complete—a third or a fourth may vacate their seats annually or biennially, and may or may not be eligible for re-election. So, too, there are various modes by which the executive may originate consistently with the representative principle. It may be simple or it may be compound; and if compound, the members of it may be changeable separately or altogether. The political head may be elected directly by the whole community, or by its local governing bodies, or by one or by both of its central representative bodies; and may be so elected for a term or for life. His assistants or ministers may be chosen by himself; or he may choose one who chooses the rest; or they may be chosen separately or bodily by one or other legislature, or by the two united. And the members of the ministry may compose a group apart from both chambers, or may be members of one or the other.

Concerning these, and many other possible arrangements which may be conceived as arising by modification and complication of them (all apparently congruous with the requirement that the making and administration of laws shall conform to public opinion) the choice is to be guided mainly by regard for simplicity and facility of working. But it seems likely that hereafter, as heretofore, the details of constitutional forms in each society, will not be determined on *à priori* grounds, or will be but partially so determined. We may conclude that they will be determined in large measure by the antecedents of the society; and that between societies of the industrial type, there will be differences of political organization consequent on genealogical differences. Recognizing the analogies furnished by individual organizations, which everywhere show us that structures evolved during the earlier stages of a type for functions then requisite, usually do not disappear at later stages, but become remoulded in adaptation to functions more or less different; we may suspect that the political institutions appropriate to the industrial type, will, in each society, continue to bear traces of the earlier political institutions evolved for other purposes; as we see that even now the new societies growing up in colonies, tend thus to preserve marks of earlier stages passed through by ancestral societies. Hence we may infer that societies which, in the future, have alike become completely industrial, will not present identical political forms; but that to the various possible forms appropriate to the type, they will present approximations determined partly by their own structures in the past and partly by the structures of the societies from which they have been derived. Recognizing this probability, let us now ask by what changes our own political constitution may be brought into congruity with the requirements.

Though there are some who contend that a single body of representatives is sufficient for the legislative needs of a free nation, yet the reasons above given warrant the suspicion that the habitual duality of legislatures, of which the rudiments are traceable in the earliest political differentiation, is not likely to be entirely lost in the future. That spontaneous division of the primitive group into the distinguished few and the undistinguished many, both of which take part in determining the actions of the group—that division which, with reviving power of the undistinguished many, reappears when there is formed a body representing it, which cooperates with the body formed of the distinguished few in deciding on national affairs, appears likely to continue. Assuming that as a matter of course two legislative bodies, if they exist hereafter, must both arise by representation, direct or indirect, it seems probable that an upper and a lower chamber may continue to display a contrast in some degree analogous to that which they have displayed thus far. For however great the degree of evolution reached by an industrial society, it cannot abolish the distinction between the superior and the inferior—the regulators and the regulated. Whatever arrangements for carrying on industry may in times to come be established, must leave outstanding the difference between those whose characters and abilities raise them to the higher positions, and those who remain in the lower. Even should all kinds of production and distribution be eventually carried on by bodies of cooperators, as a few are now to some extent, such bodies must still have their appointed heads and committees of managers. Either from an electorate constituted not, of course, of a permanently-privileged class, but of a class including all heads of industrial organizations, or from an electorate otherwise composed of all persons occupied in administration, a senate may perhaps eventually be formed consisting of the representatives of directing persons as distinguished from the representatives of persons directed. Of course in the general government, as in the government of each industrial body, the representatives of the class regulated must be ultimately supreme; but there is reason for thinking that the representatives of the regulating class might with advantage exercise a restraining power. Evidently the aspect of any law differs according as it is looked at from above or from below—by those accustomed to rule or by those accustomed to be ruled. The two aspects require to be coordinated. Without assuming that differences between the interests of these bodies will, to the last, make needful different representations of them, it may reasonably be concluded that the higher, experienced in administration, may with advantage bring its judgments to bear in qualifying the judgments of the lower, less conversant with affairs; and that social needs are likely to be most effectually met by laws issuing from their joint deliberations. Far from suggesting an ultimate unification of the two legislative bodies, the facts of evolution, everywhere showing advance in specialization, suggest rather that one or both of such two bodies, now characterizing developed political organizations, will further differentiate. Indeed we have at the present moment indications that such a change is likely to take place in our own House of Commons. To the objection that the duality of a legislative body impedes the making of laws, the reply is that a considerable amount of hindrance to change is desirable. Even as it is now among ourselves, immense mischiefs are done by ill-considered legislation; and any change which should further facilitate legislation would increase such mischiefs.

Concerning the ultimate executive agency, it appears to be an unavoidable inference that it must become, in some way or other, elective; since hereditary political headship is a trait of the developed militant type, and forms a part of that *régime* of status which is excluded by the hypothesis. Guided by such evidence as existing advanced societies afford us, we may infer that the highest State-office, in whatever way filled, will continue to decline in importance; and that the functions to be discharged by its occupant will become more and more automatic. There requires an instrumentality having certain traits which we see in our own executive, joined with certain traits which we see in the executive of the United States. On the one hand, it is needful that the men who have to carry out the will of the majority as expressed through the legislature, should be removable at pleasure; so that there may be maintained the needful subordination of their policy to public opinion. On the other hand, it is needful that displacement of them shall leave intact all that part of the executive organization required for current administrative purposes. In our own case these requirements, fulfilled to a considerable extent, fall short of complete fulfilment in the respect that the political head is not elective, and still exercises, especially over the foreign policy of the nation, a considerable amount of power. In the United States, while these requirements are fulfilled in the respect that the political head is elective, and cannot compromize the nation in its actions towards other nations, they are not fulfilled in the respect that far from being an automatic centre, having actions restrained by a ministry responsive to public opinion, he exercises, during his term of office, much independent control. Possibly in the future, the benefits of these two systems may be united and their evils avoided. The strong party antagonisms which accompany our state of transition having died away, and the place of supreme State-officer having become one of honour rather than one of power, it may happen that appointment to this place, made during the closing years of a great career to mark the nation's approbation, will be made without any social perturbation, because without any effect on policy; and that, meanwhile, such changes in the executive agency as are needful to harmonize its actions with public opinion, will be, **as** at present among ourselves, changes of ministries.

Rightly to conceive the natures and workings of the central political institutions appropriate to the industrial type, we must assume that along with the establishment of them there has gone that change just named in passing—the decline of party antagonisms. Looked at broadly, political parties are seen to arise directly or indirectly out of the conflict between militancy and industrialism. Either they stand respectively for the coercive government of the one and the free government of the other, or for particular institutions and laws belonging to the one or the other, or for religious opinions and organizations congruous with the one or the other, or for principles and practices that have been bequeathed by the one or the other, and survived under alien conditions. Habitually if we trace party feeling to its sources, we find on the one side maintenance of, and on the other opposition to, some form of inequity. Wrong is habitually alleged by this side against that; and there **must be** injustice either in the thing done or in the allegation concerning it. Hence as fast as the *régime* of voluntary cooperation with its appropriate ideas, sentiments, and usages, pervades the whole society—as fast as there disappear all those arrangements which in any way trench upon the equal freedom of these or those citizens, party warfare must practically die away. Such differences of opinion only can remain as concern matters of detail and minor questions of administration. Evidently there is approach to such a state in proportion as the graver injustices descending from the militant type disappear. Evidently, too, one concomitant is that increasing subdivision of parties commonly lamented, which promises to bring about the result that no course can be taken at the dictation of any one moiety in power; but every course taken, having the assent of the average of parties, will be thereby proved in harmony with the aggregate will of the community. And clearly, with this breaking up of parties consequent on growing individuality of nature, all such party-antagonisms as we now know must cease.

Concerning local government **we** may conclude that as centralization is an essential trait of the militant type, decentralization is an essential trait of the industrial type. With that independence which the *régime* of voluntary cooperation generates, there arises resistance not only to dictation by one man, and to dictation by a class, but even to dictation by a majority, when it restrains individual action in ways not necessary for maintaining harmonious social relations. One result must be that the inhabitants of each locality will object to be controlled by the inhabitants of other localities, in matters of purely local concern. In respect of such laws as equally apply to all individuals, and such laws as affect the inhabitants of each locality in their intercourse with those of other localities, the will of the majority of the community will be recognized as authoritative; but in respect of arrangements not affecting the community at large, but affecting only the members forming one part, we may infer that there will arise such tendency to resist dictation by members of other parts, as will involve the carrying of local rule to the greatest practicable limit. Municipal and kindred governments may be expected to exercise legislative and administrative powers, subject to no greater control by the central government than is needful for the concord of the whole community.

Neither these nor any other speculations concerning ultimate political forms can, however, be regarded as anything more than tentative. They are ventured here simply as foreshadowing the general nature of the changes to be anticipated; and in so far as they are specific, can be at the best but partially right. We may be sure that the future will bring unforeseen political arrangements along with many other unforeseen things. As already implied, there will probably be considerable variety in the special forms of the political institutions of industrial societies: all of them bearing traces of past institutions which have been brought into congruity with the representative principle. And here I may add that little stress need be laid on one or other speciality of form; since, given citizens having the presupposed appropriate natures, and but small differences in the ultimate effects will result from differences in the machinery used.

§ 579. Somewhat more definitely, and with somewhat greater positiveness, may we, I think, infer the political functions carried on by those political structures proper to the developed industrial type. Already these have been generally indicated; but here they must be indicated somewhat more specifically.

We have seen that when corporate action is no longer needed for preserving the society as a whole from destruction or injury by other societies, the end which remains for it is that of preserving the component members of the society from destruction or injury by one another: injury, as here interpreted, including not only immediate, but also remote, breaches of equity. Citizens whose natures have through many generations of voluntary cooperation and accompanying regard for one another's claims, been moulded into the appropriate form, will entirely agree to maintain such political institutions as may continue needful for insuring to each that the activities he carries on within limits imposed by the activities of others, shall bring to him all the directly-resulting benefits, or such benefits as indirectly result under voluntary agreements; and each will be ready to yield up such small portion of the proceeds of his labour, as may be required to maintain the agency for adjudicating in complex cases where the equitable course is not manifest, and for such legislative and administrative purposes as may prove needful for effecting an equitable division of all natural advantages. Resistance to extension of government beyond the sphere thus indicated, must eventually have a two-fold origin— egoistic and altruistic.

In the first place, it cannot be supposed that citizens having the characters indicated, will, in their corporate capacity, agree to impose on themselves individually, other restraints than those necessitated by regard for one another's spheres of action. Each has had fostered in him by the discipline of daily life carried on under contract, a sentiment prompting assertion of his claim to free action within the implied limits; and there cannot therefore arise in an aggregate of such, any sentiment which would tolerate further limits. And that any part should impose such further limits on the rest, is also contrary to the hypothesis; since it presupposes that political inequality, or status, which is excluded by the industrial type. Moreover, it is manifest that the taking from citizens of funds for public purposes other than those above specified, is negatived. For while there will ever be a unanimous desire to maintain for each and all the conditions needful for severally carrying on their private activities and enjoying the products, the probabilities are immense against agreement for any other public end. And in the absence of such agreement, there must arise resistance by the dissentients to the costs and administrative restraints required for achieving such other end. There must be dissatisfaction and opposition on the part of the minority from whom certain returns of their labours are taken, not for fulfilling their own desires, but for fulfilling the desires of others. There must be an inequality of treatment which does not consist with the *régime* of voluntary cooperation fully carried out.

At the same time that the employment of political agencies for other ends than that of maintaining equitable relations among citizens, will meet with egoistic resistance from a minority who do not desire such other ends, it will also meet with altruistic resistance from the rest. In other words, the altruism of the rest will prevent them from achieving such further ends for their own satisfaction, at the cost of dissatisfaction to those who do not agree with them. To one who is ruled by a predominant sentiment of justice, the thought of profiting in any way, direct or indirect, at the expense of another, is repugnant; and in a community of such, none will desire to achieve by public agency at the cost of all, benefits which a part do not participate in, or do not wish for. Given in all citizens a quick sense of equity, and it must happen, for example, that while those who have no children will protest against the taking away of their property to educate the children of others, the others will no less protest against having the education of their children partially paid for by forced exactions from the childless, from the unmarried, and from those whose means are in many cases less than their own. So that the eventual limitation of State-action to the fundamental one described, is insured by a simultaneous increase of opposition to other actions and a decrease of desire for them.

§ 580. The restricted sphere for political institutions thus inferred as characterizing the developed industrial type, may also be otherwise inferred.

For this limitation of State-functions is one outcome of that process of specialization of functions which accompanies organic and super-organic evolution at large. Be it in an animal or be it in a society, the progress of organization is constantly shown by the multiplication of particular structures adapted to particular ends. Everywhere we see the law to be that a part which originally served several purposes and achieved none of them well, becomes divided into parts each of which performs one of the purposes, and, acquiring specially-adapted structures, performs it better. Throughout the foregoing chapters we have seen this truth variously illustrated by the evolution of the governmental organization itself. It remains here to point out that it is further illustrated in a larger way, by the division which has arisen, and will grow ever more decided, between the functions of the governmental organization as a whole, and the functions of the other organizations which the society includes.

Already we have seen that in the militant type, political control extends over all parts of the lives of the citizens. Already we have seen that as industrial development brings the associated political changes, the range of this control decreases: ways of living are no longer dictated; dress ceases to be prescribed; the rules of class-subordination lose their peremptoriness; religious beliefs and observances are not insisted upon; modes of cultivating the land and carrying on manufactures are no longer fixed by law; and the exchange of commodities, both within the community and with other communities, becomes gradually unshackled. That is to say, as industrialism has progressed, the State has retreated from the greater part of those regulative actions it once undertook. This change has gone along with an increasing opposition of citizens to these various kinds of control, and a decreasing tendency on the part of the State to exercise them. Unless we assume that the end has now been reached, the implication is that with future progress of industrialism, these correlative changes will continue. Citizens will carry still further their resistance to State-dictation; while the tendency to State-dictation will diminish. Though recently, along with re-invigoration of militancy, there have gone extensions of governmental interference, yet this is interpretable as a temporary wave of reaction. We may expect that with the ending of the present retrograde movement and resumption of unchecked industrial development, that increasing restriction of State-functions which has unquestionably gone on during the later stages of civilization, will be resumed; and, for anything that appears to the contrary, will continue until there is reached the limit above indicated.

Along with this progressing limitation of political functions, has gone increasing adaptation of political agencies to the protecting function, and better discharge of it. During unqualified militancy, while the preservation of the society as a whole against other societies was the dominant need, the preservation of the individuals forming the society from destruction or injury by one another, was little cared for; and in so far as it was cared for, was cared for mainly out of regard for the strength of the whole society, and its efficiency for war. But those same changes which have cut off so many political functions at that time exercised, have greatly developed this essential and permanent political function. There has been a growing efficiency of the organization for guarding life and property; due to an increasing demand on the part of citizens that their safety shall be insured, and an increasing readiness on the part of the State to respond. Evidently our own time, with its extended arrangements for administering justice, and its growing wish for codification of the law, exhibits a progress in this direction; which will end only when the State undertakes to administer civil justice to the citizen free of cost, as it now undertakes, free of cost, to protect his person and punish criminal aggression on him.

And the accompanying conclusion is that there will be simultaneously carried further that trait which already characterizes the most industrially-organized societies—the performance of increasingly-numerous and increasingly-important functions by other organizations than those which form departments of the government. Already in our own case private enterprise, working through incorporated bodies of citizens, achieves ends undreamed of as so achievable in primitive societies; and in the future, other ends undreamed of now as so achievable, will be achieved.

§ 581. A corollary having important practical bearings may be drawn. The several changes making up the transformation above indicated, are normally connected in their amounts; and mischief must occur if the due proportions among them are not maintained. There is a certain right relation to one another, and a right relation to the natures of citizens, which may not be disregarded with impunity.

The days when "paper constitutions" were believed in have gone by—if not with all, still with instructed people. The general truth that the characters of the units determine the character of the aggregate, though not admitted overtly and fully, is yet admitted to some extent—to the extent that most politically-educated persons do not expect forthwith completely to change the state of a society by this or that kind of legislation. But when fully admitted, this truth carries with it the conclusion that political institutions cannot be effectually modified faster than the characters of citizens are modified; and that if greater modifications are by any accident produced, the excess of change is sure to be undone by some counter-change. When, as in France, people undisciplined in freedom are suddenly made politically free, they show by some *plébiscite* that they willingly deliver over their power to an autocrat, or they work their parliamentary system in such way as to make a popular statesman into a dictator. When, as in the United States, republican institutions, instead of being slowly evolved, are all at once created, there grows up within them an agency of wire-pulling politicians, exercising a real rule which overrides the nominal rule of the people at large. When, as at home, an extended franchise, very soon re-extended, vastly augments the mass of those who, having before been controlled are made controllers, they presently fall under the rule of an organized body that chooses their candidates and arranges for them a political programme, which they must either accept or be powerless. So that in the absence of a duly-adapted character, liberty given in one direction is lost in another.

Allied to the normal relation between character and institutions, are the normal relations among institutions themselves; and the evils which arise from disregard of the second relations are allied to those which arise from disregard of the first. Substantially there is produced the same general effect. The slavery mitigated in one direction is intensified in another. Coercion over the individual, relaxed here is tightened there. For, as we have seen, that change which accompanies development of the industrial type, and is involved by the progress towards those purely equitable relations which the *régime* of voluntary cooperation brings, implies that the political structures simultaneously become popular in their origin and restricted in their functions. But if they become more popular in their origin without becoming more restricted in their functions, the effect is to foster arrangements which benefit the inferior at the expense of the superior; and by so doing work towards degradation. Swayed as individuals are on the average by an egoism which dominates over their altruism, it must happen that even when they become so far equitable in their sentiments that they will not commit direct injustices, they will remain liable to commit injustices of indirect kinds. And since the majority must ever be formed of the inferior, legislation, if unrestricted in its range, will inevitably be moulded by them in such way as more or less remotely to work out to their own advantage, and to the disadvantage of the superior. The politics of trades'-unions exemplify the tendency. Their usages have become such that the more energetic and skilful workmen are not allowed to profit to the full extent of their capacities; because, if they did so, they would discredit and disadvantage those of lower capacities, who, forming the majority, establish and enforce the usages. In multitudinous ways a like tendency must act through a political organization, if, while all citizens have equal powers, the organization can be used for other purposes than administering justice. State-machineries worked by taxes falling in more than due proportion on those whose greater powers have brought them greater means, will give to citizens of smaller powers more benefits than they have earned. And this burdening of the better for the benefit of the worse, must check the evolution of a higher and more adapted nature: the ultimate result being that a community by which this policy is pursued, will, other things equal, fail in competition with a community which pursues the purely equitable policy, and will eventually disappear in the race of civilization.

In brief, the diffusion of political power unaccompanied by the limitation of political functions, issues in communism. For the direct defrauding of the many by the few, it substitutes the indirect defrauding of the few by the many: evil proportionate to the inequity, being the result in the one case as in the other.

§ 582. But the conclusion of profoundest moment to which all lines of argument converge, is that the possibility of a high social state, political as well as general, fundamentally depends on the cessation of war. After all that has been said it is needless to emphasize afresh the truth that persistent militancy, maintaining adapted institutions, must inevitably prevent, or else neutralize, changes in the direction of more equitable institutions and laws; while permanent peace will of necessity be followed by social ameliorations of every kind.

From war has been gained all that it had to give. The peopling of the Earth by the more powerful and intelligent races, is a benefit in great measure achieved; and what remains to be done, calls for no other agency than the quiet pressure of a spreading industrial civilization on a barbarism which slowly dwindles. That integration of simple groups into compound ones, and of these into doubly compound ones, which war has effected, until at length great nations have been produced, is a process already carried as far as seems either practicable or desirable. Empires formed of alien peoples habitually fall to pieces when the coercive power which holds them together fails; and even could they be held together, would not form harmoniously-working wholes: peaceful federation is the only further consolidation to be looked for. Such large advantage as war has yielded by developing that political organization which, beginning with the leadership of the best warrior has ended in complex governments and systems of administration, has been fully obtained; and there only remains for the future to preserve and re-mould its useful parts while getting rid of those no longer required. So, too, that organization of labour initiated by war—an organization which, setting out with the relation of owner and slave and developing into that of master and servant, has, by elaboration, given us industrial structures having numerous grades of officials, from head-directors down to foremen—has been developed quite as far as is requisite for combined action; and has to be hereafter modified, not in the direction of greater military subordination, but rather in the opposition direction. Again, the power of continuous application, lacking in the savage and to be gained only under that coercive discipline which the militant type of society establishes, has been already in large measure acquired by the civilized man; and such further degree of it as is needed, will be produced under the stress of industrial competition in free communities. Nor is it otherwise with great public works and developed industrial arts. Though, in the canal cut by the Persians across the isthmus of Athos, and again in a canal of two miles long made by the Fijians, we see both that war is the first prompter to such undertakings and that the despotic rule established by it is the needful agency for carrying them out; yet we also see that industrial evolution has now reached a stage at which commercial advantage supplies a sufficient stimulus, and private trading corporations a sufficient power, to execute works far larger and more numerous. And though from early days when flint arrowheads were chipped and clubs carved, down to present days when armour-plates a foot thick are rolled, the needs of defence and offence have urged on invention and mechanical skill; yet in our own generation steam-hammers, hydraulic rams, and multitudinous new appliances from locomotives to telephones, prove that industrial needs alone have come to furnish abundant pressure whereby, hereafter, the industrial arts will be further advanced. Thus, that social evolution which had to be achieved through the conflicts of societies with one another, has already been achieved; and no further benefits are to be looked for.

Only further evils are to be looked for from the continuance of militancy in civilized nations. The general lesson taught by all the foregoing chapters is that, indispensable as has been this process by which nations have been consolidated, organized, and disciplined, and requisite as has been the implied coercion to develop certain traits of individual human nature, yet that, beyond the unimaginable amount of suffering directly involved by the process, there has been an unimaginable amount of suffering indirectly involved; alike by the forms of political institutions necessitated, and by the accompanying type of individual nature fostered. And they show by implication that for the diminution of this suffering, not only of the direct kind but of the indirect kind, the one thing needful is the checking of international antagonisms and the diminution of those armaments which are at once cause and consequence of them. With the repression of militant activities and decay of militant organizations, will come amelioration of political institutions as of all other institutions. Without them, no such ameliorations are permanently possible. Liberty overtly gained in name and form will be unobtrusively taken away in fact.

It is not to be expected, however, that any very marked effects are to be produced by the clearest demonstration of this truth—even by a demonstration beyond all question. A general congruity has to be maintained between the social state at any time necessitated by circumstances, and the accepted theories of conduct, political and individual. Such acceptance as there may be of doctrines at variance with the temporary needs, can never be more than nominal in degree, or limited in range, or both. The acceptance which guides conduct will always be of such theories, no matter how logically indefensible, as are consistent with the average modes of action, public and private. All that can be done by diffusing a doctrine much in advance of the time, is to facilitate the action of forces tending to cause advance. The forces themselves can be but in small degrees increased; but something may be done by preventing mis-direction of them. Of the sentiment at any time enlisted on behalf of a higher social state, there is always some (and at the present time a great deal) which, having the broad vague form of sympathy with the masses, spends itself in efforts for their welfare by multiplication of political agencies of one or other kind. Led by the prospect of immediate beneficial results, those swayed by this sympathy are unconscious that they are helping further to elaborate a social organization at variance with that required for a higher form of social life, and are, by so doing, increasing the obstacles to attainment of that higher form. On a portion of such the foregoing chapters may have some effect by leading them to consider whether the arrangements they are advocating involve increase of that public regulation characterizing the militant type, or whether they tend to produce that greater individuality and more extended voluntary cooperation, characterizing the industrial type. To deter here and there one from doing mischief by imprudent zeal, is the chief proximate effect to be hoped for.

PART VI.

ECCLESIASTICAL INSTITUTIONS.

CHAPTER I.

THE RELIGIOUS IDEA.

§ 583. THERE can be no true conception of a structure without a true conception of its function. To understand how an organization originated and developed, it is requisite to understand the need subserved at the outset and afterwards. Rightly to trace the evolution of Ecclesiastical Institutions, therefore, we must know whence came the ideas and sentiments implied by them. Are these innate or are they derived?

Not only by theologians at large but also by some who have treated religion rationalistically, it is held that man is by constitution a religious being. Prof. Max Müller's speculations are pervaded by this assumption; and in such books as that by Mr. R. W. Mackay on *The Progress of the Intellect*, it is contended that man is by nature a monotheist. But this doctrine, once almost universally accepted, has been rudely shaken by the facts which psychologists and anthropologists have brought to light.

There is clear proof that minds which have from infancy been cut off by bodily defects from intercourse with the minds of adults, are devoid of religious ideas. The deaf Dr. Kitto, in his book called *The Lost Senses* (p. 200), quotes the testimony of an American lady who was deaf and dumb, but at a mature age was instructed, and who said "the idea that the world must have had a Creator never occurred to her, nor to any other of several intelligent pupils, of similar age." Similarly, the Rev. Samuel Smith, after "twenty-eight years' almost daily contact" with such, says of a deaf-mute, "he has no idea of his immortal nature, and it has not been found in a single instance, that an uneducated deaf-mute has had any conception of the existence of a Supreme Being as the Creator and Ruler of the universe."

The implication is that civilized men have no innate tendency to form religious ideas; and this implication is supported by proofs that among various savages religious ideas do not exist. Sir John Lubbock has given many of these in his *Prehistoric Times* and his *Origin of Civilization*; and others may be added. Thus of a Wedda, who, when in jail received instruction, Mr. Hartshorne writes—"he had no idea of a soul, of a Supreme Being, or of a future state." Concerning an African race Heuglin says—"the Dōr do not seem to have religious conceptions properly so called, but they believe in spirits." We learn from Schweinfurth that "the Bongo have not the remotest conception of immortality. . . . All religion, in our sense of the word religion, is quite unknown to the Bongo." It is true that in such cases there is commonly a notion, here distinct and there vague, of something supernatural associated with the dead. While now, in answer to a question, asserting that death brings annihilation, the savage at another time shows great fear of places where the dead are: implying either a half-formed idea that the dead will suddenly awake, as a sleeper does, or else some faint notion of a double. Not even this notion exists in all cases; as is well shown by Sir Samuel Baker's conversation with a chief of the Latooki—a Nile tribe.

"'Have you no belief in a future existence after death?' . .
Commoro (*loq.*).—'Existence *after* death! How can that be? Can a dead man get out of his grave unless we dig him out?'
'Do you think man is like a beast, that dies and is ended?'
Commoro.—'Certainly; an ox is stronger than a man; but he dies, and his bones last longer; they are bigger. A man's bones break quickly—he is weak.'

'Is not a man superior in sense to an ox? Has he not a mind to direct his actions?'

Commoro.—'Some men are not so clever as an ox. Men must sow corn to obtain food, but the ox and wild animals can procure it without sowing.'

'Do you not know that there is a spirit within you more than flesh? Do you not dream and wander in thought to distant places in your sleep? Nevertheless, your body rests in one spot. How do you account for this?'

Commoro, laughing.—'Well, how do *you* account for it? It is a thing I cannot understand; it occurs to me every night.'

* * *

'Have you no idea of the existence of spirits superior to either man or beast? Have you no fear of evil except from bodily causes?'

Commoro.—'I am afraid of elephants and other animals when in the jungle at night, but of nothing else.'

'Then you believe in nothing; neither in a good nor evil spirit! And you believe that when you die it will be the end of body and spirit; that you are like other animals; and that there is no distinction between man and beast; both disappear, and end at death?'

Commoro.—'Of course they do.'"

And then in response to Baker's repetition of St. Paul's argument derived from the decaying seed, which our funeral service emphasizes, Commoro said:—

"'Exactly so; that I understand. But the *original* grain does *not* rise again; it rots like the dead man, and is ended; the fruit produced is not the same grain that we buried, but the *production* of that grain: so it is with man,—I die, and decay, and am ended; but my children grow up like the fruit of the grain. Some men have no children, and some grains perish without fruit; then all are ended.'"

Clearly, then, religious ideas have not that supernatural origin commonly alleged; and we are taught, by implication, that they have a natural origin. How do they originate?

§ 584. In the first volume of this work, nearly a score chapters are devoted to an account of primitive ideas at large; and especially ideas concerning the natures and actions of supernatural agents. Instead of referring the reader back to those chapters, I think it better to state afresh, in brief, the doctrine they contain. I do this partly because that doctrine, at variance both with current beliefs and the beliefs of the mythologists, needs re-emphasizing; partly because citing a further series of illustrations will strengthen the argument; and partly because a greater effect may be wrought by bringing the several groups of facts and inferences into closer connexion.

As typifying that genesis of religious conceptions to be delineated in this chapter, a statement made by Mr. Brough Smyth in his elaborate work *The Aborigines of Victoria* may first be given. When an Australian, of mark as a hunter or counsellor, is buried, the medicine-man, seated or lying beside the grave, praising the deceased and listening for his replies, said—" The dead man had promised that if his murder should be sufficiently avenged his spirit would not haunt the tribe, nor cause them fear, nor mislead them into wrong tracks, nor bring sickness amongst them, nor make loud noises in the night." Here we may recognize the essential elements of a cult. There is belief in a being of the kind we call supernatural—a spirit. There are praises of this being, which he is supposed to hear. On condition that his injunctions are fulfilled, he is said to promise that he will not make mischievous use of his superhuman powers—will not hurt the living by pestilence, nor deceive them, nor frighten them.

Is it not manifest that from germs of this kind elaborate religions may be evolved? When, as among the ancestor-worshipping Malagasy, we find, as given by M. Réville, the prayer,—" Nyang, méchant et puissant esprit, ne fais pas gronder le tonnerre sur nos têtes. Dis à la mer de rester dans ses bords. Épargne, Nyang, les fruits qui mûrissent. Ne sèche pas le riz dans sa fleur ;" it is a conclusion scarcely to be resisted that Nyang is but the more developed form of a spirit such as that propitiated and petitioned by the Australian. On reading the Japanese sayings, " that the spirits of the dead continue to exist in the unseen world, which is everywhere about us, and that they all become gods, of varying character and degrees of influence," and also that "the gods who do harm are to be appeased, so that they may not punish those who have offended them, and all the gods are to be worshipped, so that they may be induced to increase their favours;" we are strengthened in the suspicion that these maleficent gods and beneficent gods have all been derived from "the spirits of the dead ... of varying character and influence." From the circumstance that in India as Sir Alfred Lyall tells us, "it would seem that the honours which are at first paid to all departed spirits come gradually to be concentrated, as divine honours, upon the Manes of notables," we derive further support for this view. And when by facts of these kinds we are reminded that among the Greeks down to the time of Plato, parallel beliefs were current, as is shown in the *Republic*, where Socrates groups as the "chiefest of all" requirements "the service of gods, demigods, and heroes ... and the rites which have to be observed in order to propitiate the inhabitants of the world below," proving that there still survived "that fear of the wrath of the departed which strongly possessed the early Greek mind;" we get from this kinship of beliefs among races remote in time, space, and culture, strong warrant for the inference that ghost-propitiation is the origin of all religions.

This inference receives support wherever we look. As, until lately, no traces of pre-historic man were supposed to exist, though now that attention has been drawn to them, the implements he used are found everywhere; so, once being entertained, the hypothesis that religions in general are derived from ancestor-worship, finds proofs among all races and in every country. Each new book of travels yields fresh evidence; and from the histories of ancient peoples come more numerous illustrations the more closely they are examined.

Here I will re-exemplify the chief factors and stages in this genesis of religious beliefs; citing, in large measure, books that have been published since the first volume of this work.

§ 585. The African savage Commoro, quoted above, and shown by his last reply to be more acute than his questioner, had no theory of dreams. To the inquiry how he accounted for the consciousness of wandering while asleep, he said—"It is a thing I cannot understand." And here it may be remarked in passing, that where there existed no conception of a double which goes away during sleep, there existed no belief in a double which survives after death. But with savages who are more ready to accept interpretations than Commoro, the supposition that the adventures had in dreams are real, prevails. The Zulus may be instanced. To Bishop Callaway one of them said:—

"When a dead man comes [in a dream] he does not come in the form of a snake, nor as a mere shade; but he comes in very person, just as if he was not dead, and talks with the man of his tribe; and he does not think it is the dead man until he sees on awaking, and says, 'Truly I thought that So-and-so was still living; and forsooth it is his shade which has come to me.'"

Similarly with the Andamanese (who hold that a man's reflected image is one of his souls), the belief is that "in dreams it is the soul which, having taken its departure through the nostrils, sees or is engaged in the manner represented to the sleeper."

Abnormal forms of insensibility are regarded as due to more prolonged absences of the wandering double; and this is so whether the insensibility results naturally or artificially. That originally, the accepted interpretations of these unusual states of apparent unconsciousness were of this kind, we see in the belief expressed by Montaigne, that the "souls of men when at liberty, and loosed from the body, either by sleep, *or some extasie*, divine, foretel, and see things which whilst joyn'd to the body they could not see." Then at the present time among the Waraus (Guiana Indians) to gain magical power a man takes infusion of tobacco, "and, in the death-like state of sickness to which it reduces him, his spirit is supposed to leave the body, and to visit and receive power from the yauhahu . . . the dreaded beings under whose influence he is believed to remain ever after."

From the ordinary absence of the other-self in sleep and its extraordinary absences in swoon, apoplexy, etc., the transition is to its unlimited absence at death; when, after an interval of waiting, the expectation of immediate return is given up. Still, the belief is that, deaf to entreaties though the other-self has become, it either does from time to time return, or will eventually return. Commonly, the spirit is supposed to linger near the body or revisit it; as by the Iroquois, or by the Chinooks, who "speak of the dead walking at night, when they are supposed to awake, and get up to search for food." Long surviving among superior races, in the alleged nightly wanderings of de-materialized ghosts, this belief survives in its original crude form in the vampyre stories current in some places.

One sequence of the primitive belief in the materiality of the double, is the ministering to such desires as were manifest during life. Hence the shell with "some of her own milk beside the grave" of an infant, which an Andamanese mother leaves; hence the "food and oblations to the dead" by the Chippewas, etc.; hence the leaving with the corpse all needful implements, as by the Chinooks; hence the "fire kept burning there [the grave] for many weeks," as among the Waraus; hence the immolation of wives and slaves with the chief, as still, according to Cameron, at Urua in Central Africa. Hence, in short, the universality among the uncivilized and semi-civilized of these funeral rites implying belief that the ghost has the same sensations and emotions as the living man. Originally this belief is entertained literally; as by the Zulus, who in a case named said, "the Ancestral spirits came and eat up all the meat, and when the people returned from bathing, they found all the meat eaten up." But by some peoples the ghost, conceived as less material, is supposed to profit by the spirit of the thing offered: instance the Nicaraguans, by whom food " was tied to the body before cremation;" and instance the Ahts, who " burn blankets when burying their friends," that they may not be " sent shivering to the world below."

Ministrations to the double of the deceased, habitually made at the funeral, are in many places continued—here on special occasions and here at regular intervals. For if the ghost is not duly attended to, there may come mischief. Men of various types visit their dead from time to time to carry food, drink, etc.; as the Gonds, by whom, at the graves of honoured persons, "offerings continue to be presented annually for many years." Others, as the Ukiahs and Sanéls of California, "sprinkle food about the favorite haunts of the dead." Elsewhere, ghosts are supposed to come to places where food is being prepared for them; as instance Zululand. Bishop Callaway quotes a Zulu as saying—" These dead men are fools! Why have they revealed themselves by killing the child in this way, without telling me? Go and fetch the goat, boys."

The habitats of these doubles of the dead, who are like the living in their appetites and passions, are variously conceived. Some peoples, as the Shillook of the White Nile, "imagine of the dead that they are lingering amongst the living and still attend them." Other peoples, as for instance the Santals, think that the ghosts of their ancestors inhabit the adjacent woods. Among the Sonoras and the Mohaves of North America, the cliffs and hills are their imagined places of abode. "The Land of the Blest" says Schoolcraft, "is not in the sky. We are presented rather ... with a new earth, or terrene abode." Where, as very generally, the ghost is believed to return to the region whence the tribe came, obstacles have to be overcome. Some, as the Chibchas, tell of difficult rivers to be crossed to reach it; and others of seas: the Naowe (of Australia) think that their ghosts depart and people the islands in Spencer's Gulf. With these materialistic conceptions of the other-self and its place of abode, there go similarly materialistic conceptions of its doings after death. Schoolcraft, describing the hereafter of Indian belief, says the ordinary avocations of life are carried on with less of vicissitude and hardship. The notion of the Chibchas was that "in the future state, each nation had its own particular location, so that they could cultivate the ground." And everywhere we find an approach to parallelism between the life here and the imagined life hereafter. Moreover, the social relations in the other world, are supposed, even among comparatively-advanced peoples, to repeat those of this world. "Some of them [Taouist temples] are called Kung, *palace;* and the endeavour is made in these to represent the gods of the religion in their celestial abodes, seated on their thrones in their palaces, either administering justice or giving instruction:" recalling the Greek idea of Hades. That like ideas prevailed among the early English, is curiously shown by a passage Kemble quotes from King Alfred, concerning the permission to compound for crimes by the bot in money, "except in cases of treason against a lord, to which they dared not assign any mercy; because Almighty God adjudged none to them that despised him, nor did Christ . . . adjudge any to him that sold him unto death: and he commanded that a lord should be loved like himself."

Grave-heaps on which food is repeatedly placed, as by the Woolwas of Central America, or heaps of stones such as the "obo" described by Prejevalski, which "a Mongol never passes without adding a stone, rag, or tuft of camels' hair, as an offering," and which, as in Afghanistan, manifestly arise as coverings over dead men, are by such observances made into altars. In some cases they acquire this character quite definitely. On the grave of a prince in Vera Paz, there was "a stone altar erected above all, upon which incense was burned and sacrifices were made in memory of the deceased." Various peoples make shelters for such incipient altars or developed altars. By the Mosquitos "a rude hut is constructed over the grave, serving as a receptacle for the choice food, drink," etc. In Africa the Wakhutu "usually erect small pent-houses over them [the graves], where they place offerings of food." Major Serpa Pinto's work contains a cut representing a native chief's mausoleum, in which we see the grave covered by a building on six wooden columns—a building needing but additional columns to make it like a small Greek temple. Similarly in Borneo. The drawing of "Rajah Dinda's family sepulchre," given by Bock, shows development of the grave-shed into a temple of the oriental type. A like connexion existed among the Greeks.

"The 'heroön' was a kind of chapel raised to the memory of a hero.... It was at first a funeral monument ($\sigma\tilde{\eta}\mu\alpha$) surrounded by a sacred enclosure ($\tau\acute{\epsilon}\mu\epsilon\nu o\varsigma$); but the importance of the worship there rendered to the heroes soon converted it into a real 'hieron' [temple]."

And in our own time Mohammedans, notwithstanding their professed monotheism, show us a like transformation with great clearness. A saint's mausoleum in Egypt, is a "sacred edifice." People passing by, stop and become "pious worshippers" of "our lord Abdallah." "In the corner of the sanctuary stands a wax candle as long and thick as an elephant's tusk;" and there is a surrounding court with "niches for prayer, and the graves of the favoured dead." The last quotation implies something more. Along with development of grave-heaps into altars and grave-sheds into religious edifices, and food for the ghost into sacrifices, there goes on the development of praise and prayer. Instance, in addition to the above, the old account Dapper gives, translated by Ogilby, which describes how the negroes near the Gambia erected small huts over graves, "whither their surviving Friends and Acquaintance at set-times repair, to ask pardon for any offences or injuries done them while alive."

The growth of ancestor-worship, thus far illustrated under its separate aspects, may be clearly exhibited under its combined aspects by quotations from a recent book, *Africana*, by the Rev. Duff MacDonald, one of the missionaries of the Blantyre settlement. Detached sentences from his account, scattered here and there over fifty pages, run as follows:—

"The man may be buried in his own dwelling" (p. 109). "His old house thus becomes a kind of temple" (p. 109). "The deceased is now in the spirit world, and receives offerings and adoration" (p. 110). "Now he is a god with power to watch over them, and help them, and control their destiny" (p. 61). "The spirit of a deceased man is called his Mulungu" (p. 59). The probably correct derivation of this word is "stated by Bleek [the philologist], which makes it originally mean 'great ancestor'" (p. 67). "Their god appears to them in dreams. They may see him as they knew him in days gone by" (p. 61). "The gods of the natives are nearly as numerous as their dead" (p. 68). "Each worshipper turns most naturally to the spirits of his own departed relatives" (p. 68). A chief "will present his offering to his own immediate predecessor, and say, 'Oh, father, I do not know all your relatives, you know them all, invite them to feast with you'" (p. 68). "The spirit of an old chief may have a whole mountain for his residence, but he dwells chiefly on the cloudy summit" (p. 60). "A great chief that has been successful in his wars does not pass out of memory so soon. He may become the god of a mountain or a lake, and may receive homage as a local deity long after his own descendants have been driven from the spot. When there is a supplication for rain the inhabitants of the country pray not so much to their own forefathers as to the god of yonder mountain on whose shoulders the great rain clouds repose" (p. 70). "Beyond and above the spirits of their fathers, and chiefs localised on hills, the Wayao speak of others that they consider superior. Only their home is more associated with the country which the Yao left; so that they too at one time may have been looked upon really as local deities" (p. 71). (Vol. I, pp. 59–110.)

Let us pass now to certain more indirect results of the ghost-theory. Distinguishing but confusedly between semblance and reality, the savage thinks that the representation of a thing partakes of the properties of the thing. Hence he believes that the effigy of a dead man (originally placed on the grave) becomes a habitation for his ghost. This belief spreads to effigies otherwise placed. Concerning "a rude figure of a naked man and woman" which some Land

Dyaks place on the path to their farms, St. John says "These figures are said to be inhabited each by a spirit."

Because of the indwelling doubles of the dead, such images are in many cases propitiated. Speaking of the idols made by the people west of Lake Nyassa, Livingstone says "they present pombe, flour, bhang, tobacco, and light a fire for them to smoke by. They represent the departed father or mother, and it is supposed that they are pleased with the offerings made to their representatives . . . names of dead chiefs are sometimes given to them." Bastian tells us that a negress in Sierra Leone had in her room four idols whose mouths she daily daubed with maize and palm-oil: one for herself, one for her dead husband, and one for each of her children. Often the representation is extremely rude. The Damaras have "an image, consisting of two pieces of wood, supposed to represent the household deity, or rather the deified parent," which is brought out on certain occasions. And of the Bhils we read—"Their usual ceremonies consist in merely smearing the idol, which is seldom anything but a shapeless stone, with vermilion and red lead, or oil; offering, with protestations and a petition, an animal and some liquor."

Here we see the transition to that form of fetichism in which an object having but a rude likeness to a human being, or no likeness at all, is nevertheless supposed to be inhabited by a ghost. I may add that the connexion between development of the ghost-theory and development of fetichism, is instructively shown by the absence of both from an African people described by Thomson:—

"The Wahebe appear to be as free from superstitious notions as any tribe I have seen . . . there was an entire absence of the usual signs of that fetichism, which is so prevalent elsewhere. They seem, however, to have no respect for their dead; the bodies being generally thrown into the jungle to be eaten by the hyenas."

And just the same connexion of facts is shown in the account of the Masai more recently given by him.

In several ways there arises identification of ancestors

with animals, and consequent reverence for the animals: now resulting in superstitious regard, and now in worship. Creatures which frequent burial places or places supposed to be haunted by spirits, as well as creatures which fly by night, are liable to be taken for forms assumed by deceased men. Thus the Bongo dread—

"Ghosts, whose abode is said to be in the shadowy darkness of the woods. Spirits, devils, and witches have their general appellation of 'bitaboh;' wood-goblins being specially called 'ronga.' Comprehended under the same term are all the bats ... as likewise are owls of every kind."

Similarly, the belief that ghosts often return to their old homes, leads to the belief that house-frequenting snakes are embodiments of them. The negroes round Blantyre think that "if a dead man wants to frighten his wife he may persist in coming as a serpent;" and "when a man kills a serpent thus belonging to a spirit, he goes and makes an apology to the offended god, saying, 'Please, please, I did not know that it was your serpent.'" Moreover, "serpents were regarded as familiar and domestic divinities by a multitude of Indo-European peoples;" and "in some districts of Poland [in 1762] the peasants are very careful to give milk and eggs to a species of black serpent which glides about in their ... houses, and they would be in despair if the least harm befel these reptiles." Beliefs of the same class, suggested in other ways, occur in North America. The Apaches "consider the rattlesnake as the form to be assumed by the wicked after death." By the people of Nayarit it was thought that "during the day they [ghosts] were allowed to consort with the living, in the form of flies, to seek food:" recalling a cult of the Philistines and also a Babylonian belief expressed in the first Izdubar legend, in which it is said that "the gods of Uruk Suburi (the blessed) turned to flies."

Identification of the doubles of the dead with animals— now with those which frequent houses or places which the doubles are supposed to haunt, and now with those which are like certain of the dead in their malicious or beneficent natures

—is in other cases traceable to misinterpretation of names. We read of the Ainos of Japan that "their highest eulogy on a man is to compare him to a bear. Thus Shinondi said of Benri the chief 'He is as strong as a bear,' and the old Fate praising Pipichari called him 'The young bear.'" Here the transition from comparison to metaphor illustrates the origin of animal names. And then on finding that the Ainos worship the bear, though they kill it, and that after killing it at the bear-festival they shout in chorus—"We kill you, O bear! come back soon into an Aino," we see how identification of the bear with an ancestral Aino, and consequent propitiation of the bear, may arise. Hence when we read "that the ancestor of the Mongol royal house was a wolf," and that the family name was Wolf; and when we remember the multitudinous cases of animal-names borne by North American Indians, with the associated totem-system; this cause of identification of ancestors with animals, and consequent sacredness of the animals, becomes sufficiently obvious. Even without going beyond our own country we find significant evidence. In early days there was a tradition that Earl Siward of Northumbria had a grandfather who was a bear in a Norwegian forest; and "the bear who was the ancestor of Siward and Ulf had also, it would seem, known ursine descendants." Now Siward was distinguished by "his gigantic stature, his vast strength and personal prowess;" and hence we may reasonably conclude that, as in the case of the Ainos above given, the supposed ursine descent had arisen from misinterpretation of a metaphor applied to a similarly powerful progenitor. In yet other cases, sacredness of certain animals results from the idea that deceased men have migrated into them. Some Dyaks refuse to eat venison in consequence of a belief that their ancestors "take the form of deer after death;" and among the Esquimaux "the Angekok announces to the mourners into what animal the soul of the departed has passed." Thus there are several ways in which respect for, and sometimes worship of, an animal arises: all of them, however, implying identification of it with a human being.

A pupil of the Edinburgh institution for deaf-mutes said, "before I came to school, I thought that the stars were placed in the firmament like grates of fire." Recalling, as this does, the belief of some North Americans, that the brighter stars in the Milky Way are camp-fires made by the dead on their way to the other world, we are shown how naturally the identification of stars with persons may occur. When a sportsman, hearing a shot in the adjacent wood, exclaims—"That's Jones," he is not supposed to mean that Jones is the sound; he is known to mean that Jones made the sound. But when a savage, pointing to a particular star originally thought of as the camp-fire of such or such a departed man, says—"There he is," the children he is instructing naturally suppose him to mean that the star itself is the departed man: especially when receiving the statement through an undeveloped language. Hence such facts as that the Californians think ghosts travel to "where earth and sky meet, to become stars, chiefs assuming the most brilliant forms." Hence such facts as that the Mangaians say of certain two stars that they are children whose mother "was a scold and gave them no peace," and that going to "an elevated point of rock," they "leaped up into the sky;" where they were followed by their parents, who have not yet caught them. In ways like these there arises personalization of stars and constellations; and remembering, as just shown, how general is the identification of human beings with animals in primitive societies, we may perceive how there also originate animal-constellations; such as Callisto, who, metamorphosed into a she-bear, became the bear in heaven. That metaphorical naming may cause personalization of the heavens at large, we have good evidence. A Hawaiian king bore the name Kalani-nui-Liho Liho, meaning "the heavens great and dark;" whence it is clear that (reversing the order alleged by the mythologists) Zeus may naturally have been at first a living person, and that his identification with the sky resulted from his metaphorical name.

There are proofs that like confusion of metaphor with fact leads to Sun-worship. Complimentary naming after the Sun occurs everywhere; and, where it is associated with power, becomes inherited. The chiefs of the Hurons bore the name of the Sun; and Humboldt remarks that "the 'sun-kings' among the Natches recall to mind the Heliades of the first eastern colony of Rhodes." Out of numerous illustrations from Egypt, may be quoted an inscription from Silsilis —"Hail to thee! king of Egypt! Sun of the foreign peoples. . . . Life, salvation, health to him! he is a shining sun." In such cases, then, worship of the ancestor readily becomes worship of the Sun. The like happens with other celestial appearances. "In the Beirût school" says Jessup, " are and have been girls named . . . Morning Dawn, Dew, Rose. . . . I once visited a man in the village of Brummana who had six daughters, whom he named *Sun, Morning, Zephyr breeze*," &c. Another was named *Star*. Here, again, the superiority, or good fortune, or remarkable fate, of an individual thus named, would originate propitiation of a personalized phenomenon. That personalization of the wind had an origin of this kind is indicated by a Bushman legend. "The wind" it says "was formerly a person. He became a feathered thing. And he flew, while he no longer walked as formerly; for he flew, and he dwelt in the mountain . . . he inhabited a mountain-hole." Here, too, we are reminded that in sundry parts of the world there occurs the notion that not only the divine ancestors who begat the race came out of caves, but that Nature-gods also did. A legend of the Mexicans tells of the Sun and Moon coming out of caves; and in the conception of a cave inhabited by the wind, the modern Bushman does but repeat the ancient Greek. As descending from the traditions of cave-dwellers, stories of this kind, with accompanying worship, are natural;

but otherwise they imply superfluous absurdities which cannot be legitimately ascribed even to the most unintelligent. That in primitive times names are used in ways showing such lack of discrimination as leads to the confusions here alleged, we have proof. Grote says of the goddess Atē,—"the same name is here employed sometimes to designate the person, sometimes the attribute or event not personified." And again, it has been remarked that "in Homer, Aïdes is invariably the name of a god; but in later times it was transferred to his house, his abode or kingdom." Nature-worship, then, is but an aberrant form of ghost-worship.

In their normal forms, as in their abnormal forms, all gods arise by apotheosis. Originally, the god is the superior living man whose power is conceived as superhuman. From uncivilized peoples at present, and from civilized peoples during their past, evidence is derived. Mr. Selous says —'the chief of these kraals, 'Situngweesa,' is considered a very powerful 'Umlimo,' or god, by the Amandebele." So, too, among existing Hindus, "General Nicholson . . . was adored as a hero in his lifetime, in spite of his violent persecution of his own devotees." The *Rig Veda* shows that it was thus with the ancient people of India. Their gods are addressed—"Thou, Agni, the earliest and most Angiras-like sage" (R. V., i, 31). "Thou Agni, the most eminent rishi" (iii, 21, 3). "Thou [Indra] art an anciently-born rishi" (viii, 6, 41). "Indra is a priest, Indra is a rishi" (viii, 16, 7). That Achilles was apotheosized, and that according to tradition the Pythian priestess preferred to address Lykurgus as a god, are examples sufficiently reminding us of man-derived deities among the Greeks. It is a familiar fact, too, that with the Romans and subject peoples emperor-worship became a developed cult. In "every one of the Gaulish cities," "a large number of men, who belonged to the highest as well as to the middle classes, were priests and flamens of Augustus, flamens of Drusus, priests of Vespasian or Marcus Aurelius." "The statues of the emperors were real idols, to which they offered incense, victims, and prayers." And how natural to other European peoples in those days were conceptions leading to such cults, is curiously shown by an incident in the campaign of Tiberius, then a prince, carried on in Germany in A.D. 5, when Romans and Teutons were on opposite sides of the Elbe.

"One of the barbarians, an aged man, powerfully built and, to judge from his attire, of high rank, got into an excavated trunk (such as they use for boats) and rowed his vessel to the middle of the river. There he asked and obtained leave to come safely to our side and to see the prince. Having come to shore, he first for a long time silently looked at the prince and finally broke out into these words: 'Mad, indeed, are our young men. For if you are far, they worship you as gods, and if you approach, they rather fear your weapons than do you homage. But I, by thy kind permission, O prince, to day have seen the gods of whom before I had heard."

That some of our own ancestors regarded gods simply as superior men is also clear. If the Norseman "thought himself unfairly treated, even by his gods, he openly took them to task and forsook their worship;" and, reminding us of some existing savages, we read of a Norse warrior "wishing ardently that he could but meet with Odin, that he might attack him."

As, in primitive thought, divinity is thus synonymous with superiority; and as at first a god may be either a powerful living person (commonly of conquering race) or a dead person who has acquired supernatural power as a ghost; there come two origins for semi-divine beings—the one by unions between the conquering god-race and the conquered race distinguished as men, and the other by supposed intercourse between living persons and spirits. We have seen that dream-life in general is at first undistinguished from waking life. And if the events of ordinary dreams are regarded as real, we may infer that the concomitants of dreams of a certain kind create a specially strong belief in their reality. Once having become established in the popular mind, this belief in their reality is, on occasion, taken advantage of. At Hamóa (Navigator's Islands) "they have an idea which is very convenient to the reputation of the females, that some of these *hotooa pow* [mischievous spirits] molest them in their sleep, in consequence of which there are many supernatural conceptions." Among the Dyaks it is the same. We are told both by Brooke and St. John of children who were begotten by certain spirits. Of like origin and nature was the doctrine of the Babylonians concerning male and female spirits and their offspring. And the beliefs in incubi and succubi lasted in European history down to comparatively late times: sometimes giving rise to traditions like that of Robert the Devil. Of course the statement respecting the nature of the supernatural parent is variable—he is demoniacal or he is divine; and consequently there now and then result such stories as those of the Greeks about god-descended men.

Thus Comparative Sociology discloses a common origin for each leading element of religious belief. The conception of the ghost, along with the multiplying and complicating ideas arising from it, we find everywhere—alike in the arctic regions and in the tropics; in the forests of North America and in the deserts of Arabia; in the valleys of the Himalayas and in African jungles; on the flanks of the Andes and in the Polynesian islands. It is exhibited with equal clearness by races so remote in type from one another, that competent judges think they must have diverged before the existing distribution of land and sea was established— among straight-haired, curly-haired, woolly-haired races; among, white, tawny, copper-coloured, black. And we find it among peoples who have made no advances in civilization as well as among the semi-civilized and the civilized. Thus we have abundant proofs of the natural genesis of religions.

§ 586. To give to these proofs, re-inforcing those before

given, a final re-inforcement, let me here, however, instead of taking separately each leading religious conception as similarly exhibited by different peoples, take the whole series of them as exhibited by the same people.

That belief in the reality of dream-scenes and dream-persons, which, as we before saw (§ 530), the Egyptians had in common with primitive peoples at large, went along with the belief, also commonly associated with it, that shadows are entities. A man's shadow was "considered an important part of his personality;" and the Book of the Dead treats it "as something substantial." Again, a man's other-self, called his *ka*, accompanied him while alive; and we see "the Egyptian king frequently sculptured in the act of propitiating his own *ka*," as the Karen does at the present day. "The disembodied personality" had "a material form and substance. The soul had a body of its own, and could eat and drink." But, as partially implied by this statement, each man was supposed to have personalities of a less material kind. After death "the soul, though bound to the body, was at liberty to leave the grave and return to it during the daytime in any form it chose;" and a papyrus tells of mummies who "converse in their catacomb about certain circumstances of their past life upon earth." Having desires, the *ka* must be ministered to; and, as M. Maspero says, "le double des pains, des liquides, de la viande, passait dans l'autre monde et y nourrissait le *Double* de l'homme." Along with this belief that the bodily desires and satisfactions continued in the second life, there naturally went a conception of the second life as substantially like the first; as is shown by the elaborate delineations of it contained in ancient tombs, such as the tomb of Ti.

Along with ministrations to the appetites of the supposed material or semi-material dead, resulting from these beliefs, there went ministrations to desires of other kinds. In the richly-adorned sepulchral chamber of king Mycerinus's daughter, there was a daily burning of incense; and at night a lamp was "kept burning in the apartment." Habitually there were public praises of the dead; and to tempt back to Egypt a valued subject, a king promises that "the poor shall make their moan at the door of thy tomb. Prayers shall be addressed to thee." Such sacrifices, praises, and prayers, continued from festival to festival, and, eventually, from generation to generation, thus grew into established worships. "The monuments of the time of the building of the pyramids mention priests and prophets which were devoted to the service of Kheops, Chabryes, and other rulers, and who offered them sacrifices"—priests who had successors down even to the 26th dynasty. Such priesthoods were established for worship not of the royal dead only, but for worship of other dead. To ensure sacrifices to their statues, great landowners made "contracts with the priests of their town," prescribing the kinds of food and drink to be offered. So far was this system carried that Hapi Tefa, the governor of a district, to maintain services to himself "for all time . . . provides salaries for the priests." As implied in some of the foregoing extracts, there arose an idol-worship by differentiation from worship of the dead. The *ka*, expected eventually to return and re-animate the mummy, could enter also a statue of wood or stone representing the deceased. Hence some marvellous elaborations. In the Egyptian tomb, sometimes called the "house of the double," there was a walled-up space having but a small opening, which contained images of the dead, more or less numerous; so that if re-animation of the mummy was prevented by destruction of it, any one of these might be utilized in its place.

The proofs thus furnished that their idolatry was developed from their ancestor-worship, are accompanied by proofs that their animal-worship was similarly developed. The god Ammon Ra is represented as saying to Thothmes III—

"I have caused them to behold thy majesty, even as it were the star Seschet (the evening star) . . . I have caused them to behold thy majesty as it were a bull young and full of spirit . . . I have caused them to behold thy majesty as it were a crocodile [and similarly with a lion, an eagle, and a jackal] . . . It is I who protecteth thee, oh my cherished son ! Horus, valiant bull, reigning over the Thebaid."

Here, in the first place, we are shown, as we were shown by the Ainos, that there takes place a transition from simile to metaphor: "thy majesty, as it were a bull," presently becomes "Horus, valiant bull." This naturally leads in subsequent times to confusion of the man with the animal, and consequent worship of the animal. We may further see that complimentary comparisons to other animals, similarly passing through metaphors into identifications, are likely to generate belief in a deified individual who had sundry forms. Another case shows us how, from what was at first eulogistic naming of a local ruler, there may grow up the adoption of an animal-image for a known living person. We read of "the Ram, who is the Lord of the city of Mendes, the Great God, the Life of Ra, the Generator, the Prince of young women." We find the king speaking of himself as "the image of the divine Ram, the living portrait of him . . . the divine efflux of the prolific Ram . . . the eldest son of the Ram." And then, further, we are told that the king afterwards deified the first of his consorts, and "commanded that her Ram-image should be placed in all temples."

So, too, literal interpretation of metaphors leads to worship of heavenly bodies. As above, the star Seschet comes to be identified with an individual; and so, continually, does the Sun. Thus it is said of a king—"My lord the Sun, Amenhotep III, the Prince of Thebes, rewarded me. He is the Sun-god himself;" and it is also said of him "no king has done the like, since the time of the reign of the Sun-god Ra, who possessed the land." In kindred manner we are told of the sarcophagus provided for another king, Amenemhat, that "never the like had been provided since the time of the god Ra." These quotations show that this complimentary metaphor was used in so positive a way as to cause acceptance of it as fact; and thus to generate a belief that the Sun had been actual ruler over Egypt.

The derivation of all these beliefs from ancestor-worship, clear as the above evidence makes it, becomes clearer still when we observe, on the one hand, how the name "god" was applied to a superior living individual, and, on the other hand, how completely human in all their attributes were the gods, otherwise so-called. The relatively small difference between the conceptions of the divine and the human, is shown by the significant fact that in the hieroglyphics, one and the same "determinative" means, according to the context, god, ancestor, august person. Hence we need not wonder on finding king Sahura of the 5th dynasty called "God, who strikes all nations, and reaches all countries with his arm;" or on meeting with like deifications of other historical kings and queens, such as Mencheres and Nofert-Ari-Aähmes. And on finding omnipotence and omnipresence ascribed to a living king, as to Ramses II., we see little further scope for deification. Indeed we see no further scope; since along with these exalted conceptions of certain men there went low conceptions of gods.

"The bodies of the gods are spoken of as well as their souls, and they have both parts and passions; they are described as suffering from hunger and thirst, old age, disease, fear and sorrow. They perspire, their limbs quake, their head aches, their teeth chatter, their eyes weep, their nose bleeds, 'poison takes possession of their flesh.' . . . All the great gods require protection. Osiris is helpless against his enemies, and his remains are protected by his wife and sister."*

---

* It is strange how impervious to evidence the mind becomes when once pre-possessed. One would have thought that such an accumulation of proofs, congruous with the proofs yielded by multitudinous other societies, would have convinced everyone that the Egyptian religion was a developed ancestor-worship. But such proofs appear to have no effects in the minds of the theologians and the mythologists. Though the ancient Egyptian tradition is that "the land of Punt was the original seat of the gods," whence "the holy ones had travelled to the Nile valley, at their head Amon, Horus, Hathor;" though there is also the tradition that "during the first age a Dynasty of the Gods reigned in the land; this was followed by the age of the Demigods; and the dynasty of the mysterious Manes closed the prehistoric time;" though

The saying that one half the world does not know how the other half lives, may be paralleled by the saying that one half the world has no idea what the other half thinks, and what it once thought itself. Habitually at a later mental stage, there is a forgetting of that which was familiar at an earlier mental stage. Ordinarily in adult life many thoughts and feelings of childhood have faded so utterly that there is an incapacity for even imagining them; and, similarly, from the consciousness of cultured humanity there have so completely disappeared certain notions natural to the consciousness of uncultured humanity, that it has become almost incredible they should ever have been entertained. But just as certain as it is that the absurd beliefs at which parents laugh when displayed in their children, were once their own; so certain is it that advanced peoples to whom primitive conceptions seem ridiculous, had forefathers who held these primitive conceptions. Their own theory of things has arisen by slow modification of that original theory of things in which, from the supposed reality of dreams, there resulted the supposed reality of ghosts; whence developed all kinds of supposed supernatural beings.

§ 587. Is there any exception to this generalization? Are we to conclude that amid the numerous religions, varying in

---

these traditions are congruous with that deification of kings, priests, minor potentates, and, in a sense, even ordinary persons, which Egyptian history at large shows us; yet all this evidence is disregarded from the desire to ascribe a primitive monotheism or a primitive nature-worship. For these the sole authorities are statements made by the later Egyptian priests or contained in certain of the inscriptions—statements, written or spoken, which were necessarily preceded by a long period during which the art of recording did n t exist, and a further long period of culture—statements which naturally embodied relatively advanced conceptions. It would be about as wise to deny that the primitive Hebrew worship was that prescribed in Leviticus because such worship is denounced by Amos and by Hosea. It would be about as wise to take the conception of Zeus entertained by Socrates as disproving the gross anthropomorphism of the primitive Greeks. It would be about as wise to instance some refined modern version of Christianity, like that of Maurice, as showing what mediæval Christians believed.

their forms and degrees of elaboration, which have this common origin, there exists one which has a different origin? Must we say that while all the rest are natural, the religion possessed by the Hebrews which has come down to us with modifications, is supernatural?

If, in seeking an answer, we compare this supposed exceptional religion with the others, we do not find it so unlike them as to imply an unlike genesis. Contrariwise, we find it presenting throughout remarkable likenesses to them. We will consider these in groups.

In the first place, the plasma of superstitions amid which the religion of the Hebrews evolved, was of the same nature with that found everywhere. Though, during the early nomadic stage, the belief in a permanently-existing soul was undeveloped, yet there was shown belief in the reality of dreams and of the beings seen in dreams. At a later stage we find that the dead were supposed to hear and sometimes to answer; there was propitiation of the dead by gashing the body and cutting the hair; there was giving of food for the dead; spirits of the dead were believed to haunt burial-places; and demons entering into men caused their maladies and their sins. Much given, like existing savages, to amulets, charms, exorcisms, etc., the Hebrews also had functionaries who corresponded to medicine men—men having "familiar spirits," "wizards" (Isaiah viii, 19), and others, originally called seers but afterwards prophets (1 Sam. ix, 9); to whom they made presents in return for information, even when seeking lost asses. And Samuel, in calling for thunder and rain, played the part of a weather-doctor—a personage still found in various parts of the world.

Sundry traditions they held in common with other peoples. Their legend of the deluge, besides being allied to that of the Accadians, was allied to that of the Hindus; among whom the Śathapatha-brāhmaṇa tells how Manu was instructed by Vishnu to make an ark to escape the coming flood, which came as foretold and "swept away all living creatures; Manu alone was left." The story of Moses' birth is paralleled by an Assyrian story, which says—"I am Sargina the great King . . . my mother . . . in a secret place she brought me forth: she placed me in an ark of bulrushes . . . she threw me into the river . . ." etc. Similarly with the calendar and its entailed observances. "The Assyrian months were lunar . . . the seventh, fourteenth, twenty-first, and twenty-eighth days, being the sabbaths. On these sabbath days, extra work and even missions of mercy were forbidden . . . The enactments were similar in character to those of the Jewish code."

So again is it with their Theology. Under the common title *Elohim*, were comprehended distinguished living persons, ordinary ghosts, superior ghosts or gods. That is to say, with the Hebrews as with the Egyptians and numerous other peoples, a god simply meant a powerful being, existing visibly or invisibly. As the Egyptian for god, *Nutar*, was variously used to indicate strength; so was *Il* or *El* among the Hebrews, who applied it to heroes and also "to the gods of the gentiles." Out of these conceptions grew up, as in other cases, the propitiation or worship of various supernatural beings—a polytheism. Abraham was a demi-god to whom prayers were addressed. "They sacrificed unto devils, not to God; to gods whom they knew not, to new *gods that came newly up*, whom your fathers feared not" (Deut. xxxii, 17). That the belief in other gods than Jahveh long survived, is shown by Solomon's sacrifices to them, as well as by the denunciations of the prophets. Moreover, even after Jahveh had become the acknowledged great-god, the general conception remained essentially polytheistic. For just as in the *Iliad* (bk. v, 1000–1120) the gods and goddesses are represented as fighting with sword and lance the battles of the mortals whose causes they espoused; so the angels and archangels of the Hebrew pantheon are said to fight in Heaven when the peoples they respectively patronize fight on earth: both ideas being paralleled by those of some existing savages.

Seeing then that Jahveh was originally one god among many—the god who became supreme; let us ask what was his nature as shown by the records. Not dwelling on the story of the garden of Eden (probably accepted from the Accadians) where God walked and talked in human fashion; and passing by the time when "the Lord came down to see the city and the tower, which the children of men builded;" we may turn to such occasions as those on which Jacob wrestled with him, and on which "the Lord spake unto Moses face to face, as a man speaketh unto his friend." These, and many kindred statements, show that by the Hebrews in early days, Jahveh, "the strong one," "a man of war," having been originally a local potentate (like those who even now are called gods by the Bedouins), was, in after times, regarded as the most powerful among the various spirits worshipped: the places where sacrifices to him were made, being originally high places (2 Kings xii, 3), such as those habitually used for the burials of superior persons; as they are still in the same regions. Says Burkhardt of the Bedouins—"the saints' tombs are generally placed on the summits of mountains," and "to him [a saint] all the neighbouring Arabs address their vows." Here we see parallelism to the early religious ideas of Greeks, Scandinavians, and others; among whom gods, indistinguishable from men in appearance, sometimes entered into conflicts with them, not always successfully. Moreover, this "God of battles," whose severe punishments, often inflicted, were for insubordination, was clearly a local god—"the god of Israel." The command "thou shalt have none other gods but me," did not imply that there were none other, but that the Israelites were not to recognize their authority. The admission that the Hebrew god was not the only god is tacitly made by the expression "our" god as used by the Hebrews to distinguish Jahveh from others. And though with these admissions that

Jahveh was one god among many, there were assertions of universality of rule ; these were paralleled by assertions concerning certain gods of the Egyptians—nay, by assertions concerning a living Pharaoh, of whom it is said "no place is without thy goodness. Thy sayings are the law of every land. . . . Thou hast millions of ears. . . . Whatsoever is done in secret, thy eye seeth it." Along with the limitations of Jahveh's authority in range, went limitations of it in degree. There was no claim to omnipotence. Not forgetting the alleged failure of his attempt personally to slay Moses, we may pass on to the defeats of the Israelites when they fought by his advice, as in two battles with the Benjaminites, and as in a battle with the Philistines when "the ark of God was taken" (1 Sam. iv, 3–10). And then, beyond this, we are told that though "the Lord was with Judah," he "could not drive out the inhabitants of the valley, because they had chariots of iron." (Judges i, 19.) That is, there were incapacities equalling those attributed by other peoples to their gods. Similarly with intellectual and moral nature. Jahveh receives information ; he goes to see whether reports are true ; he repents of what he has done—all implying anything but omniscience. Like Egyptian and Assyrian kings, he continually lauds himself ; and while saying "I will not give my glory to another" (Isai. xlviii, 11), he describes himself as jealous, as revengeful, and as a merciless destroyer of enemies. He sends a lying spirit to mislead a king, as Zeus does to Agamemnon (2 Chron. xviii, 20-2) ; by his own account he will deceive a prophet that he may prophesy falsely, intending then to destroy him (Ezekiel xiv, 9) ; he hardens men's hearts that he may inflict evils on them for what they then do ; and, as when he prompts David to number Israel, suggests a supposed sin that he may afterwards punish those who have not committed it. He acts as did the Greek gods ; from whom bad impulses were supposed to come, and who were similarly indiscriminate in their revenges.

The forms of worship show us like parallelisms. Not dwelling on the intended or actual human sacrifices (though by grouping the sacrifice of a son with sacrifices of rams and calves, as methods of propitiation to be repudiated, Micah implies in ch. vi, 6-9 that the two had been associated in the Hebrew mind), it suffices to point out that the prescribed ceremonies in temples, had the characters usual everywhere. Called in sundry places the "bread of God," the offerings, like those to Egyptian gods and mummies, included bread, meat, fat, oil, blood, drink, fruits, etc.; and there was maintained, as by other peoples, a constant fire, as well as burnings of incense: twice daily by the Hebrews, and four times daily by the Mexicans. Jahveh was supposed to enjoy the "sweet savour" of the burnt offerings, like the idol-inhabiting gods of the negroes (§ 161). Associated with the belief that "the blood is the life," this, either poured on the ground or on the altar, according to circumstances, was reserved for Jahveh; as with the ancient Mexican and Central American gods, to whom was continually offered up the blood alike of sacrificed men and animals: now the image of the god being anointed with it, and now the cornice of the doorway of the temple. As the Egyptians and as the Greeks, so did the Hebrews offer hecatombs of oxen and sheep to their god; sometimes numbering many thousands (1 Kings viii, 62-64). To the Hebrews, it was a command that unblemished animals only should be used for sacrifices; and so among the Greeks a "law provided that the best of the cattle should be offered to the Gods," and among the Peruvians it was imperative that "all should be without spot or blemish." A still more remarkable likeness exists. Those orders made in Leviticus, under which certain parts of animals are to be given to Jahveh while other parts are left to the priests, remind us of those endowment-deeds, by which Egyptian landowners provided that for their ghosts should be reserved certain joints of the sacrificed animals, while the remaining parts were made over to the *ka*-priests. Again, just as we have seen that the gods of the Wayao, who were ghosts of ancient great chiefs, dwelt on the cloudy summits of certain adjacent mountains; and just as the residence of "cloud-compelling Jove" was the top of Olympus, where storms gathered; so the Hebrew god "descended in the cloud" on the summit of Mount Sinai, sometimes with thunder and lightning. Moreover, the statement that from thence Moses brought down the tables of the commands, alleged to be given by Jahveh, parallels the statement that from Mount Ida in Crete, from the cave where Zeus was said to have been brought up (or from the connected Mount Iuktas reputed in ancient times to contain the burial place of Zeus), Rhadamanthus first brought down Zeus' decrees, and Minos repaired to obtain re-inforced authority for his laws.*

Various other likenesses may be briefly noted. With the account of the council held by Jahveh when compassing Ahab's destruction, may be compared the account of the council of the Egyptian gods assembled to advise Ra, when contemplating the destruction of the world, and also the accounts of the councils of the Greek gods held by Zeus. Images of the gods, supposed to be inhabited by them, have been taken to battle by various peoples; as by the Hebrews was the ark of the covenant, which was a dwelling place of Jahveh. As by many savages, who even when living dislike

* It matters not to the argument whether this was or was not the Olympian Zeus. It suffices that he was a king, whose mountain-dwelling ghost became a god giving commands. But that the two personages were originally one is a tenable conclusion. Having a belief in a god inhabiting a neighbouring mountain where the clouds gathered, a migrating people, settling elsewhere, near a mountain similarly distinguished as an originator of storms, would naturally infer that their god had come with them. A recently published work, *Africana*, has yielded me some evidence supporting this conclusion; in so far that the Wayaos regard as superior, certain gods originally localized in the country they left, and who yet must, in a sense, be present with them if they are regarded as their superior gods. The different genealogy of the Olympian Zeus goes for little, considering what differences there were among the genealogies of historical persons among the Greeks.

their names to be known) it is forbidden to call a dead man by his real name, especially if distinguished; and as among the early Romans, it was a "deeply cherished belief that the name of the proper tutelary spirit of the community ought to remain for ever unpronounced;" so was it with the Hebrews in early days: their god was not named. Dancing was a form of worship among the Hebrews as it was among the Greeks and among various savages: instance the Iroquois. Fasts and penances like those of the Hebrews exist, or have existed, in many places; especially in ancient Mexico, Central America, and Peru, where they were extremely severe. The fulfilments of prophecies alleged by the Hebrews were paralleled by fulfilments of prophecies alleged by the Greeks; and the Greeks in like manner took them to be evidence of the truth of their religion. Nay we are told the same even of the Sandwich Islanders, who said that Captain Cook's death "fulfilled the prophecies of the priests, who had foretold this sad catastrophe." The working of miracles alleged of the Hebrew god as though it were special, is one of the ordinary things alleged of the gods of all peoples throughout the world. The translation of the living Elijah recalls the Chaldean legend of Izdubar's "translated ancestor, Hasisadra or Xisuthrus;" and in New World mythologies, there are the cases of Hiawatha, who was carried living to heaven in his magic canoe, and the hero of the Arawâks, Arawanili. As by the Hebrews, Jahveh is represented as having in the earliest times appeared to men in human shape, but not in later times; so by the Greeks, the theophany frequently alleged in the *Iliad*, becomes rare in traditions of later date. Nay, the like happened with the ancient Central Americans. Said an Indian in answer to Fr. Bobadilla—"For a long time our gods have not come nor spoken to them [the devotees]. But formerly they used to do so, as our ancestors told us."

Nor do parallelisms fail us when we turn to the more developed form of the Hebrew religion. That the story of a god-descended person should be habitually spoken of by Christians as though it were special to their religion, is strange considering their familiarity with stories of god-descended persons among the Greeks,—Æsculapius, Pythagoras, Plato. But it is not the Greek religion only which furnished such parallels. The Assyrian king Nebuchadnezzar asserted that he had been god-begotten. It is a tradition among the Mongols that Alung Goa, who herself "had a spirit for her father," bore three sons by a spirit. In ancient Peru if any of the virgins of the Sun "appeared to be pregnant, she said it was by the Sun, and this was believed, unless there was any evidence to the contrary." And among the existing inhabitants of Mangaia it is the tradition that "the lovely Ina-ani-vai" had two sons by the great god Tangaroa. The position, too, of mediator held by the god-descended son, has answering positions elsewhere. Among the Fijian gods, "*Tokairambe* and *Tui Lakemba Randinandina* seem to stand next to Ndengei, being his sons, and acting as mediators by transmitting the prayers of suppliants to their father."

Once more we have, in various places, observances corresponding to the eucharist. All such observances originate from the primitive notion that the natures of men, inhering in all their parts, inhere also in whatever becomes incorporated with them; so that a bond is established between those who eat of the same food. As furnishing one out of many instances, I may name the Padam, who "hold inviolate any engagement cemented by an interchange of meat as food." Believing that the ghosts of the dead, retaining their appetites, feed either on the material food offered or on the spirit of it, this conception is extended to them. Hence arise, in various parts of the world, feasts at which living and dead are supposed to join; and thus to renew the relation of subordination on the one side and friendliness on the other. And this eating with the ghost or the god, which by the Mexicans, was transformed into "eating the god" (symbolized by a cake made up with the blood of a victim), was associated with a bond of service to the god for a specified period. Briefly stringing together minor likenesses, we may note that the Christian crusades to get possession of the holy sepulchre, had their prototype in the sacred war of the Greeks to obtain access to Delphi; that as, among Christians, part of the worship consists in reciting the doings of the Hebrew god, prophets, and kings, so worship among the Greeks consisted partly in reciting the great deeds of the Homeric gods and heroes; that Greek temples were made rich by precious gifts from kings and wealthy men to obtain divine favour or forgiveness, as Christian cathedrals have been; that St. Peter's at Rome was built by funds raised from various catholic countries, as the temple of Delphi was rebuilt by contributions from various Grecian states; that the doctrine of special providences, general over the world, was as dominant among the Greeks as it has been among Christians, so that, in the words of Grote, "the lives of the Saints bring us even back to the simple and ever-operative theology of the Homeric age;" and lastly that various religions, alike in the new and old worlds, show us, in common with Christianity, baptism, confession, canonization, celibacy, the saying of grace, and other minor observances.

§ 588. What are we to conclude from all this evidence? What must we think of this unity of character exhibited by religions at large? And then, more especially, what shall we say of the family likeness existing between the creed of Christendom and other creeds? Observe the facts.

Alike in those minds among the civilized which, by defective senses, have been cut off from instruction, and in the minds of various primitive peoples, religious conceptions do not exist. Wherever the rudiments of them exist, they have, as their form, a belief in, and sacrifices to, the doubles of the dead. The ghost-theory, with resulting propitiation of ordinary ghosts, habitually survives along with belief in, and propitiation of, supernatural beings of more powerful kinds; known at first by the same generic name as ordinary ghosts, and differentiating by small steps. And the worships of the supposed supernatural beings, up even to the highest, are the same in nature, and differ only in their degrees of elaboration. What do these correspondences imply? Do they not imply that in common with other phenomena displayed by human beings as socially aggregated, religions have a natural genesis?

Are we to make an exception of the religion current among ourselves? If we say that its likenesses to the rest hide a transcendant unlikeness, several implications must be recognized. One is that the Cause to which we can put no limits in Space or Time, and of which our entire Solar System is a relatively infinitesimal product, took the disguise of a man for the purpose of covenanting with a shepherd-chief in Syria. Another is that this Energy, unceasingly manifested everywhere, throughout past, present, and future, ascribed to himself under this human form, not only the limited knowledge and limited powers which various passages show Jahveh to have had, but also moral attributes which we should now think discreditable to a human being. And a third is that we must suppose an intention even more repugnant to our moral sense. For if these numerous parallelisms between the Christian religion and other religions, do not prove likeness of origin and development, then the implication is that a complete simulation of the natural by the supernatural has been deliberately devised to deceive those who examine critically what they are taught. Appearances have been arranged for the purpose of misleading sincere inquirers, that they may be eternally damned for seeking the truth.

On those who accept this last alternative, no reasonings will have any effect. Here we finally part company with them by accepting the first; and, accepting it, shall find that Ecclesiastical Institutions are at once rendered intelligible in their rise and progress.

CHAPTER II.

MEDICINE-MEN AND PRIESTS.

§ 589. A SATISFACTORY distinction between priests and medicine-men is difficult to find. Both are concerned with supernatural agents, which in their original forms are ghosts; and their ways of dealing with these supernatural agents are so variously mingled, that at the outset no clear classification can be made.

Among the Patagonians the same men officiate in the "three-fold capacity of priests, magicians, and doctors;" and among the North American Indians the functions of "sorcerer, prophet, physician, exorciser, priest, and rain-doctor," are united. The Pe-i-men of Guiana "act as conjurors, soothsayers, physicians, judges, and priests." So, too, Ellis says that in the Sandwich Islands the doctors are generally priests and sorcerers. In other cases we find separation beginning; as witness the New Zealanders, who, in addition to priests, had at least one in each tribe who was a reputed sorcerer. And with advancing social organization there habitually comes a permanent separation.

In point of time the medicine-man takes precedence. Describers of the degraded Fuegians, speak only of wizards; and even of the relatively-advanced Mapuchés on the adjacent continent, we read that they have no priests, though they have diviners and magicians. In Australian tribes the only men concerned with the supernatural are the *boyala-men* or doctors; and the like is alleged by Bonwick of the

Tasmanians. Moreover, in many other instances, those who are called priests among uncivilized peoples, do little else than practise sorcery under one or other form. The *pajé* or priest of the Mundurucús " fixes upon the time most propitious for attacking the enemy; exorcises evil spirits, and professes to cure the sick;" and the like is the case with the Uaupés. In various tribes of North America, as the Clallums, Chippewayans, Crees, the priests' actions are simply those of a conjuror.

How shall we understand this confusion of the two functions, and the early predominance of that necromantic function which eventually becomes so subordinate?

§ 590. If we remember that in primitive thought the other world repeats this world, to the extent that its ghostly inhabitants lead similar lives, stand in like social relations, and are moved by the same passions; we shall see that the various ways of dealing with ghosts, adopted by medicine-men and priests, are analogous to the various ways men adopt of dealing with one another; and that in both cases the ways change according to circumstances.

See how each member of a savage tribe stands towards other savages. There are first the members of adjacent tribes, chronically hostile, and ever on the watch to injure him and his fellows. Among those of his own tribe there are parents and near relatives from whom, in most cases, he looks for benefit and aid; and towards whom his conduct is in the main amicable, though occasionally antagonistic. Of the rest, there are some inferior to himself over whom he habitually domineers; there are others proved by experience to be stronger and more cunning, of whom he habitually stands in fear, and to whom his behaviour is propitiatory; and there are many whose inferiority or superiority is so far undecided, that he deals with them now in one way and now in another as the occasion prompts—changing from bullying to submission or from submission to bullying, as he finds one or other answer. Thus to the living around him, he variously adapts his actions—now to conciliate, now to oppose, now to injure, according as his ends seem best subserved.

Men's ghosts being at first conceived as in all things like their originals, it results that the assemblage of them to which dead members of the tribe and of adjacent tribes give rise, is habitually thought of by each person as standing to him in relations like those in which living friends and enemies stand to him. How literally this is so, is well shown by a passage from Bishop Callaway's account of the Zulus, in which an interlocutor describes his relations with the spirit of his brother.

"You come to me, coming for the purpose of killing me. It is clear that you were a bad fellow when you were a man: are you still a bad fellow under the ground?"

Ghosts and ghost-derived gods being thus thought of as repeating the traits and modes of behaviour of living men, it naturally happens that the modes of treating them are similarly adjusted—there are like efforts, now to please, now to deceive, now to coerce. Stewart tells us of the Nagas that they cheat one of their gods who is blind, by pretending that a small sacrifice is a large one. Among the Bouriats, the evil spirit to whom an illness is ascribed, is deluded by an effigy —is supposed "to mistake the effigy for the sick person," and when the effigy is destroyed thinks he has succeeded. In Kibokwé, Cameron saw a "sham devil," whose "functions were to frighten away the devils who haunted the woods." Believing in spirits everywhere around, the Kamtschatkans "adored them when their wishes were fulfilled, and insulted them when their affairs went amiss." The incantations over a sick New Zealander were made "with the expectation of either propitiating the angry deity, or of driving him away:" to which latter end threats to "kill and eat him," or to burn him, were employed. The Wáralís, who worship Wághiá, on being asked—"Do you ever scold Wághiá?" replied—"To be sure, we do. We say, You fellow, we have given you a chicken, a gnat, and yet you strike us! What more do you want?" And then to cases like these, in which the conduct towards certain ghosts and ghost-derived gods, is wholly or partially antagonistic, have to be added the cases, occurring abundantly everywhere, in which those ghosts who are supposed to stand in amicable relations with the living, are propitiated by gifts, by praises, and by expressions of subordination, with the view of obtaining their good offices—ghosts who receive extra propitiations when they are supposed to be angry, and therefore likely to inflict evils.

Thus, then, arises a general contrast between the actions and characters of men who deal antagonistically with supernatural beings and men who deal sympathetically. Hence the difference between medicine-men and priests; and hence, too, the early predominance of medicine-men.

§ 591. For in primitive societies relations of enmity, both outside the tribe and inside the tribe, are more general and marked than relations of amity; and therefore the doubles of the dead are more frequently thought of as foes than as friends.

As already shown at length in §§ 118, 119, one of the first corollaries drawn from the ghost-theory is, that ghosts are the causes of disasters. Numerous doubles of the dead supposed to haunt the neighbourhood, are those of enemies to the tribe. Of the rest, the larger number are those with whom there have been relations of antagonism or jealousy. The ghosts of friends, too, and even of relatives, are apt to take offence and to revenge themselves. Hence, accidents, misfortunes, diseases, deaths, perpetually suggest the agency of malevolent spirits and the need for combating them. Modes of driving them away are devised; and the man who gains repute for success in using such modes becomes an important personage. Led by the primitive conception of ghosts as like their originals in their sensations, emotions, and ideas, he tries to frighten them by threats, by grimaces, by horrible noises; or to disgust them by stenches and by things to which they are averse; or, in cases of disease, to make the body a disagreeable habitat by subjecting it to intolerable heat or violent ill-usage. And the medicine-man, deluding himself as well as others into the belief that spirits have been expelled by him, comes to be thought of as having the ability to coerce them, and so to get supernatural aid: as instance a pagé of the Uaupés, who is "believed to have power to kill enemies, to bring or send away rain, to destroy dogs or game, to make the fish leave a river, and to afflict with various diseases."

The early predominance of the medicine-man as distinguished from the priest, has a further cause. At first the only ghosts regarded as friendly are those of relatives, and more especially of parents. The result is that propitiatory acts, mostly performed by descendants, are relatively private. But the functions of the medicine-man are not thus limited in area. As a driver away of malicious ghosts, he is called upon now by this family and now by that; and so comes to be a public agent, having duties co-extensive with the tribe. Such priestly character as he occasionally acquires by the use of propitiatory measures, qualifies but little his original character. He remains essentially an exorcist.

It should be added that the medicine-man proper, has some capacity for higher development as a social factor, though he cannot in this respect compare with the priest. Already in § 474, instances have been given showing that repute as a sorcerer sometimes conduces to the attainment and maintenance of political power; and here is another.

"The King of *Great Cassan* [Gambea] call'd *Magro* . . . was well skill'd in *Necromantick Arts*. . . . One time to shew his Art, he caused a strong Wind to blow. . . . Another time desiring to be resolved of some questioned particular, after his Charms a smoke and flame arose out of the Earth, by which he gathered the answer to his demand."

We also saw in § 198 that the medicine-man, regarded with fear, occasionally becomes a god.

§ 592. In subsequent stages when social ranks, from head ruler downwards, have been formed, and when there has evolved a mythology having gradations of supernatural beings—when, simultaneously, there have grown up priesthoods ministering to those superior supernatural beings who cannot be coerced but must be propitiated; a secondary confusion arises between the functions of medicine-men and priests. Malevolent spirits, instead of being expelled directly by the sorcerer's own power, are expelled by the aid of some superior spirit. The priest comes to play the part of an exorcist by calling on the supernatural being with whom he maintains friendly relations, to drive out some inferior supernatural being who is doing mischief.

This partial usurpation by the priest of the medicine-man's functions, we trace alike in the earliest civilizations and in existing civilizations. At the one extreme we have the fact that the Egyptians "believed . . . in the incessant intervention of the gods; and their magical literature is based on the notion of frightening one god by the terrors of a more powerful divinity;" and at the other extreme we have the fact that in old editions of our *Book of Common Prayer*, unclean spirits are commanded to depart "in the name of the Father, of the Son, and of the Holy Ghost."

There may be added the evidence which early records yield, that the superior supernatural beings invoked to expel inferior supernatural beings, had been themselves at one time medicine-men. Summarizing a tablet which he translates, Smith says—

"It is supposed in it that a man was under a curse, and Merodach, one of the gods, seeing him, went to the god Hea his father and enquired how to cure him. Hea, the god of Wisdom, in answer related the ceremonies and incantations, for effecting his recovery, and these are recorded in the tablet for the benefit of the faithful in after times."

§ 593. Thus, after recognizing the fact that in primitive

belief the doubles of the dead, like their originals in all things, admit of being similarly dealt with, and may therefore be induced to yield benefits or desist from inflicting evils, by bribing them, praising them, asking their forgiveness, or by deceiving and cajoling them, or by threatening, frightening, or coercing them; we see that the modes of dealing with ghosts, broadly contrasted as antagonistic and sympathetic, initiate the distinction between medicine-man and priest.

It is needless here to follow out the relatively unimportant social developments which originate from the medicine-man. Noting, as we have done, that he occasionally grows politically powerful, and sometimes becomes the object of a cult after his death, it will suffice if we note further, that during civilization he has varieties of decreasingly-conspicuous descendants, who, under one or other name, using one or other method, are supposed to have supernatural power or knowledge. Scattered samples of them still survive under the forms of wise women and the like, in our rural districts.

But the other class of those who are concerned with the supernatural, becoming, as it does, conspicuous and powerful, and acquiring as society develops an organization often very elaborate, and a dominance sometimes supreme, must be dealt with at length.

CHAPTER III.

PRIESTLY DUTIES OF DESCENDANTS.

§ 594. As we have before seen (§ 87), it is in some cases the custom to destroy corpses for the purpose of preventing resurrection of them and consequent annoyance by them; and in other cases where no such measure of protection is taken, the dead are, without discrimination between relatives and others, dreaded as causers of misfortunes and diseases. Illustrations of this belief as existing among various savages were given in Part I, Chaps. XVI, XVII. Here is another from New Britain.

The Matukanaputa natives "bury their dead underneath the hut which was lately inhabited by the deceased, after which the relatives go for a long canoe journey, staying away some months . . . they say . . . the spirit of the departed stays in his late residence for some time after his death, and eventually finding no one to torment goes away for good; the surviving relatives then return and remain there as formerly."

Even where ghosts are regarded as generally looking on their descendants with goodwill, they are apt to take offence and to need propitiation. We read of the Santāls that from the silent gloom of the adjacent grove—

"the byegone generations watch their children and children's children playing their several parts in life, not altogether with an unfriendly eye. Nevertheless the ghostly inhabitants of the grove are sharp critics, and deal out crooked limbs, cramps and leprosy, unless duly appeased."

But while recognizing the fact that ghosts in general are usually held to be more or less malicious, we find, as might be expected, that the smallest amount of enmity and the greatest amount of amity are supposed to be felt by the ghosts of relatives. Indeed by some races such ghosts are considered purely beneficent; as by the Karens, who think their meritorious ancestors "exercise a general watch care over their children on earth."

Though among various peoples there is propitiation chiefly of bad spirits, while good spirits are ignored as not likely to do mischief; yet wherever ancestor-worship preserves its original lineaments, we find the chief attention paid to the spirits of kindred. Prompted as offerings on graves originally are by affection for the deceased, and called forth as praises are by actual regrets for his or her departure, it naturally happens that these propitiations are made more by relatives than by others.

§ 595. Hence then the truth, everywhere illustrated, that those who perform the offices of the primitive cult are, at the outset, children or other members of the family. Hence then the fact that in Samoa—

"Prayers at the grave of a parent or brother or *chief* were common. Some, for example, would pray for health in sickness and might or might not recover."

Hence the fact that the people of Banks' Island, setting out on a voyage, would say—

"'Uncle! Father! plenty of pigs for you, plenty of money, kava for your drinking, twenty bags of food for your eating in the canoe. I pray you look upon me; let me go safe on the sea.'"

And hence once more the fact that among the Blantyre negroes—

"If they pray for a successful hunting expedition and return laden with venison or ivory, they know that it is their old relative that has done it, and they give him a thank offering. If the hunting party get nothing, they may say 'the spirit has been sulky with us,' . . . and refuse the thank-offering."

Unquestionably these cases, re-inforcing many before given, show us the beginnings of a family-religion. Along with that fear of a supernatural being which forms the central element of every religion, we see sacrifice and prayer, gratitude and hope, as well as the expectation of getting benefits proportionate to propitiations.

§ 596. An interpretation is thus furnished of the fact that in undeveloped societies the priestly function is generally diffused.

We find this to be the case at present among the uncivilized; as in New Caledonia, where "almost every family has its priest;" as in Madagascar, where other worships have arisen "long subsequently to the prevalence of the worship of household gods;" and as among the aborigines of India, who, though they propitiate ancestors, have not "in general, a regular and established priesthood." So, too, was it with the people who made the first advances in civilization—the Egyptians. Each family maintained the sacrifices to its own dead; and the greater deities had a semi-private worship, carried on by actual or nominal descendants. The like held of the Greeks and Romans, who joined sacrifices made to their public gods, chiefly by priests, with sacrifices made by private persons to their household gods who were dead relatives. And it is the same at the present time in China, where priesthoods devoted to wider worships, have not supplanted the primitive worship of departed progenitors by their offspring.

Having thus observed that in the earliest stage, propitiation of the double of a dead man by offerings, praises, etc., is carried on by surviving relatives, we have now to observe that this family-cult acquires a more definite form by the devolution of its functions on one member of the family.

CHAPTER IV.

ELDEST MALE DESCENDANTS AS QUASI-PRIESTS.

§ 597. THOUGH in the earliest stages sacrifices to the ghost of the dead man are made by descendants in general, yet in conformity with the law of the instability of the homogeneous, an inequality soon arises: the propitiatory function falls into the hands of one member of the group. Of the Samoans we read that "the father of the family was the *high-priest*." The like was true of the Tahitians: "in the family . . . the father was the priest." Of Madagascar, Drury says—"Every man here . . . is a Priest for himself and Family." Similarly in Asia. Among the Ostyaks "the father of a family was the sole priest, magician, and god maker;" and among the Gonds religious rites are "for the most part performed by some aged relative." With higher races it is, or has been, the same. By existing Hindoos the daily offering to ancestors is made by the head of the family. While "every good Chinaman regularly, every day, burns incense before the tablet to his father's memory," on important occasions the rites are performed by the head of the brotherhood. That family-headship brought the like duties in respect of manes-worship among Greeks and Romans, needs no showing. Speaking of primitive Sabæans, Palgrave says—"presidence in worship was, it seems, the privilege merely of greater age or of family headship;" and even among the Jews, to whom propitiation of the dead had been forbidden, there long survived the usage which had resulted from it. Kuenen remarks that though, up to David's time, "the competence of every Israelite to offer sacrifice was not doubted," yet "it was the kings and the heads of the tribes and families especially who made use of this privilege."

In the course of evolution under all its forms, differentiations tend ever to become more definite and fixed; and the differentiation above indicated is no exception. Eventually the usage so hardens, that the performance of sacrificial rites to ancestors is restricted to particular descendants. Speaking of the ancient Aryans, Sir Henry Maine says—"not only must the ancestor worshipped be a male ancestor, but the worshipper must be the male child or other male descendant."

§ 598. Hence certain sequences which we must note before we can rightly understand the institutions which eventually become established. In ancient Egypt "it was most important that a man should have a son established in his seat after him who should perform the due rites [of sacrifice to his *ka*, or double] and see that they were performed by others." Still more strongly was the need felt by the ancient Aryans. Says Duncker, "according to the law [of the Brahmans] every man ought to marry; he must have a son who may one day pour for him the libations for the dead." And we further read concerning them:—

"But the chief reason [for allowing polygamy] was that a son must necessarily be born to the father to offer libations for the dead to him. If the legitimate wife was barren, or brought forth daughters only, the defect must be remedied by a second wife. Even now, Hindoo wives, in a similar case, are urgent with their husbands to associate a second wife with them, in order that they may not die without male issue. How strongly the necessity was felt in ancient times is shown by an indication of the Rigveda, where the childless widow summons her brother-in-law to her bed, and by the narrative in the Epos of the widows of the king who died without a son, for whom children are raised up by a relation, and these children pass for the issue of the dead king (p. 85, 101). The law shows that such a custom did exist, and is not a poetic invention. It permits a son to be begotten by the brother of the husband, or the nearest of kin after him; in any case by a man of the same race (*gotra*), even in the life-time of the husband with his consent."

Among the Jews, too, though interdicted by their law from making material sacrifices to the dead, there survived the need for a son to utter the sacrificial prayer.

"Part of this extreme desire for sons is rooted in the fact that men alone can really pray, that men only can repeat the Kaddish, a prayer that has become almost a corner-stone of Hebraism, for there is deemed inherent in it a marvellous power. It is held that this prayer spoken by children over their parents' graves releases their souls from purgatory, that it is able to penetrate graves, and tell the dead parents that their children remember them."

So is it too in China, where a chief anxiety during life is to make provision for proper sacrifices after death. Failure of a first wife to bear a male child who may perform them, is considered a legitimate reason for taking a second wife; and in the Corea, where the funeral ceremonies are so elaborate that the mourners have cues to weep or cease weeping, we are shown the quasi-priestly function of the son, and also get an indication of the descent of this function. After a death "a man must be at once appointed *Shangjoo*, or male Chief Mourner. The eldest son, if living, or, failing him, his son rather than his brother, is the proper Shangjoo. . . . When these friends arrive, they mourn altogether, with the Shangjoo at their head." And among the Shangjoo's duties is that of putting food into the deceased's mouth: performing, at the same time, the reverential obeisance—baring his left shoulder.

§ 599. The primitive and long-surviving belief in a second life repeating the first in its needs—a belief which, as we see, prompted surprising usages for procuring an actual or nominal son who should minister to these needs—prompted, in other cases, a usage which, though infrequent among ourselves, has been and still is frequent in societies less divergent from early types: so frequent as to cause surprise until we understand its origin. Says Satow—"The practice of adoption, which supplies the childless with heirs, is common all over the East, but its justification in Japan is the necessity of keeping up the ancestral sacrifices." Accounts of Greeks and Romans show us that a kindred custom had among them a kindred motive. Though, as indicated in §§ 319 and 452, the practice of adoption had, among these people, survived from the times when its chief purpose was that of strengthening the patriarchal group; yet it is clear that the more special form of adoption which grew up had another purpose. Such a ceremony as that of a mock birth, whereby a fictitious son was made to simulate as nearly as might be a real son, could not have had a political origin, but must have had a domestic origin; and this origin was the one above indicated. As is pointed out by Prof. Hunter, Gaius speaks of " the great desire of the ancients to have vacant inheritances filled up, in order that there might be some one to perform the sacred rites, which were specially called for at the time of death." And since the context shows that this was the dominant reason for easy legalization of inheritance, it becomes clear that it was not primarily in the interest of the son, or the fictitious son, or the adopted son, that heirship was soon settled; but in the interest of the departed person. Just as, in ancient Egypt, men made bequests and endowed priests for the purpose of carrying on sacrifices in the private shrines erected to them; so did Roman fathers secure to themselves dutiful heirs, artificial when not natural, to minister to their ghosts out of the transmitted property.

Further significant evidence is supplied by the fact that heirship involved sacrifice. It was thus with the Eastern Aryans. Sir Henry Maine, speaking of the "elaborate liturgy and ritual" for ancestor-worship among the Hindus, says—"In the eye of the ancient Hindu sacerdotal lawyer, the whole law of Inheritance is dependent on its accurate observance." Or as Prof. Hunter remarks of these people— "The earliest notions of succession to deceased persons are connected with duties rather than with rights, with sacrifices rather than with property." And it was so with the Western

Aryans. Sir Henry Maine quotes the appeal of a Greek orator on behalf of a litigant—" Decide between us, which of us should have the succession and make the sacrifices at the tomb." And he points out that " the number, costliness, and importance of these ceremonies and oblations [to the dead] among the Romans," were such that even when they came to be less regarded, " the charges for them were still a heavy burden on Inheritances." Nay, even in mediæval Christendom there survived the same general conception in a modified form. Personal property was held to be " primarily a fund for the celebration of masses to deliver the soul of the owner from purgatory."

That these obligations to the dead had a religious character, is shown by the fact that where they have survived down to our own day, they take precedence of all other obligations. In India " a man may be pardoned for neglecting all his social duties, but he is for ever cursed if he fails to perform the funeral obsequies of his parents, and to present them with the offerings due to them."

§ 600. That we may the better comprehend early ideas of the claim supposed to be made by the double of the dead man on his property and his heir, it will be well to give some ancient examples of the way in which a son, or one who by a fiction stands in the position of a son, speaks of, or speaks to, his actual or nominal father who has died.

In Egypt, at Beni-hassan, an inscription by Chnumhotep says—" I made to flourish the name of my father, and I built the chapels for his *ka*. I caused my statues to be conveyed to the holy dwelling, and distributed to them their offerings in pure gifts. I instituted the officiating priest, to whom I gave donations in lands and peasants." Similarly at Abydos, Rameses II says concerning the worship of his father, Seti I :—

"I dedicated to thee the lands of the South for the service of thy temple, and the lands of the North, they bring to thee their gifts before thy beautiful countenance . . . I fixed for thee the number of the fields . . . great is their number according to their valuation in acres. I provided thee with land-surveyors and husbandmen, to deliver the corn for thy revenues."

Both which extracts exhibit the successor as being, in some sort, a steward for the deceased, administering on his behalf.

So was it in an adjacent empire. Assyria's "first rulers were called Patesi or 'Viceroys' of Assur;" and an inscription of Tiglath-Pileser says :—

"Ashur (and) the great gods, the guardians of my kingdom, who have government and laws to my dominions, and ordered an enlarged frontier to their territory, having committed to (my) hand their valiant and warlike servants, I have subdued the lands and the peoples and the strong places, and the Kings who were hostile to Ashur."

If now we remember that in Egypt the *ka*, or double of the dead man, was expected to return after a long period to re-animate his mummy and resume his original life—if we recall, too, the case of the Peruvians, who, similarly providing elaborately for the welfare of departed persons, similarly believed that they would eventually return—if we find ourselves thus carried back to the primitive notion that death is simply a long-suspended animation ; we may suspect the original conception to be that when he revives, a man will reclaim whatever he originally had ; and that therefore whoever holds his property, holds it subject to his prior claim—holds it as a kind of tenant who may be dispossessed by the owner, and whose sacred duty meanwhile is to administer it primarily for the owner's benefit.

§ 601. Be this so or not, however, the facts grouped as above, clearly show how, among the progenitors of the civilized peoples of the Old World, as well as among peoples who still retain early institutions, there arose those arrangements of the family-cult which existed, or still exist.

What has happened where descent in the female line obtains, is not clear. I have met with no statements showing that in societies characterized by this usage, the duty of ministering to the double of the dead man devolved on one of his children rather than on others. But the above facts show that, where the system of counting kinship through males has been established, the descent of the priestly function follows the same law as the descent of property; and there are other facts showing it more directly.

At the present time the connexion between the two is well displayed in China, where "it is regarded as indispensable that there should be some one to *burn incense to the manes of the dead*, from the eldest son down to posterity in the direct line of the eldest son, either by an own child or an adopted child;" and where the eldest son, who inherits more than other sons, has to bear the cost of the offerings. So, too, is it in the Corea, where, as already pointed out, the Shangjoo, or chief mourner, is either the eldest son or the eldest son of the eldest. When the corpse is buried, "if there are graves of ancestors in that place already, the Shangjoo sacrifices before them also, informing them of the new arrival."

These facts, along with foregoing ones, show that devolution of the sacrificial office accompanies devolution of property, because the property has to bear the costs of the sacrifices. We see that in societies characterized by the patriarchal form of organization, a son, who alone was capable of inheriting, could alone have due means of ministering to the deceased, and therefore could alone be priest. Whence obviously resulted the necessity for having a male descendant, as indicated above.

At the same time we are shown how, under the patriarchal type of society in its first stages, the domestic, the political, and the ecclesiastical, are undistinguished. These sacrifices made to the departed head of a family-group are primarily domestic. As the family-group develops into the compound group, the patriarch at its head acquires a quasi-political character; and these offerings made to him after death are in the nature of tribute, while fulfilment of the commands he left, disobedience to which may bring punishment when he returns, implies civil subordination. At the same time, in so far as these actions are performed to propitiate a being distinguished as supernatural, those who perform them acquire a quasi-ecclesiastical character.

CHAPTER V.

THE RULER AS PRIEST.

§ 602. In Chapters XIV and XV of Part I, we saw that according to the primitive Theory of Things, this life and this world stand in close relations with the other life and the other world. As implied at the end of the last chapter, one of the many results is that throughout early stages of social evolution, the secular and the sacred are but little distinguished.

Speaking of religion and politics, Huc remarks that "in the Eastern regions of Asia they were formerly one and the same thing, if we may judge from tradition. . . . The name of *heaven* was given to the Empire, the sovereign called himself *God*." How intimately blended were conceived to be the affairs of the material and spiritual worlds by the ancient Ethiopians, is well shown in Maspero's translation of a tablet describing the choice of a king by them.

"Then said each of them [the assembled host] unto his mate: 'It is true! since the time heaven was, since the royal crown was, . . . Ra decreed to give it unto his son whom he loves, so that the king be an image of Ra amongst the living; and has not Ra put himself in this land, that this land may be in peace?' Then said each of them unto his mate: 'But Ra has he not gone away to heaven, and is not his seat empty without a king . . . ?' So this whole host mourned, saying: 'There is a Lord standing amongst us without our knowing him!'" [The host eventually agrees to go to Amen-Ra, "who is the god of Kush," and ask him to give them their "Lord to vivify" them. Amen-Ra selects one of the Royal Brothers. The new king makes his obeisance to Amen-Ra, "and smelt the earth very much, very much, saying: 'Come to me, Amen-Ra, Lord of the seats of both worlds.'"]

Again of the ancient Peruvians we read that—
"If the estates of the King were not sufficient to provide for the excessive cost of a war, then those of the Sun were made available, which the Ynca considered to be his, as the legitimate child and heir of the Deity."

If from the primitive belief that the double of the dead man will presently return and resume his life, there results the conception that the son who holds his property and ministers to him from its proceeds is but a deputy, then this fusion of the sacred with the secular is a corollary. When we read of the New Caledonians that in Tokelau, while "the king, Tui Tokelau, is high priest as well," "their great god is called Tui Tokelau, or king of Tokelau," we have a typical instance of the union which results from this supposed vicegerency.

§ 603. While the growth of the family into the cluster of families, ending in the formation of the village-community, which often includes affiliated strangers, involves that the patriarch ceases to have the three-fold character of domestic, political, and ecclesiastical head, his character remains two-fold: he habitually retains, as in the case just named, the functions of ruler and priest. This connexion of offices we everywhere find in early stages of social evolution; and we observe it continuing through later stages.

In Tanna, "the chief acts as high-priest;" and the like is true in other islands of the group. The kings of Mangaia "were 'te ara pia o Rongo' *i.e.*, 'the mouth-pieces, or priests, of Rongo.'" Among the New Zealanders "the offices of chief and priest were generally united and hereditary." "The king of Madagascar ... is high-priest of the realm." In the Sandwich Islands the king "uttered the responses of the oracle, from his concealment in a frame of wicker-work." Of Humphrey's Island we read that the king "was high priest as well." Similarly with rude peoples in America. "The Pueblo chiefs seem to be at the same time priests," says Bancroft; and we learn the like from Ross concerning the Chinooks, and from Hutchison concerning the Bolivian

Indians. Of various semi-civilized peoples, past and present, we have similar accounts. The traditional "founders of the Maya civilization, united in their persons the qualities of high-priest and king." In ancient Peru, the Ynca was high-priest: "as the representative of the Sun, he stood at the head of the priesthood, and presided at the most important of the religious festivals." Of Siam, Thomson writes—"the King himself is High Priest." We are told by Crawfurd that the Javanese king is "the first minister of religion." In China the ritual laws give to the Emperor-Pontiff "the exclusive privilege of worshipping the Supreme, and prohibit subjects from offering the great sacrifices." And in Japan, the Mikado was "chief of the national religion." The early records of Old World peoples show us the same connexion. The Egyptian king, head of the priesthood, was everywhere represented in their monuments as sacrificing to a god. The Assyrian king was similarly represented; and the inscriptions show that Tiglath Pileser was "high-priest of Babylon." So, too, in the Hebrew records we read of David officiating as priest. It was the same with Aryan peoples in ancient days. Among the Greeks, as described by Homer, acts of public devotion "are everywhere performed by the chiefs without the intervention of a priest." The Spartan kings were priests of Zeus; and they received the perquisites due to priests. So "at Athens, the archon-king . . . embraced in his functions all that belonged to the State-religion. He was a real *rex sacrorum*." And that the like was the case among the Romans, "we know from the fact that the 'rex sacrificulus' was appointed on the abolition of the monarchy to perform such sacrifices as could only be performed by a king." Nor did the Aryans who spread northwards fail to furnish illustrations. Among the primitive Scandinavians the head man was "minister and magistrate in one:" in early days "each chief, as he settled, built his own hof or temple, and assumed the functions of priest himself."

This connexion long continued in a modified form throughout mediæval Europe. King Gontran was "like a priest among priests." Charlemagne, too, had a kind of high-priestly character: on solemn occasions he bore relics on his shoulders and danced before relics. Nor indeed is the connexion entirely broken even now.*

§ 604. In illustrating this primitive identity of ruler and priest, and in tracing out the long-continued connexion between the two, I have been unavoidably led away from the consideration of this double function as seen at the outset. Fully to understand the genesis of the priest properly so called, we must return for a moment to early stages.

At first the priestly actions of the chief differ in nothing from the priestly actions of other heads of families. The heads of all families forming the tribe, severally sacrifice to their departed ancestors; and the chief does the like to his departed ancestors. How, then, does his priestly character become more decided than theirs?

Elsewhere I suggested that besides propitiating the ghosts of dead relatives, the members of a primitive community will naturally, in some cases, think it prudent to propitiate the ghost of a dead chief, regarded as more powerful than other ghosts, and as not unlikely to do them mischief if friendly

* The fact that most people on reading that Melchizedek was priest and king, are struck by the connexion as anomalous, well exemplifies the quality of current education. When, as I have just learned, a clergyman examining young ladies at their confirmation, names as remarkable this combination of characters, which is the normal combination, we may judge how widely prevalent is the ignorance of cardinal truths in the histories of societies: an ignorance which goes along with knowledge of those multitudinous trivialities that make up primers of history and figure on examination papers. But our many-headed political pope, which is as fit to prescribe a system of education as was the ecclesiastical pope to tell Galileo the structure of the Solar System, thinks well that children should learn (even though the lessons add to that strain which injures health) what woman this or that king married, who commanded at this or that battle, what was the punishment of this rebel or that conspirator, &c.; while they are left in utter darkness respecting the early stages of leading institutions under which they live.

relations are not maintained by occasional offerings. I had not, when making the suggestion, any evidence; but conclusive evidence has since been furnished by the Rev. Duff MacDonald's *Africana*. The following three extracts show the transition from priestly actions of a private character to those of a public character, among the Blantyre negroes.

"On the subject of the village gods opinions differ. Some say that everyone in the village, whether a relative of the chief or not, must worship the forefathers of the chief. Others say that a person not related to the chief must worship his own forefathers, otherwise their spirits will bring trouble upon him. To reconcile these authorities we may mention that nearly everyone in the village is related to its chief, or if not related is, in courtesy, considered so. Any person not related to the village chief would be polite enough on all public occasions to recognise the village god: on occasions of private prayer . . . he would approach the spirits of his own forefathers."

"The chief of a village has another title to the priesthood. It is his relatives that are the village gods."

"Apart from the case of dreams and a few such private matters, it is not usual for anyone to approach the gods except the chief of the village. He is the recognised high priest who presents prayers and offerings on behalf of all that live in his village."

Here, then, we see very clearly the first stage in the differentiation of the chief into the priest proper—the man who intercedes with the supernatural being not on his own behalf simply, nor on behalf only of members of his family, but on behalf of unrelated persons. This is, indeed, a stage in which, as shown by the disagreement among the people themselves, the differentiation is incomplete. In another part of Africa, we find it more definitely established. At Onitsha on the Niger, "the people reverence him [the king] as the mediator between the gods and themselves, and salute him with the title of *Igue*, which in Ēbo means supreme being." A kindred state of things is illustrated among remote and unallied peoples. In Samoa, where the chiefs were priests, "every village had its god, and everyone born in that village was regarded as the property of that god." And among the ancient Peruvians, more advanced though they were in their social organization, a like primitive arrangement was traceable. The *huacas* were adored by the entire village; the *canopas* by particular families, and only the priests spoke to, and brought offerings to, the *huacas*.

These few out of many cases, while they sufficiently exemplify the incipient parting of the sacred function from the secular function, also illustrate the truth which everywhere meets us, that the political and religious obligations are originally both obligations of allegiance, very little distinguished from one another—the one being allegiance to the living chief and the other allegiance to the ghost of the dead chief.

To prevent misapprehension a parenthetic remark must be made. This growth of a distinction between the public worship of his ancestor by a chief, and the private worship of their ancestors by other men, which makes the chief's priestly character relatively decided, is apt to be modified by circumstances. Where allegiance to the ghost of a deceased patriarch or founder of the tribe, has become so well established through generations that he assumes the character of a god; and where, by war or migration, the growing society is so broken up that its members are separated from their chief and priest; it naturally results that while continuing to sacrifice to the doubles of their dead relatives, these separated members of the society begin to sacrifice on their own account to the traditional god. Among the ancient Scandinavians "every father of a family was a priest in his own house," where he sacrificed to Odin. Similarly among the Homeric Greeks. While chiefs made public sacrifices to the gods, sacrifices and prayers were made to them by private persons, in addition to the sacrifices made to their own ancestors. The like was the case with the Romans. And even among the Hebrews, prohibited from worshipping ancestors, the existence of public propitiators of Jahveh did not exclude "the competence of every Israelite" to perform propitiatory rites: the nomadic habits preventing concentration of the priestly function.

Phenomena of this kind, however, manifestly belong to a more advanced stage and not to that first stage in which, as we see, the genesis of the god and the priest are concurrent.

§ 605. Thus, then, the ghost-theory, which explains the multitudinous phenomena of religion in general, explains also the genesis of the priestly function, and the original union of it with the governing function.

Propitiations of the doubles of dead men, made at first by all their relatives and afterwards by heads of families, come to be somewhat distinguished when made by the head of the most powerful family. With increased predominance of the powerful family, and conception of the ghost of its deceased head as superior to other ghosts, there arises the wish, at first in some, then in more, and then in all, to propitiate him. And this wish eventually generates the habit of making offerings and prayers to him through his ruling descendant, whose priestly character thus becomes decided.

We have now to observe how, with the progress of social evolution, the sacerdotal function, though for a long time retained and occasionally exercised by the political head, comes to be performed more and more by proxy.

CHAPTER VI.

THE RISE OF A PRIESTHOOD.

§ 606. In §§ 480 and 504, I have drawn conclusions from the fact, obvious *a priori* and illustrated everywhere, that with increase of a chief's territory, there comes an accumulation of business which necessitates the employment of assistants; whence follows the habit of frequently, and at length permanently, deputing one or other of his functions, such as general, judge, etc. Among the functions thus deputed, more or less frequently, is that of priest.

That such deputation takes place under pressure of affairs, civil or military, we see in the case of the Romans. As the kings could not always attend to the sacrifices, having often to make war, Numa (who performed, according to Livy, the majority of the sacerdotal offices) "instituted flamens to replace the kings when the latter were absent;" and, adds M. Coulanges, "thus the Roman priesthood was only an emanation from the primitive royalty." How causes of this kind operate in simple societies, we are shown by a sentence in Mr. MacDonald's account of the Blantyre negroes. He says :—" If the chief is from home his wife will act [as priest], and if both are absent, his younger brother." As occurring in a ruder society where the blood-relationship of the chief to the god is still recognized, this case shows us, better than that of the Romans, how a priesthood normally originates.

This vicarious priest-ship of the younger brother, here arising temporarily, in other cases becomes permanent. Of the New Zealanders, who have in many cases chiefs who are at the same time priests, we read that in other cases the brother of the chief is priest. In the Mexican empire "the high-priest in the kingdom of Acolhuacan [and in that of Tlacupan] was, according to some historians, always the second son of the king." So, too, in ancient Peru "they had a high priest, who was an uncle or brother of the king, or at least a legitimate member of the royal family." As this last case shows, when the ruling man, still exercising the priestly function on great occasions, does not invariably make his younger brother his deputy on ordinary occasions, the office of high-priest still habitually falls to some blood-relation. Thus of the Khonds we read that "the chief civil and sacerdotal offices appear originally to have been united, or, at least, to have been always held by members of the chief patriarchal family." In Tahiti, where the king frequently personified the god, receiving the offerings brought to the temple and the prayers of the supplicants, and where he was sometimes the priest of the nation, "the highest sacerdotal dignity was often possessed by some member of the reigning family." Dupuis tells us that one of the priests of Ashantee belonged to the "king's own family." Among the Maya nations of America "the high-priests were members of the royal families." And in ancient Egypt there existed a kindred connexion. The king himself being high-priest, it was natural that the priesthood should include some of his relatives; and Brugsch, speaking of the high-priests of Ptah, says—"We find among their number princes of the blood royal. As an example we may name the prince Khamus, a favourite son of Ramses II."

In some cases the priestly functions of the head man are performed by a female relative. Among the Damaras the chief's daughter is priestess; and, "besides attending to the sacrifices, it is her duty to keep up the 'holy fire.'" On appointed occasions among the Dahomans, sacrifices are brought to the tomb (presumably of a king) and "before the tomb, a Tansi-no priestess, of blood-royal, offers up to the Ghost a prayer." Similarly in ancient Peru, a chief priestess who was one of the virgins of the Sun, and who was regarded as his principal wife, "was either the sister or the daughter of the ruler." On reading that among the Chibchas, with the priests "as with the caziques, the sister's son inherited," we may suspect that usages of this kind were consequent on descent in the female line. Among the Damaras this law of descent is still in force; it was manifestly at one time the law among the Peruvians; and the high political position of women among the Dahomans suggests that it was once the law with them also. Further reason for assuming this cause is supplied by the fact that in Dahomey and Peru, the priestly organization in general is largely officered by women; and that in Madagascar too, where descent is in the female line, there are women-priests. Obviously the transition from the usage of tracing descent through females to that of tracing descent through males, or the mixture of peoples respectively recognizing these unlike laws of descent, will cause anomalies; as instance that shown us by the Karens, whose village priests are males, but who, in their family ancestor-worship, "require that the officiating priest shall be a woman, the oldest of the family."

This deputation of priestly functions to members of a ruling family, usual in early stages, may be considered the normal differentiation; since the god being the apotheosized ancestor, the sacrifices made to him continue to be the sacrifices made by descendants. Even where descent is not real, or has ceased to be believed, it is still pretended; as in Egypt, where the king habitually claimed kinship with a god, and where, by consequence, members of his family were hypothetically of divine descent.

§ 607. But while this is distinguishable as the usual origin of a priesthood, there are other origins. In a preceding chapter we saw that there is at the outset no clear distinction between the medicine-man and the priest. Though the one is a driver away of spirits rather than a propitiator of them, while the other treats them as friends rather than enemies, yet either occasionally adopts the policy of the other. The priest sometimes plays the part of exorcisor and the medicine-man endeavours to appease: instance the Australian medicine-man described in § 584. Among the Ostyaks the shamans, who are medicine-men, are also "intermediators between the people and their gods." The business of a Gond medicine-man is "to exorcise evil spirits, to interpret the wishes of the fetish, to compel rain, and so on." And the same men who, among the Kukis, have to pacify a god who is angry and has caused disease, are often supposed to abuse "the influence they possess with supernatural agents." Evidently there is here indicated another origin of a priesthood.

Especially in cases where the medicine-man is supposed to obtain for the tribe certain benefits by controlling the weather through the agency of supernatural beings, does he participate in the character of priest. On recalling the case of Samuel, who while a judge over Israel also offered sacrifice to Jahveh as a priest and also controlled the weather by his influence with Jahveh (thus uniting the offices of ruler, priest and weather-doctor), we are shown how a kindred union of functions may in other cases similarly arise. Such facts as that among the Obbo the chief is also the rain-maker, and that Sechele, king of the Bechuanas, practises "rain-magic," besides re-inforcing the evidence given in § 474 that supposed power over supernatural beings strengthens the hands of political heads, shows also that, as having the function of obtaining from the supernatural beings benefits for the society, they in so far fulfil the priestly office.

In other cases there arise within the tribe the worships of apotheosized persons who were not related to the apotheosized chief; but who, for some reason or other, have left behind awe-inspiring reputations. Hislop tells us of a Gond who boasts of miraculous powers, and who "has erected a sacred mound to the manes of his father, who was similarly gifted, and he uses the awe which attaches to this spot as a means of extorting money from the deluded Queen"—money partly spent in offerings to "his deified ancestor:" the rest being appropriated by himself. And Sir Alfred Lyall in his *Asiatic Studies*, variously illustrates this sporadic origin of new deities severally apt to originate priesthoods.

Hence it seems inferable that in early stages there occasionally arise men not descended from the chief's ancestor, who acquire quasi-priestly characters, and may even succeed in supplanting priests of normal origin. Especially is such usurpation likely to happen where by migration or by war, there have been produced fragments of the society which do not contain within themselves descendants of the traditional god.

§ 608. So long as there continues undivided, a community of which the deceased founder has become the village god, propitiated on behalf of his descendants by the nearest of kin among them, who also serves as intermediator for other heads of families respectively worshipping their ancestors, no advance in the development of a priesthood is likely to take place. But when increase of numbers necessitates parting, there comes a further differentiation. How this arises we are well shown by a statement of Andersson concerning the Damaras:—"A portion of such fire [sacred fire] is also given to the head man of a kraal, when about to remove from that of the chief. The duties of a vestal then devolve upon the daughter of the emigrant." Evidently where a dead ruler, or other remarkable member of the tribe, has become a traditional god, so well established that propitiation of him has become imperative, migrating portions of the tribe, carrying their cult with them, must have someone to perform the rites on their behalf. Always the probability is that the detached group contains men akin to the chief of the parent tribe, and therefore descendants, direct or collateral, of the worshipped god; and on one of these, in virtue of greatest age or nearest relationship, the function is likely to fall. And since the reasons which determine this choice tend also to determine inheritance of the function, the genesis of a priestly caste becomes intelligible. Light is thrown on the matter by Hislop's statement that though the Gonds are without priests, there are "some men who, from supposed superior powers, or in consequence of their hereditary connection with a sacred spot, are held to be entitled to take the lead in worship." The course which change in some cases takes is shown us by the Sāntals. Hunter says—

"Two of the tribes have more especially devoted themselves to religion, and furnish a large majority of the priests. One of these represents the state religion, founded on the family basis, and administered by the descendants of the fifth son, the original family priest. . . . In some places, particularly in the north, the descendants of the second son . . . are held to make better priests than those of the fifth. . . . They are for the most part prophets, diviners, and officiating Levites of forest or other shrines, representing demon-worship; and in only a few places do they take the place of the fifth tribe."

Not only by the spread of a growing tribe into new habitats, are there thus produced conditions which further the growth of a priesthood; but kindred conditions are produced by the spread of a conquering tribe, and the establishment of its members as rulers over subordinate tribes. While it has to establish local governments, it has also to establish local ministrations of the cult it brings with it. The case of the Peruvians may be taken as typical. The Ynca-race, over-running indigenous races and leaving their religions intact, simply superposed their own religion. Hence the need for dispersed representatives of it. "The principal priest (or bishop) in each province was an Ynca, who took care that the sacrifices and ceremonies should be in conformity with those of the metropolitan." Now since the Ynca-religion was a worship of the Sun, regarded as ancestor; and since his supposed most direct descendant, the king himself, was high-priest on important occasions, while the other chief priests were "all Yncas of the blood royal;" it becomes clear that this establishment of a local priesthood of Ynca-blood, illustrates the development of a priestly caste from the ancestor-worshipping members of a conqueror's family.

§ 609. In verification of the foregoing conclusions, some evidence might be added showing that in tribes which lead peaceful lives, and in which considerable advances have been made without the establishment of strong personal governments, and therefore without the rise of apotheosized chiefs serving as village gods, there is but a feeble marking off of the priest-class. Among the Bodo and Dhimáls, for example, the priestly office is not hereditary, and is participated in by the elders of the people.

It is scarcely practicable, however, and would not be very profitable, to trace further this rise of a priesthood. Influences of sundry kinds tend everywhere to complicate, in one way or other, the primitive course of development. While we see that worshipping the spirit of the dead chief, at first carried on by his heir, is in his heir's absence deputed to a younger brother—while we see that temporary assumption of the function by a brother or other member of the family, tends to become permanent where the business of the chief increases—while we see that migrating parts of a tribe, are habitually accompanied by some of the village god's direct or collateral descendants, who carry with them the cult and perform its rites, and that where conquest of adjacent communities leads to an extension of rule, political and ecclesiastical, members of the ruling family become local priests; we find at work sundry causes which render this process irregular Besides the influence which the chief or his priestly relative is supposed to have with powerful supernatural beings, there is the competing influence ascribed to the sorcerer or rain-maker. Occasionally, too, the tribe is joined by an immigrant stranger, who, in virtue of superior knowledge or arts, excites awe; and an additional cult may result either from his teachings, or from his own apotheosis. Moreover, a leader of a migrating portion of the tribe, if in some way specially distinguished, is likely at death to become himself the object of a worship competing with the traditional worship, and perhaps initiating another priesthood. Fluctuating conditions are thus apt, even in early stages, to produce various modifications in ecclesiastical organization.

But the complications thus resulting are small compared with others which they foreshadow, and to which we may now turn our attention.

CHAPTER VII.

POLYTHEISTIC AND MONOTHEISTIC PRIESTHOODS.

§ 610. ALREADY in the preceding chapters the rudimentary form of a polytheistic priesthood has been exhibited. For wherever, with the worship of an apotheosized founder of the tribe, there co-exist in the component families of the tribe, worships of their respective ancestors, there is an undeveloped polytheism and an incipient priesthood appropriate to it. In the minds of the people there is no contrast in kind between the undistinguished ghosts and the distinguished ghosts; but only a contrast in power. In the first stage, as in later and higher stages, we have a greater supernatural being amid a number of lesser supernatural beings; all of them propitiated by like observances.

The rise of that which is commonly distinguished as polytheism, appears to result in several ways; of which two may be named as the more important.

The first of them is a concomitant of the division and spreading of tribes which outgrow their means of subsistence. Within each separated sub-tribe eventually arises some distinguished chief or medicine-man, whose greatly-feared ghost, propitiated not by his descendants only but by other members of the sub-tribe, becomes a new local god; and where there survives the cult which the sub-tribe brought with it, there will, in addition to the worship of the more ancient god common to the spreading cluster of sub-tribes, grow up in each sub-tribe the worship of a more modern god

peculiar to it. Traces of this process we find in many places. What we read of the Malagasy may be instanced as typical. They have gods who belong "respectively to different tribes or divisions of the natives, and are supposed to be the guardians and benefactors, or the titular gods, of these particular clans or tribes. Four of these are considered superior to all others"—are public or national gods. And Ellis adds that the gods of one province have little weight or authority with people of another province. As a case remote in time may be named that of the ancient Egyptians. The nomes, or original divisions of which Egypt was composed, were "of the highest antiquity": their limits being very exactly defined in inscriptions borne by the most ancient monumental structures. "Each district had a chief place where the [hereditary] governor resided, and enjoyed the protection and the cult of a special divinity, the sanctuary of which formed the centre of the religious worship of the district." That kindred evidence is furnished by accounts of other ancient peoples needs no showing. Of course along with this process goes the rise of priesthoods devoted some to the local and some to the general cults, with consequent differences in dignity. Thus of Egyptian priests we read:—
"Some also, who were attached to the service of certain divinities, held a rank far above the rest; and the priests of the great gods were looked upon with far greater consideration than those of the minor deities. In many provinces and towns, those who belonged to particular temples were in greater repute than others."

A genesis of polytheism, and of polytheistic priesthoods, equally important with, or perhaps more important than, the foregoing, but frequently, as in the last case, scarcely distinguishable from it, accompanies conquest. The over-runnings of tribe by tribe and nation by nation, which have been everywhere and always going on, have necessarily tended to impose one cult upon another; each of them already in most cases made composite by earlier processes of like kind. Not destroying the worships of the conquered, the conquerors bring in their own worships—either carrying them on among themselves only, or making the conquered join in them; but in either case multiplying the varieties of priests. The survival of cults that were of Pelasgian origin amid those of the Greeks, supplies an early instance in Europe; and later instances are supplied by the Romans. "As a conquering state Rome was constantly absorbing the religions of the tribes it conquered. On besieging a town, the Romans used solemnly to evoke the deities dwelling in it." The process was illustrated in ancient American societies. " The high-priests of Mexico were the heads of their religion only among the Mexicans, and not with respect to the other conquered nations: these . . . maintaining their priesthood independent." Similarly in Peru.

"The Yncas did not deprive the chiefs of their lordship, but his delegate lived in the valley, and the natives were ordered to worship the sun. Thus a temple was built, and many virgins and priests to celebrate festivals resided in it. But, notwithstanding that this temple of the sun was so pre-eminently established, the natives did not cease to worship also in their ancient temple of Chinchaycama."

Of additional but less important causes of complication, three may be named. The spreading reputations of local deities, and the consequent establishment of temples to them in places to which they do not belong, is one of these causes. A good example is that of Æsculapius; the worship of whom, as a local ancestor and medicine-man, originated in Pergamon, but, along with his growth into a deity, spread East and West, and eventually became established in Rome. Another additional cause, well illustrated in ancient Egypt, is the deification of powerful persons who establish priesthoods to minister to their ghosts. And a third is the occasional apotheosis of those who, for some reason or other strike the popular imagination as remarkable. This is even now active in India. Sir Alfred Lyall has exemplified it in his *Asiatic Studies*.

§ 611. The frequent genesis of new worships and continued co-existence of many worships, severally having their priesthoods, though quite normal as we here see, appears to many persons anomalous. Carrying back modern ideas to the interpretation of ancient usages, writers comment on the "tolerance" shown by the Romans in leaving intact the religions of the peoples conquered by them. But considered from their point of view instead of from our point of view, this treatment of local gods and their priests was quite natural. If everywhere, from ancestor-worship as the root, there grew up worships of known founders of tribes and traditional progenitors of entire local races, it follows that conquerors will, as a matter of course, recognize the local worships of the conquered while bringing in their own. The corollary from the universally-accepted belief is that the gods of the vanquished are just as real as those of the victors.

Sundry interpretations are yielded. Habitually in the ancient world, conquerors and settlers took measures to propitiate the local gods. All they heard about them fostered the belief that they were powerful in their respective localities, and might be mischievous if not prayed to or thanked. Hence, probably, the fact that the Egyptian Nekôs sacrificed to Apollo on the occasion of his victory over Josiah, king of Judah. Hence, to take a case from a remote region, the fact that the Peruvian Yncas, themselves Sun-worshippers, nevertheless provided sacrifices for the various *huacas* of the conquered peoples, "because it was feared that if any were omitted they would be enraged and would punish the Ynca."

Co-existence of different cults is in some cases maintained by the belief that while the allegiance of each man to his particular deity or deities is obligatory, he is not required, or not permitted, to worship the deities belonging to fellow-citizens of different origin. Thus in early times in Greece, "by the combination of various forms of religious worship Athens had become the capital, and Attica one united whole. But . . . Apollo still remained a god of the nobility, and his religion a wall of separation. . . . According to the plan of Solon this was to be changed. . . . To every free Athenian belonged henceforth the right and the duty of sacrificing to Apollo."

All which facts make it clear that not only the genesis of polytheism but the long survival of it, and consequent persistence of priesthoods devoted to different gods, are sequences of primitive ancestor-worship.

§ 612. But while, during early stages of polytheism, overt efforts at subjugation of one cult by another are not conspicuous, there habitually arises a competition which is the first step towards subjugation.

A feeling like that occasionally displayed by boys, boasting of the strengths of their respective fathers, prompts men in early stages to exaggerate the powers of their ancestors, as compared with the powers which the ancestors of others displayed; and concerning the relative greatness of the deified progenitors of their tribes, there are certain to arise disputes. This state of things was exemplified in Fiji when first described by missionaries: "each district contending for the superiority of its own divinity." Evidently among the Hebrews an implied belief, opposed to the beliefs of adjacent peoples, was—our god is greater than your god. Without denying the existence of other gods than their own, the superiority of their own was asserted. In Greece, too, the religious emulation among cities, and the desire to excite envy by the numbers of men who flocked to sacrifice to their respective deities, implied a struggle between cults—a struggle conducive to inequality. Influences such as those which caused supremacy of the Olympian festivals above kindred festivals, were ever tending among the Greeks to give some gods and their ministers a higher *status* than others. Religion being under its primary aspect the expression of allegiance—an allegiance shown first to the living patriarch or conquering hero and afterwards to his ghost; it is to be expected that causes which modify the degree and extent of allegiance to the head man while alive, will similarly modify the allegiance to his ghost after his death. How closely connected are the two kinds of fealty we see in such a fact as that at a Santal marriage, the bride must give up her clan and its gods for those of her husband : reminding us of the representation made by Naomi to Ruth—" thy sister-in-law is gone back unto her people, and unto her gods; " and the rejoinder of Ruth—" thy people shall be my people, and thy god my god."

So understanding the matter, we see how it naturally happens that just as the subjects of a living chief, for one reason or another dissatisfied with his rule, will some of them desert him and attach themselves to a neighbouring chief (§ 452); so, among a polytheistic people, this or that motive may prompt decrease in the number of devotees at one god's temple and increase those at the temple of another. Disappointments like those which lead to the beating of their idols by savages, when in return for sacrifices the idols have not given what was wanted, will, among peoples somewhat more advanced, cause alienation from a deity who has proved obstinate, and propitiation of a deity who it is hoped will be more conceding. Even at the present day, we are shown by the streams of pilgrims to Lourdes, how the spread of belief in some alleged marvel may initiate a new worship, or reinforce an old one. As with saints so with gods—there result gradations. Political influences, again, occasionally conduce to the elevation of some cults above others. Speaking of Greece, Curtius says :—

" Another religious worship which the Tyrants raised to a new importance was that of Dionysus. This god of the peasantry is everywhere opposed to the gods of the knightly houses, and was therefore favoured by all rulers who endeavoured to break the power of the aristocracy."

Chiefly, however, inequalities among the ascribed powers of gods, where many co-exist, are due to conquests. Militant activities, which establish gradations of rank among the living, also establish gradations of rank among the worshipped dead. Habitually mythologies tell of victories achieved by the gods;

habitually they describe fights among the gods themselves; and habitually they depict the chief god as the one who acquired supremacy by force. These are just the traits of a pantheon resulting from the apotheosis of conquering invaders and from the usurpations now and then witnessed among their leaders. And evidently the subjugation of peoples one by another, and consequent elevation of one pantheon above another, must be a chief cause of differences among the powers of the major and minor deities, and of contrasts in importance among their respective cults and priesthoods.

§ 613. Eventually there results under favouring conditions a gravitation towards monotheism. It is true that for a long time there may continue in the minds of a polytheistic people, a fluctuating conflict among the beliefs respecting the relative powers of their gods. Of the ancient Aryans, Professor Max Müller writes—" It would be easy to find, in the numerous hymns of the Veda, passages in which almost every single god is represented as supreme and absolute. . . . Agni is called the ruler of the universe; . . . Indra is celebrated as the strongest god, . . . and the burden of one of the songs . . . is . . . Indra is greater than all. Of Soma it is said that . . . he conquers every one." Of the Egyptian gods too, a like fact is stated. The exaggerated language of worshippers attributes now to this of them and now to that, and sometimes to a living king, a greatness so transcendent that not only all other things but all other gods exist through him.

But the position of "father of gods and men" becomes eventually settled in the minds of believers; and if subsequently usurped, the usurpation does not diminish the tendency towards monotheism but increases it; since there results the idea of a divinity more powerful than was before believed in. How recognition of superiority in a conquering people, and by implication in their gods, tends to dwarf the gods of the conquered, the ancient Peruvians show. Garcilasso tells us that Indian tribes are said to have sometimes submitted from admiration of the higher culture of the Yncas: the obligation to join in the Yncas' worship being one of the concomitants. Then of the Yncas themselves, Herrera says—

"When they saw the *Spaniards* make Arches on Centers, and take them away when the Bridge was finish'd, they all ran away, thinking the Bridge would fall; but when they saw it stand fast, and the *Spaniards* walk on it, a Cacique said, It is but Justice to serve these Men, who are the Children of the Sun."

Evidently the attitude thus displayed conduced to acceptance of the Spaniards' beliefs and worship. And such mental conquests often repeated in the evolution of societies, tend towards the absorption of local and minor conceived supernatural agents in greater and more general ones.

Especially is such absorption furthered when one who, as a living ruler, was distinguished by his passion for subjugating adjacent peoples, leaves at death unfulfilled projects of conquest, and then has his ghost propitiated by extending his dominion. As shown by a preceding extract, this was the case with the Assyrian god Ashur (§ 600); and it was so, too, with the Hebrew god Jahveh: witness Deut. xx, 10—18.

"When thou comest nigh unto a city to fight against it, then proclaim peace unto it. And it shall be, if it make thee answer of peace, and open unto thee, then it shall be, that all the people that is found therein shall be tributaries unto thee, and they shall serve thee. And if it will make no peace with thee, but will make war against thee, then thou shalt besiege it: and when the Lord thy God hath delivered it into thine hands, thou shalt smite every male thereof with the edge of the sword. . . . But of the cities of these people, which the Lord thy God doth give thee for an inheritance, thou shalt save alive nothing that breatheth: But thou shalt utterly destroy them."

From the beginning we are shown that, setting out with the double of the ordinary dead man, jealousy is a characteristic ascribed to supernatural beings at large. Ghosts not duly sacrificed to are conceived as malicious, and as apt to wreak vengeance on survivors; gods whose shrines have been neglected and whose festivals do not bring due offerings, are said to be angry, and are considered the causers of disasters; while if one of them is derived from a ruler whose love of power was insatiable, and whose ghost is considered a jealous god, tolerating no recognition of others, he tends, if his devotees become predominant, to originate a worship which suppresses other worships.

Of course with such an advance towards monotheism there goes an advance towards unification of priesthoods. The official propitiators of minor deities dwindle away and disappear; while the official propitiators of the deity who has come to be regarded as the most powerful, or as the possessor of all power, become established everywhere.

§ 614. These influences conspiring to evolve monotheism out of polytheism are reinforced by one other—the influence of advancing culture and accompanying speculative capacity. Molina says that the Ynca Yupanqui "was of such clear understanding" as to conclude that the Sun could not be the creator, but that there must be "someone who directs him;" and he ordered temples to be erected to this inferred creator. So again in Mexico, "Nezahuatl, lord of Tezcuco," disappointed in his prayers to the established idols, concluded that "there must be some god, invisible and unknown, who is the universal creator;" and he built a nine-storied temple "to the Unknown God, the Cause of Causes." Here, among peoples unallied to them, we find results like those shown us by the Greeks. In the Platonic dialogues, along with repudiation of the gross conceptions current among the uncultured, there went arguments evidently implying an advance towards monotheism. And on comparing the ideas of the Hebrew prophets with those of primitive Hebrews, and those of most co-existing Hebrews, it becomes clear that mental progress operated as a part cause of Jewish monotheism.

It may be observed, too, that once having been set up, the change towards monotheism goes on with increasing momentum among the highest intelligences. A supremacy of one supernatural agent having become established, there follows the thought that what power other supernatural agents exercise is exercised by permission. Presently they come to be conceived as deputies, entrusted with powers not their own; and in proportion as the Cause of Causes grows more predominant in thought, the secondary causes fade from thought.

§ 615. Rightly to conceive the evolution of monotheism and its accompanying ecclesiastical institutions, we must take note of several influences which qualify it.

The earlier tendencies towards the rise of a supreme deity are apt to prove abortive. Just as during the first stages of social integration, a predominant headship is often but temporary, and the power acquired by a conquering chief is frequently lost by his successor; so an ascribed headship among the gods is commonly not lasting. For this we may see more reasons than one. The double of a dead man, at first conceived as existing temporarily, becomes conceived as permanently existing only where circumstances favour remembrance of him; and in like manner supremacy among ghosts or gods, requires for its maintenance that traditions shall be well preserved, and the social state lend itself to orderly observances. In many places these conditions are inadequately fulfilled. Remarking upon the fading of traditions among the Comanches, Schoolcraft says—" I question if the names of any of their chiefs of the fourth generation ascending are retained among them;" and when, in 1770, Cook touched on the shores of New Zealand within fifteen miles of the place visited by Tasman a hundred and twenty-eight years before, he found no tradition of the event. So that though everywhere the original tendency is for the oldest known progenitor to become the chief god; yet, as we are shown by the Unkulunkulu of the Zulus, this headship of the supernatural beings is apt to fade from memory, and later headships only to be regarded. A further cause militating against an unchanged pantheon, is the rise of usurpers, or of men who, by their successes in war or other achievements, so impress themselves on the popular mind as to make relatively weak the impressions derived from traditions of earlier deified men. The acquirement of supremacy by Kronos over Uranus, and again by Zeus over Kronos, serve as illustrations. And during times in which apotheosis is an ordinary process, there is an evident tendency to such substitutions. Yet another analogy between the changes of celestial headships and the changes of terrestrial headships, may be suspected. When dealing with political institutions, we saw that power is apt to lapse from the hands of a supreme ruler into the hands of a chief minister, through whom all information comes and all orders are issued. Similarly, a secondary supernatural being regarded as intercessor with a chief supernatural being, and constantly appealed to by worshippers in that capacity, seems liable to become predominant. Among Roman Catholics the Virgin, habitually addressed in prayers, tends to occupy the foreground of consciousness; the title "Mother of God" dimly suggests a sort of supremacy; and now in the Vatican may be seen a picture in which she is represented at a higher elevation than the persons of the trinity.

Another fact to be noted respecting the evolution of monotheisms out of polytheisms—a fact congruous with the hypothesis that they are thus evolved, but not congruous with other hypotheses—is that they do not become complete; or, at least, do not maintain their purity. Already I have referred to the truth, obvious enough though habitually ignored, that the Hebrew religion, nominally monotheistic, retained a large infusion of polytheism. Archangels exercising powers in their respective spheres, and capable even of rebellion, were practically demi-gods; answering in fact, if not in name, to the inferior deities of other pantheons. Moreover, of the derived creeds, that distinguished as trinitarian is partially polytheistic; and in the mystery plays of the Middle Ages marks of polytheism were still more distinct. Nay, even belief in a devil, conceived as an independent supernatural being, implies surviving polytheism. Only by unitarians of the advanced type, and by those who are called theists, is a pure monotheism accepted.

Further, we may remark that where polytheism under its original form has been suppressed by a monotheism more or less complete, it habitually revives under a new form. Though the followers of Mahomet shed their own blood and the blood of others, to establish everywhere the worship of one god, the worship of minor gods has grown up afresh among them. Not only do the Bedouins make sacrifices at saints' tombs, but among more civilized Mahometans there is worship of their deceased holy men at shrines erected to them. Similarly, throughout mediæval Christendom, canonized priests and monks formed a new class of minor deities. As now in Fiji "nearly every chief has a god in whom he puts special trust;" so, a few centuries back, every knight had a patron saint to whom he looked for succour.

That modifications of Ecclesiastical Institutions result from causes of this kind, is sufficiently shown by the fact, so familiar that we do not observe its significance, that churches are named after, or dedicated to, saints; and that such churches "as were built over the grave of any martyr, or called by his name to preserve the memory of him, had usually the distinguishing title of *Martyrium*, or *Confessio*, or *Memoria*, given them for that particular reason." It may, indeed, be alleged that these usages were rather survivals than revivals; since, as Mosheim says, the early Christian bishops deliberately adopted them, believing that "the people would more readily embrace Christianity" if they "saw that *Christ* and the martyrs were worshipped in the same manner as formerly their gods were." But taken either way the facts show that monotheism, and the sacerdotal arrangements proper to it, did not become complete.

CHAPTER VIII.

ECCLESIASTICAL HIERARCHIES.

§ 616. THE component institutions of each society habitually exhibit kindred traits of structure. Where the political organization is but little developed, there is but little development of the ecclesiastical organization; while along with a centralized coercive civil rule there goes a religious rule no less centralized and coercive. Qualifications of this statement required to meet changes caused in the one case by revolutions and in the other case by substitutions of creeds, do not seriously affect it. Along with the restoration of equilibrium the alliance begins again to assert itself.

Before contemplating ecclesiastical hierarchies considered in themselves, let us, then, note more specifically how these two organizations, originally identical, preserve for a long time a unity of nature consequent on their common origin.

§ 617. As above implied, this relation is primarily illustrated by the cases in which, along with unsettled civil institutions there go unsettled religious institutions. The accounts given of the Nagas by Stewart and by Butler, which are to the effect that they "have no kind of internal government," and have apparently no priesthood, show also that along with their disregard of human authority, they show extremely little respect to such gods as they recognize after a fashion: dealing with beings in the spirit-world as defiantly as they do with living men. Of the Comanches, again, Schoolcraft, saying that "the authority of their chiefs is rather nominal than positive," also says—" I perceived no order of priesthood . . . if they recognise any ecclesiastical authority whatever, it resides in their chiefs." Evidently in the absence of established political headship, there cannot habitually arise recognition of a deceased political head; and there is consequently no place for an official propitiator.

With the rise of the patriarchal type of organization, both of these governmental agencies assume their initial forms. If, as in early stages, the father of a family, while domestic ruler, is also the one who makes offerings to the ancestral ghost—if the head of the clan, or chief of the village, while exercising political control also worships the spirit of the dead chief on behalf of others, as well as on his own behalf; it is clear that the ecclesiastical and political structures begin as one and the same: the co-existing medicine-man being, as already shown, not a priest properly so-called. When, for instance, we read of the Eastern Slavs that "it was customary among them for the head of the family or the tribe to offer sacrifices on behalf of all beneath a sacred tree," we see that the civil and religious functions and their agents are at first undifferentiated. Even where something like priests have arisen, yet if there is an undeveloped ruling agency they are but little distinguished from others, and they have no exclusive powers: instance the Bodo and Dhimáls, whose village heads have "a general authority of voluntary rather than coercive origin," and among whom elders "participate the functions of the priesthood." Nomadic habits, while they hinder the development of a political organization, also hinder the development of a priesthood; even when priests are distinguishable as such. Tiele says of the primitive Arabs that "the sanctuaries of the various spirits and fetishes had their own hereditary ministers, who, however, formed no priestly caste." So, too, such physical characters of a habitat, and such characters of its occupants as impede the massing of small groups into large ones, maintain simplicity of the ecclesiastical structure, as of the political. Witness the Greeks, of whom Mr. Gladstone, remarking that the priest was never "a significant personage in Greece," adds "nor had the priest of any one place or deity, so far as we know, any organic connection with the priest of any other; so that if there were priests, yet there was not a priesthood."

Conversely, along with that development of civil government which accompanies social integration, there usually goes a development of ecclesiastical government. From Polynesia we may take, as an instance, Tahiti. Here, along with the ranks of king, nobility, land-owners, and common people, there went such distinctions among the priests that each officiated in that rank only to which he belonged; and "the priests of the national temples were a distinct class." In Dahomey and Ashantee, along with a despotic government and a civil organization having many grades there go orders of priests and priestesses divided into several classes. The ancient American states, too, exhibited a like union of traits. Their centralized and graduated political systems were accompanied by ecclesiastical systems which were analogous in complexity and subordination. And that in more advanced societies there has been something approaching to parallelism between the developments of the agencies for civil rule and religious rule, needs not to be shown in detail.

To exclude misapprehension it may be as well to add that establishment of an ecclesiastical organization separate from the political organization, but akin to it in structure, appears to be largely determined by the rise of a decided distinction in thought between the affairs of this world and those of a supposed other world. Where the two are conceived as existing in continuity, or as intimately related, the organizations appropriate to their respective administrations remain either identical or imperfectly distinguished. In ancient Egypt, where the imagined ties between dead and living were very close, and where the union of civil and religious functions in the king remained a real union, "a chief priest, surrounded by a numerous priesthood, governed each city." The Japanese, too, yield an instance. Along with the belief that Japan was "the land of spiritual beings or kingdom of spirits," and along with the assumption by the Mikado of power to promote deceased persons to higher ranks in their second lives (§ 347), there went the trait that the Mikado's court had six grades of ecclesiastical ranks, and in this chief centre of rule, sacred and secular functions were originally fused: "among the ancient Japanese, government and religion were the same." Similarly in China, where the heavenly and the earthly are, as Huc points out, so little separated in conception, and where there is one authority common to the two, the functions of the established religion are discharged by men who are, at the same time, administrators of civil affairs. Not only is the emperor supreme priest, but the four prime ministers "are lords spiritual and temporal." If, as Tiele says, "the Chinese are remarkable for the complete absence of a priestly caste," it is because, along with their universal and active ancestor-worship, they have preserved that inclusion of the duties of priest in the duties of ruler, which ancestor-worship in its simple form shows us.

§ 618. Likeness between the ecclesiastical and political organizations where they have diverged, is largely due to their community of origin in the sentiment of reverence. Ready obedience to a terrestrial ruler is naturally accompanied by ready obedience to a supposed celestial ruler; and the nature which favours growth of an administration enforcing the one, favours growth of an administration enforcing the other.

This connexion was well illustrated by the ancient American societies. In Mexico, along with an "odious despotism" and extreme submissiveness of the people, making possible a governmental organization so ramified that there was a sub-sub-ruler for every twenty families, there went an immensely developed priesthood. Torquemada's estimate of 40,000 temples is thought by Clavigero to be greatly under the mark; and Clavigero says—" I should not think it rash to affirm, that there could not be less than a million of priests throughout the empire:" an estimate made more credible by Herrera's statement that "every great Man had a Priest, or Chaplain." Similarly in Peru; where, with an unqualified absolutism of the Ynca, and a political officialism so vast and elaborate that one out of every ten men had command of the others, there was a religious officialism no less extensive. Says Arriaga—" If one counts all the higher and lower officers, there is generally a minister for ten Indians or less." Obviously in the moral natures of the Mexicans and Peruvians, lies the explanation of these parallelisms. People so politically servile as those ruled over by Montezuma, who was " always carry'd on the Shoulders of Noblemen," and whose order was that "no Commoner was to look him in the Face, and if he did, dy'd for it," were naturally people content to furnish the numberless victims annually sacrificed to their gods, and ready continually to inflict on themselves propitiatory bloodlettings. And of course the social appliances for maintenance of terrestrial and celestial subordination developed among them with little resistance in corresponding degrees; as they have done, too, in Abyssinia. In the words of Bruce, "the kings of Abyssinia are above all laws;" and elsewhere he says "there is no country in the world in which there are so many churches as in Abyssinia."

Proof of the converse relation need not detain us. It will suffice to indicate the contrast presented, both politically and ecclesiastically, between the Greek societies and contemporary societies, to suggest that a social character unfavourable to the growth of a large and consolidated regulative organization of the political kind, is also unfavourable to the growth of a large and consolidated regulative organization of the ecclesiastical kind.

§ 619. Along with increase of a priesthood in size, there habitually go those specializations which constitute it a hierarchy. Integration is accompanied by differentiation.

Let us first note how the simultaneous progress of the two is implied by the fact that while the ecclesiastical organization is at first less sharply marked off from the political than it afterwards becomes, its own structures are less definitely distinguished from one another. Says Tiele—

"That the Egyptian religion, like the Chinese, was originally nothing but an organised animism, is proved by the institutions of worship. Here, too, existed no exclusive priestly caste. Descendants sacrificed to their ancestors, the officers of state to the special local divinities, the king to the deities of the whole country. Not till later did an order of scribes and a regular priesthood arise, and even these as a rule were not hereditary."

Again, we read that among the ancient Romans—

"The priests were not a distinct order from the other citizens. The Romans, indeed, had not the same regulations with respect to public employments as now obtain with us. With them the same person might regulate the police of the city, direct the affairs of the empire, propose laws, act as a judge or priest, and command an army."

And though in the case of an adopted religion the circumstances are different, yet we see that in the development of an administrative organization the same essential principle displays itself. M. Guizot writes—

"In the very earliest period, the Christian society presents itself as a simple association of a common creed and common sentiments. . . . We find among them [the first Christians] no system of determinate doctrines, no rules, no discipline, no body of magistrates. . . . In proportion as it advanced . . . a body of doctrines, of rules, of discipline, and of magistrates, began to appear ; one kind of magistrates were called πρεσβυτεροι, or *ancients*, who became the priests; another, επισκοποι, or inspectors, or superintendents, who became bishops ; a third διακονοι, or deacons, who were charged with the care of the poor, and with the distribution of alms. . . . It was the body of the faithful which prevailed, both as to the choice of functionaries, and as to the adoption of discipline, and even doctrine. The church government and the Christian people were not as yet separated."

In which last facts, while we see the gradual establishment of an ecclesiastical structure, we also see how, in the Church as in the State, there went on the separation of the small ruling part from the greater part ruled, and a gradual loss of power by the latter.

In the ecclesiastical body as in the political body, several causes, acting separately or jointly, work out the establishment of graduated authorities. Even in a cluster of small societies held together by kinship only, there tend, where priests exist, to arise differences among their amounts of influence: resulting in some subordination when they have to co-operate. Thus we read of the priests among the Bodo and Dhimáls, that "over a small circle of villages one Dhámi presides and possesses a vaguely defined but universally recognised control over the Deóshis of his district." Still more when small societies have been consolidated into a larger one by war, is the political supremacy of the conquering chief usually accompanied by ecclesiastical supremacy of the head priest of the conquering society. The tendency to this is shown even where the respective cults of the united societies remain intact. Thus it appears that "the highpriests of Mexico were the heads of their religion only among the Mexicans, and not with respect to the other conquered nations;" but we also read that the priesthood of Huitzilopochtli was that of the ruling tribe, and had, accordingly, great political influence. The Mexicatlteohuatzin had authority over other priesthoods than his own. Still more in ancient Peru, where the subjugation of the united peoples by the conquering people was absolute, a graduated priesthood of the conqueror's religion was supreme over the priesthoods of the religions professed by the conquered. After an account of the priesthood of the Sun in Cuzco, we read that—

"In the other provinces, where there were temples of the Sun, which were numerous, the natives were the priests, being relations of the local chiefs. But the principal priest (or bishop) in each province was an Ynca, who took care that the sacrifices and ceremonies should be in conformity with those of the metropolitan."

And then we are told by another writer that—

In the great temple of Cuzco, "the Ingas plac'd the Gods of all the Provinces they conquer'd, each Idol having its peculiar Altar, at which those of the Province it belong'd to offer'd very expensive Sacrifices; the Ingas thinking they had those Provinces secure, by keeping their Gods as Hostages."

In short the ancient Peruvian priesthood consisted of a major hierarchy posed on many minor hierarchies.

But besides these subordinations of one sacerdotal system to another caused by conquest, there are, as implied in the cases given, subordinations which arise within the organization of each cult. Such differences of rank and function existed in Egypt. Besides the high priests there were the *prophetæ*, the *justophori*, the *stolistes*, the *hierogrammateis*, and some others. Similarly among the Accadians. "On comptait à Babylone," says Maury, "divers ordres de prêtres ou interprètes sacrés, les *hakimim* ou savants, peut-être les médecins; les *khartumim*, ou magiciens, les *asaphim*, ou théologiens; et enfin les *kasdim* et les *gazrim*, c'est-à-dire les Chaldéens, les astrologues proprement dits." Rome, too, "had a very rich and complicated religious establishment" (1) the Pontiffs, Augurs, etc.; (2) the Rex Sacrificulus, the Sacrificers, and the Vestal Virgins; (3) Salii and Fetiales; (4) Curiones; (5) Brotherhoods. And it was so with the Mexican priests. "Some were the sacrificers, others the diviners; some were the composers of hymns, others those who sung. . . . Some priests had the charge of keeping the temple clean, some took care of the ornaments of the altars; to others belonged the instructing of youth, the correcting of the calendar, the ordering of festivals, and the care of mythological paintings."

Where, instead of coexisting religions with their priesthoods which we find in most compound societies produced by war in early stages, we have an invading religion which, monotheistic in theory, cannot recognize or tolerate other religions, there still, as it spreads, arises an organization similar in its centralization and specialization to those just contemplated. Describing the development of Church-government in Europe, M. Guizot says:—

"The bishop was, originally, the inspector, the chief of the religious congregation of each town. . . . When Christianity spread into the rural districts, the municipal bishop no longer sufficed. Then appeared the chorepiscopi, or rural bishops . . . the rural districts once Christian, the chorepiscopi in their turn no longer sufficed . . . each Christian agglomeration at all considerable became a parish, and had a priest for its religious head . . . originally parish priests acted absolutely only as representatives, as delegates of the bishops, and not in virtue of their own right. The union of all the agglomerated parishes around a town, in a circumscription for a long time vague and variable, formed the diocese. After a certain time, and in order to bring more regularity and completeness into the relations of the diocesan clergy, they formed a small association of many parishes under the name of the *rural chapter*. . . . At a later period many rural chapters were united . . . under the name of *district*, which was directed by an archdeacon . . . the diocesan organization was then complete. . . . All the dioceses in the civil province formed the ecclesiastical province, under the direction of the metropolitan or archbishop."

Fully to understand this development of ecclesiastical organization, it is needful to glance at the process by which it was effected, and to observe how the increasing integration necessitated the increasing differentiation.

"During a great part of this [the second] century, the Christian churches were independent on each other, nor were they joined together by association, confederacy, or any other bonds, but those of charity . . . But, in process of time, all the Christian churches of a province were formed into one large ecclesiastical body, which, like confederate states, assembled at certain times in order to deliberate about the common interests of the whole. . . . These *councils* . . . changed the whole face of the church, and gave it a new form; for by them the ancient privileges of the people were considerably diminished, and the power and authority of the bishops greatly augmented. The humility, indeed, and prudence of these pious prelates prevented their assuming all at once the power with which they were afterward invested. . . . But they soon changed this humble tone, imperceptibly extended the limits of their authority, turned their influence into dominion, and their counsels into laws. . . . Another effect of these councils was, the gradual abolition of that perfect equality, which reigned among all bishops in the primitive times. For the order and decency of these assemblies required, that some one of the provincial bishops met in council, should be invested with a superior degree of power and authority; and hence the rights of Metropolitans derive their origin. . . . The universal church had now the appearance of one vast republic formed by a combination of a great number of little states. This occasioned the creation of a new order of ecclesiastics, who were appointed, in different parts of the world, as heads of the church. . . . Such was the nature and office of the *patriarchs*, among whom, at length, ambition, being arrived at its most insolent period, formed a new dignity, investing the bishop of *Rome*, and his successors, with the title and authority of prince of Patriarchs."

To complete the conception it needs only to add that, while there was going on this centralization of the higher offices, there was going on a minuter differentiation of the lower. Says Lingard, speaking of the Anglo-Saxon clergy—

"These ministers were at first confined to the three orders of bishops, priests, and deacons: but in proportion as the number of proselytes increased, the services of additional but subordinate officers were required: and we soon meet, in the more celebrated churches, with subdeacons, lectors or cantors, exorcists, acolythists, and ostiarii or door-keepers. . . . All these were ordained, with appropriate forms, by the bishop."

§ 620. Among leading traits in the development of ecclesiastical institutions, have to be added the rise and establishment of monasticism.

For the origin of ascetic practices, we must once more go back to the ghost-theory, and to certain resulting ideas and acts common among the uncivilized (§§ 103 and 140). There are the mutilations and blood-lettings at funerals; there are the fastings consequent on sacrifices of animals and food at the grave; and in some cases there are the deficiencies of clothing which follow the leaving of dresses (always of the best) for the departed. Pleasing the dead is therefore inevitably associated in thought with pain borne by the living. This connexion of ideas grows most marked where the ghost to be propitiated is that of some ruling man, notorious for his greediness, his love of bloodshed, and, in many cases, his appetite for human flesh. To such a ruling man, gaining power by conquest, and becoming a much-feared god after his decease, there arise propitiatory ceremonies which entail severe sufferings. Hence where, as in ancient Mexico, we find cannibal deities to whom multitudes of human victims were sacrificed; we also find that there were, among priests and others, self-mutilations of serious kinds, frequent self-bleedings, self-whippings, prolonged fasts, etc. The incidental but conspicuous trait of such actions, usurped in men's minds the place of the essential but less obtrusive trait. Sufferings having been the concomitants of sacrifices made to ghosts and gods, there grew up the notion that submission to these concomitant sufferings was itself pleasing to ghosts and gods; and eventually, that the bearing of gratuitous sufferings was pleasing. All over the world, ascetic practices have thus originated.

This, however, is not the sole origin of ascetic practices. They have been by all peoples adopted for the purpose of bringing on those abnormal mental states which are supposed to imply either possession by spirits, or communion with spirits. Savages fast that they may have dreams, and obtain the supernatural guidance which they think dreams give to them; and especially among medicine-men, and those in training to become such, there is abstinence and submission to various privations, with the view of producing the maniacal excitement which they, and those around, mistake for inspiration. Thus arises the belief that by persistent self-mortifications, there may be obtained an in-dwelling divine spirit; and the ascetic consequently comes to be regarded as a holy man.*

* It is curious to observe how this primitive idea still holds its ground. In Blunt's *Ecclesiastic Dictionary* there is a laudatory description of the prophet Daniel, as "using his ascetic practices as a special means of attaining Divine light:" the writer being apparently ignorant that medicine-men al over the world, have ever been doing the same thing with the same intent.

Led into his mode of life by the two-fold belief that voluntary submission to pain pleases God, and that mortifications of the flesh bring inspiration, the ascetic makes his appearance among the devotees of every religion which reaches any considerable development. Though there is little reference to permanent anchorites in ancient American societies, we are told of temporary religious retirements; as in Guatemala, where the high-priest, who was in some cases the king, fasted " four, or even eight, months in seclusion;" and as in Peru, where the Yncas occasionally lived in solitude and fasted. Among the religions of the old world, Buddhism, Judaism, Christianity, and Mohammetanism, have all furnished numerous examples. Biblical history shows that " in times anterior to the Gospel, prophets and martyrs 'in sheepskins and goatskins,' wandered over mountains and deserts, and dwelt in caves." This discipline of separateness and abstinence, indicated as early as the days of Moses in the " vow of a Nazarite," and shown by the Essenes to be still existing in later times, reappeared in the discipline of the Christian hermits, who were the first monks or solitaries: the two words being originally equivalent. These grew numerous during the persecutions of the third century, when their retreats became refuges.

"From that time to the reign of Constantine, monachism was confined to the hermits, or anchorets, living in private cells in the wilderness. But when Pachomius had erected monasteries in Egypt, other countries presently followed the example, and so the monastic life came to its full maturity in the church."

Or, as Lingard describes the process:—

"Wherever there dwelt a monk [a recluse] of superior reputation for sanctity, the desire of profiting by his advice and example induced others to fix their habitations in his neighbourhood: he became their Abbas or spiritual father, they his voluntary subjects: and the group of separate cells which they formed around him was known to others by the name of his monastery."

Thus, beginning as usual in a dispersed unorganized form, and progressing to small clusters such as those of the Cœnobites in Egypt, severally governed by a superior with a steward, monastic bodies, growing common, at the same time acquired definite organizations; and by-and-by, as in the case of the Benedictines, came to have a common rule or mode of government and life. Though in their early days monks were regarded as men more holy than the clergy, they did not exercise clerical functions; but in the fifth and sixth centuries they acquired some of these, and in so doing became subject to bishops: the result being a long struggle to maintain independence on the one side and to enforce authority on the other, which ended in practical incorporation with the Church.

Of course there thus arose a further complication of the ecclesiastical hierarchy, which it will be sufficient just to note without describing in detail.

§ 621. For present purposes, indeed, no further account of ecclesiastical hierarchies is needed. We are here concerned only with the general aspects of their evolution.

Examination discloses a relation between ecclesiastical and political governments in respect of degree. Where there is but little of the one there is but little of the other; and in societies which have developed a highly coercive secular rule there habitually exists a highly coercive religious rule.

It has been shown that growing from a common root, and having their structures slightly differentiated in early societies, the political and ecclesiastical organizations long continue to be distinguished very imperfectly.

This intimate relationship between the two forms of regulation, alike in their instrumentalities and in their extents, has a moral origin. Extreme submissiveness of nature fosters an extreme development of both the political and religious controls. Contrariwise the growth of the agencies effecting such controls, is kept in check by the sentiment of independence; which while it resists the despotism of living rulers is unfavourable to extreme self-abasement in propitiation of deities.

While the body which maintains the observances of a cult grows in mass, it also increases in structure; and whether the cult is an indigenous or an invading one, there hence results a hierarchy of sacerdotal functionaries analogous in its general principles of organization to the graduated system of political functionaries. In the one case as in the other the differentiation, setting out from a state in which power is distributed with approximate uniformity, advances to a state in which, while the mass becomes entirely subordinate, the controlling agency displays within itself a subordination of the many to the few and to the one.

CHAPTER IX.

AN ECCLESIASTICAL SYSTEM AS A SOCIAL BOND.

§ 622. ONCE more we must return to the religious idea and the religious sentiment in their rudimentary forms, to find an explanation of the part played by ecclesiastical systems in social development.

Though ancestor-worship has died out, there survive among us certain of the conceptions and feelings appropriate to it, and certain resulting observances, which enable us to understand its original effects, and the original effects of those cults immediately derived from it. I refer more especially to the behaviour of descendants after the death of a parent or grand-parent. Three traits, of which we shall presently see the significance, may be noted.

When a funeral takes place, natural affection and usage supporting it, prompt the assembling of the family or clan: of children especially, of other relations to a considerable extent, and in a measure of friends. All, by taking part in the ceremony, join in that expression of respect which constituted the original worship and still remains a qualified form of worship. The burial of a progenitor consequently becomes an occasion on which, more than on any other, there is a revival of the thoughts and feelings appropriate to relationship, and a strengthening of the bonds among kindred.

An incidental result which is still more significant, not unfrequently occurs. If antagonisms among members of the family exist, they are not allowed to show themselves.

Being possessed by a common sentiment towards the dead, and in so far made to sympathize, those who have been at enmity have their animosities to some extent mitigated; and not uncommonly reconciliations are effected. So that beyond a strengthening of the family-group by the gathering together of its members, there is a strengthening of it caused by the healing of breaches.

One more co-operative influence exists. The injunctions of the deceased are made known; and when these have reference to family-differences, obedience to them furthers harmony. Though it is true that directions concerning the distribution of property often initiate new quarrels, yet in respect of pre-existing quarrels, the known wish of the dying man that they should be ended, is influential in causing compromise or forgiveness; and if there has been a desire on his part that some particular course or policy should be pursued after his death, this desire, even orally expressed, tends very much to become a law to his descendants, and so to produce unity of action among them.

If in our days these influences still have considerable power, they must have had great power in days when there was a vivid conception of ancestral ghosts as liable to be made angry by disregard of their wishes, and able to punish the disobedient. Evidently the family-cult in primitive times, must have greatly tended to maintain the family bond : alike by causing periodic assemblings for sacrifice, by repressing dissensions, and by producing conformity to the same injunctions.

Rising as we do from the ordinary father to the patriarch heading numerous families, propitiation of whose ghost is imperative on all of them, and thence to some head of kindred clans who, leading them to conquest, becomes after death a local chief god, above all others feared and obeyed; we may expect to find in the cults everywhere derived from ancestor-worship, the same influence which ancestor-worship in its simple original form shows us. We shall not be disappointed. Even concerning peoples so rude as the Ostyaks, we find the remark that "the use of the same consecrated spot, or the same priest, is also a bond of union;" and higher races yield still clearer evidence. Let us study it under the heads above indicated.

§ 623. The original tribes of the Egyptians, inhabiting areas which eventually became the *nomes*, were severally held together by special worships. The central point in each "was always, in the first place, a temple, about which a city became formed." And since "some animals, sacred in one province, were held in abhorrence in another"—since, as we have seen, the animal-naming of ancestral chiefs, revered within the tribe but hated beyond it, naturally originated this; we have reason for concluding that each local bond of union was the worship of an original ancestor-god.

Early Greek civilization shows like influences at work; and records enable us to trace them to a higher stage. Grote writes—

"The sentiment of fraternity, between two tribes or villages, first manifested itself by sending a sacred legation or Theôria to offer sacrifice at each other's festivals and to partake in the recreations which followed." . . . "Sometimes this tendency to religious fraternity took a form called an Amphiktyony, different from the common festival. A certain number of towns entered into an exclusive religious partnership, for the celebration of sacrifices periodically to the god of a particular temple, which was supposed to be the common property and under the common protection of all."

Then concerning the most important of these unions, we read in Curtius—

"All Greek collective national names attach themselves to particular sanctuaries: these are the centres of union, and the starting-points of history. . . . In this respect Apollo, as the god of the Thessalian Amphictyony, may be said to be the founder of the common nationality of the Hellenes, and the originator of Hellenic history."

If with this we join the further significant fact that "the Dorians . . . even called Dorus, the ancestor of their race, an so of Apollo, and recognized in the spread of the worship of the latter their proper mission in history;" the filiation of this religious development upon ancestor-worship becomes manifest. And since the periodic gatherings for sacrifice initiated the Amphictyonic council, the statutes of which "had their origin in the Apolline religion," and were regarded with respect by the separate Grecian states "in all matters touching on rights common to all;" we have clear proof that the federal bond originated in a common worship.

The like happened in Italy. Concerning the Etruscans, Mommsen says—"Each of these leagues consisted of twelve communities, which recognized a metropolis, especially for purposes of worship, and a federal head or rather a high-priest." It was thus with the Latins too. Alba was the chief place of the Latin league; and it was also the place at which the tribes forming the league assembled for their religious festivals: such union as existed among them was sanctified by a cult in which all joined. A kindred fact is alleged of ancient Rome. "The oldest constitution of Rome is religious throughout," says Seeley. "Institutions suggested by naked utility come in later, and those which they practically supersede are not abolished, but formally retained on account of their religious character."

Though generally in such cases the need for joint defence against external enemies is the chief prompter to federation; yet in each case the federation formed is determined by that community of sacred rites which from time to time brings the dispersed divisions of the same stock together, and keeps alive in them the idea of a common origin as well as the sentiment appropriate to it.

Though Christendom has not exemplified in any considerable degree a like consolidating effect—though its worship, being an adopted one has not supplied that bond which results where the worship is of some great founder of the tribe or traditional god of the race; yet it can hardly be questioned that unity of creed and ceremony has to some extent served as an integrating principle. Though Christian brotherhood has not been much displayed among Christian peoples, still, it has not been absolutely a mere name. Indeed it is manifest that since similarity of thought and sympathy of feeling must further harmony by diminishing reasons for difference, agreement in religion necessarily favours union.

§ 624. Still more clearly shown is the parallelism between suspension of family animosities at funerals, and temporary cessation of hostilities between clans on occasions of common religious festivals.

Already in § 144 I have pointed out that among some of the uncivilized, burial places of chiefs become sacred, to the extent that fighting in them is forbidden: one of the results being the initiation of sanctuaries. Naturally an interdict against quarrels at burial-places, or sacred places where sacrifices are to be made, tends to become an interdict against quarrels with those who are going there to sacrifice. The Tahitians would not molest an enemy who came to make offerings to the national idol; and among the Chibchas pilgrims to Iraca (Sogamoso) were protected by the religious character of the country even in time of war. These cases at once recall cases from ancient European history. Of the tribes which originated the Roman civilization, we read— "There are, however, indications that during the Latin festival [sacrifices to Jupiter], just as was the case during the festivals of the Hellenic leagues, 'a truce of God' was observed throughout all Latium." And the instance with which Mommsen here makes a comparison, being much more specific, is particularly instructive. First serving to regulate the worship of a deity common to all, and to maintain a temporary peace among worshippers, the Amphictyonic council served to guarantee "a safe and inviolate transit even through hostile Hellenic states" to the sacrifices and to the games which became associated with them. And here from the temporary suspensions of antagonisms came secondary effects furthering union.

"The festivals of the gods thus worshipped in common were national festivals. From the system of festivals it was only a step to a common calendar. A common purse was needed for the preservation of the buildings in which the worship was carried on, and for furnishing sacrifices; this made a common coinage necessary. The common purse and temple-treasures required administrators, for whose choice it was requisite to assemble, and whose administration of their office had to be watched by a representation of the federated tribes. In case of dispute between the Amphictyones, a judicial authority was wanted to preserve the common peace, or punish its violation in the name of the god. Thus the insignificant beginning of common annual festivals gradually came to transform the whole of public life; the constant carrying of arms was given up, intercourse was rendered safe, and the sanctity of temples and altars recognized. But the most important result of all was, that the members of the Amphictyony learnt to regard themselves as one united body against those standing outside it; out of a number of tribes arose a nation, which required a common name to distinguish it, and its political and religious system, from all other tribes."

And that, little as it operated, acceptance of a common creed tended somewhat towards consolidation of European peoples, we see alike in the weekly suspensions of feudal fights under the influence of the Church, in the longer suspensions of larger quarrels under promise to the pope during the crusades and in the consequent combined action of kings who at other times were enemies; as shown by the fighting of Philip Augustus and Richard I. under the same banners.

And then beyond these various influences indirectly aiding consolidation, come the direct influences of judgments supposed to come from God through an inspired person—Delphian oracle or Catholic high-priest. "As men of a privileged spiritual endowment" the priests of Delphi were "possessed of the capacity and mission of becoming in the name of their god the teachers and counsellors, in all matters, of the children of the land;" and obviously, in so far as their judgments concerning inter-tribal questions were respected, they served to prevent wars. In like manner belief in the pope as a medium through whom the divine will was communicated, tended in those who held it to cause subordination to his decisions concerning international disputes, and in so far to diminish the dissolving effects of perpetual conflicts: instance the acceptance of his arbitration by Philip Augustus and Richard I. under threat of ecclesiastical punishment; instance the maintenance of peace between the kings of Castile and Portugal by Innocent III. under penalty of excommunication; instance Eleanor's invocation—" has not God given you the power to govern nations;" instance the formal enunciation of the theory that the pope was supreme judge in disputes among princes.

§ 625. No less clearly do the facts justify the analogy above pointed out between the recognized duty of fulfilling a deceased parent's wishes, and the imperative obligation of conforming to a divinely-ordained law.

Twice in six months within my own small circle of friends, I have seen exemplified the subordination of conduct to the imagined dictate of a deceased person: the first example being yielded by one who, after long hesitation, decided to alter a house built by his father, but only in such way as he thought his father would have approved; the second being yielded by one who, not himself objecting to play a game on Sunday, declined because he thought his late wife would not have liked it. If in such cases supposed wishes of the dead become transformed into rules of conduct, much more must expressed injunctions tend to do this. And since maintenance of family-union is an end which such expressed injunctions are always likely to have in view—since the commands of the dying patriarch, or the conquering chief, naturally aim at prosperity of the clan or tribe he governed; the rules or laws which ancestor-worship originates, will usually be of a kind which, while intrinsically furthering social cohesion, further it also by producing ideas of obligation common to all.

Already in §§ 529—30 I have pointed out that, among primitive men, the customs which stand in place of laws, embody the ideas and feelings of past generations; and, religiously conformed to as they are, exhibit the rule of the dead over the living. From usages of the Veddahs, the Scandinavians, and the Hebrews, I there drew evidence that in some cases the ghosts of the dead are appealed to for guidance in special emergencies; and I gave proof that, more generally, apotheosized men or gods are asked for directions: instances being cited from accounts of Egyptians, Peruvians, Tahitians, Tongans, Samoans, Hebrews, and sundry Aryan peoples. Further, it was shown that from particular commands answering special invocations, there was a transition to general commands, passing into permanent laws: there being in the bodies of laws so derived, a mingling of regulations of all kinds—sacred, secular, public, domestic, personal. Here let me add evidence reinforcing that before given.

"Agriculture was inculcated as a sacred duty upon the follower of Zoroaster, and he was taught that it was incumbent upon all who worshipped Ahuramasda to lead a settled life. . . . Everything that the Nomad was enjoined to avoid was thus inculcated, as a religious duty, upon the followers of Zoroaster. . . . The principles of Zoroaster, and of similar teachers, led to the federation of settled tribes, out of which arose the mighty empires of antiquity."

Evidently bodies of laws regarded as supernaturally given by the traditional god of the race, originating in the way shown, habitually tend to restrain the anti-social actions of individuals towards one another, and to enforce concerted action in the dealings of the society with other societies: in both ways conducing to social cohesion.

§ 626. The general influence of Ecclesiastical Institutions is conservative in a double sense. In several ways they maintain and strengthen social bonds, and so conserve the social aggregate; and they do this in large measure by conserving beliefs, sentiments, and usages which, evolved during earlier stages of the society, are shown by its survival to have had an approximate fitness to the requirements, and are likely still to have it in great measure. Elsewhere (*Study of Sociology*, Chap. V) I have, for another purpose, exemplified the extreme resistance to change offered by Ecclesiastical Institutions, and this more especially in respect of all things pertaining to the ecclesiastical organization itself. Here let me add a further series of illustrations.

The ancient Mexicans had "flint knives used in the sacrifices." In San Salvador, the sacrificer had "a knife of flint, with which he opened the breast of the victim." Among the Chibchas, again, when a boy was sacrificed, "they killed him with a reed knife;" and at the present time among the Karens, the sacrificial hog offered to deified ancestors, "is not killed with a knife or spear; but a sharpened bamboo is forced into it." In many other cases the implements used for sacred purposes are either surviving tools of the most archaic types, or else of relatively ancient types; as in pagan Rome where "down to the latest times copper alone might be used, *e.g.* for the sacred plough and the shear-knife of the priests," and where also an ancient dress was used during religious ceremonies. Among the Nagas, the fire for roasting a sacrificed animal is "freshly kindled by means of rubbing together two dry pieces of wood;" and on like occasions among the Todas, "although fire may be readily procured from the Mand, a *sacred* fire is created by the rubbing of sticks." The Damaras keep a sacred fire always burning; and should this be accidentally extinguished "the fire is re-lit in the primitive way—namely, by friction." Even in Europe there long continued a like connexion of ideas and practices. Says Peschel, speaking of the fire-drill, "this mode of kindling fire was retained till quite recently in Germany, for popular superstition attributed miraculous power to a fire generated by this ancient method;" and in the Western Isles of Scotland at the end of the seventeenth century, they still obtained fire for sacrificial purposes by the friction of wood in cases of plague and murrain. So is it with the form of speech. Beyond such examples as the use of extinct tongues by Jews and by Roman Catholics for religious services, and the retention of an ancient language as a sacred language by the Copts, and the like use by the Egyptian priests of an archaic type of writing, we have illustrations furnished by the uncivilized. Schoolcraft says of the Creeks that their old language (the Seminole) is "taught by women to the children as a kind of religious duty." In Dahomey, too, the priest " pronounces an allocution in the unintelligible hierarchic tongue." And the origin of Japanese Buddhism " is shown to this day in the repetition of prayers in an unknown language, and the retention of an Indian alphabet and writing—the Sanscrit or Devanagari— in all the religious works of Japan." This same tendency was variously exemplified among the Hebrews; as we see in the prescription of unhewn stone for altars (Exod. xx, 25-6), the use of unleavened bread for offerings (Judges, vi, 19-21), and the interdict on building a temple in place of the primitive tent and tabernacle alleged to have been the divine habitation in earlier days (2 Sam. vii, 4-6). And a like persistence was shown in Greece. Religious institutions, says Grote, " often continued unaltered throughout all the political changes."

Of course while thus resisting changes of usage, ecclesiastical functionaries have resisted with equal or greater strenuousness, changes of beliefs; since any revolution in the inherited body of beliefs, tends in some measure to shake all parts of it, by diminishing the general authority of ancestral teaching. This familiar aspect of ecclesiastical conservatism, congruous with the aspects above exemplified, it is needless to illustrate.

§ 627. Again, then, the ghost-theory yields us the needful clue. As, before, we found that all religious observances may be traced back to funeral observances; so here, we find these influences which ecclesiastical institutions exert, have their germs in the influences exerted by the feelings entertained towards the dead. The burial of a late parent is an occasion on which the members of the family gather together and become bound by a renewed sense of kinship; on which any antagonism among them is temporarily or permanently extinguished; and on which they are further united by being subject in common to the deceased man's wishes, and made, in so far, to act in concert. The sentiment of filial piety thus manifesting itself, enlarges in its sphere when the deceased man is the patriarch, or the founder of the tribe, or the hero of the race. But be it in worship of a god or funeral of a parent, we ever see the same three influences— strengthening of union, suspension of hostilities, reinforcement of transmitted commands. In both cases the process of integration is in several ways furthered.

Thus, looking at it generally, we may say that ecclesiasticism stands for the principle of social continuity. Above all other agencies it is that which conduces to cohesion; not only between the coexisting parts of a nation, but also between its present generation and its past generations. In both ways it helps to maintain the individuality of the society. Or, changing somewhat the point of view, we may say that ecclesiasticism, embodying in its primitive form the rule of the dead over the living, and sanctifying in its more advanced forms the authority of the past over the present, has for its function to preserve in force the organized product of earlier experiences *versus* the modifying effects of more recent experiences. Evidently this organized product of past experiences is not without credentials. The life of the society has, up to the time being, been maintained under it; and hence a perennial reason for resistance to deviation. If we consider that habitually the chief or ruler, propitiation of whose ghost originates a local cult, acquired his position through successes of one or other kind, we must infer that obedience to the commands emanating from him, and maintenance of the usages he initiated, is, on the average of cases, conducive to social prosperity so long as conditions remain the same; and that therefore this intense conservatism of ecclesiastical institutions is not without a justification.

Even irrespective of the relative fitness of the inherited cult to the inherited social circumstances, there is an advantage in, if not indeed a necessity for, acceptance of traditional beliefs, and consequent conformity to the resulting customs and rules. For before an assemblage of men can become organized, the men must be held together, and kept ever in presence of the conditions to which they have to become adapted; and that they may be thus held, the coercive influence of their traditional beliefs must be strong. So great are the obstacles which the anti-social traits of the savage (§§ 33–38) offer to that social cohesion which is the first condition to social progress, that he can be kept within the needful bonds only by a sentiment prompting absolute submission—submission to secular rule reinforced by that sacred rule which is at first in unison with it. And hence, as I have before pointed out, the truth that in whatever place arising—Egypt, Assyria, Peru, Mexico, China—social evolution throughout all its earlier stages has been accompanied not only by extreme subordination to living kings, but also by elaborate worships of the deities originating from dead kings.

CHAPTER X.

THE MILITARY FUNCTIONS OF PRIESTS.

§ 628. AMONG the many errors which result from carrying back advanced ideas and sentiments to the interpretation of primitive institutions, few are greater than that of associating priestly functions with actions classed as high in kind, and dissociating them from brutal and savage actions. Did not men's prepossessions render them impervious to evidence, even their Bible readings might raise doubts; and wider readings would prove that among mankind at large, priests have displayed and cultivated not the higher but rather the lower passions of humanity.

We at once see that this must be so, when we remember that instead of deities conceived as possessing all perfections, moral and intellectual, most peoples have had deities conceived as possessing ferocious natures, often in no way distinguished from the diabolical. Of the ancient Mexicans we read that their "Princes sent to one another to prepare for War, because their Gods demanded something to eat;" and that their armies "fought, only endeavouring to take Prisoners, that they might have Men to feed those Gods." According to Jackson, the Fijian priests told those around "that bloodshed and war, and everything connected with them, were acceptable to their gods." Though Pindar repudiates the ascription of cannibalism to the Greek gods, yet the narrative of Pausanias shows that even in his day, human victims were occasionally sacrificed to Zeus; and the *Iliad* tacitly ascribes to the Greek gods natures lower than it ascribes to men: lying, treachery, blood-thirstiness, adultery, are without palliation attributed to them. The fact that they took part in the battles of the men with whom they respectively sided, reminds us of the Assyrians, among whom also direct divine aid in fighting was alleged. Says an inscription of Esarhaddon:—

"Ishtar queen of war and battle, who loves my piety, stood by my side. She broke their bows. Their line of battle in her rage she destroyed. To their army she spoke thus: 'An unsparing deity am I.'"

And kindred traits are directly or tacitly ascribed to the primitive Hebrew god. I do not refer only to sacrifices of human victims, or to such phrases as "the Lord is a man of war," and "God himself is with us for our captain" (2 Chron. xiii, 12); but I refer more particularly to the indiscriminate slaughter said to be ordered by God, and to the fact that a religious war is assumed to be naturally a bloody war: instance the statement in 1 Chron. v, 22—"there fell down many slain, because the war was of God." All which divine traits, attributed by early historic peoples as well as by existing barbarians, are accounted for when we remember that mythologies, which habitually describe battles among the gods for supremacy, are but transfigured accounts of struggles among primitive rulers, in which the stronger, more blood-thirsty, and more unscrupulous, usually prevailed.

Fully to understand the original connexion between military deeds and religious duties, we must recollect that when gods are not supposed to be active participators in the battles commanded or countenanced by them, they are supposed to be present in representative idols, or in certain equivalents for idols. Everywhere we find parallels to the statement made by Cook, that the Sandwich Islanders carry their war-gods with them to battle. Among the ancient Mexicans when meeting a foe, "the priests with their idols marched in the front." Certain of the Yucatanese had "idols, which they adored as gods of battles. .. They carried these when they went to fight the Chinamitas, their neighbours and mortal foes." Of the Chibchas, Herrera, referring to private idols, says—"So great was their Devotion, that whithersoever they went, the Idol was carry'd, holding it with one Arm and fighting with the other in their Battles." Nor has it been otherwise in the old world. The account in 2 Samuel, v, 21, shows that the Philistines carried their images of the gods with them when fighting; and the ark, regarded by the Hebrews as a residence of Jahveh, was taken out to war not unfrequently (2 Samuel, xi). Indeed in 1 Samuel, iv, we read that the Hebrews, having been defeated by the Philistines, sent for the ark that it might save them; "and when the ark of the covenant of the Lord came into the camp, all Israel shouted with a great shout, so that the earth rang again. . . . And the Philistines were afraid, for they said, God is come into the camp." Moreover, on calling to mind the sacrifices habitually made before and after, and sometimes during, battles by uncivilized and semi-civilized peoples, we are further shown how close has been the connexion between killing enemies and pleasing deities.

Priests being the official propitiators of deities, the corollary is obvious. While often restrainers from wars with those of the same blood, they are originally stimulators to wars with those of other bloods worshipping other deities. Thus, concerning the Mexicans above referred to, who fought to provide victims for their gods, we read that "when the Priests thought fit, they went to the Kings, and told them, they must remember the Idols who were starving with Hunger." The Assyrian priests had further motives. "They lived on the revenues of the temples . . . were directly interested in war, as a portion of the spoil was dedicated to the temples." But without multiplying instances, it will suffice to recall the fact that even among the Hebrews, while king and people were in some cases inclined to show clemency, priests insisted upon *cherem*—

merciless indiscriminate slaughter; and Samuel "cried unto the Lord all night" because Saul, though he had "utterly destroyed" the Amalekites, had not killed their king and all their cattle: reminding us of the Fijian who, not having done his utmost in slaying, worked himself into a "religious frenzy," calling out continually "the god is angry with me."

This preliminary brief survey prepares us to find that in early stages of social evolution along with sacerdotal functions go military functions. Let us look at these under their leading aspects.

§ 629. The truth that in the normal order the chief, who is originally the greatest warrior, is also the primitive priest, implies union of military and sacerdotal functions in the same person. At first the head fighter is the head propitiator of the gods. The frescoes and inscriptions of Egypt and Assyria, presenting the king as at once leader in war and leader in worship, illustrate a connexion habitually found.

This connexion is even closer than at first appears; for among the most important sacrifices made by kings to gods, are those made on the eve of battle to gain divine favour, or after victory in token of thanks. That is to say, the king discharges his function of religious propitiator in the most conspicuous way, at the time when his military headship is exercised in the most conspicuous way.

With but small modification, this connexion of functions is occasionally shown where the leadership in war is not exercised by the ruling man or body, but by an appointed general; for in such cases generals assume priestly functions. The Mexicans furnished an instance. The office of high-priest "involved, almost always, the duties of Tlacochcalcatl, or commander-in-chief of the army." So was it with the ancient civilized peoples of Europe. At Rome, "before setting out on an expedition, the army being assembled, the general repeated prayers and offered a sacrifice. The custom was the same at Athens and at Sparta." To which we may add that, among the Romans, " the army in the field was the image of the city, and its religion followed it:" the sacred hearth was perpetually burning, there were augurs and diviners, and king or commander sacrificed before and after battle. And, indeed, the priestly function of the Roman commander was such that in some cases he paid more attention to sacrificing than to fighting.

Nor does the community end here. Beyond this union of military functions with sacerdotal functions in leaders, there occur among the uncivilized, cases in which active parts in fighting are taken by priests. Concerning the Tahitians, whose " chiefs and priests were often among the most famous boxers and wrestlers," Ellis says that " the priests were not exempted from the battle, they bore arms, and marched with the warriors to the combat." Presently we shall have to note that parallels have been furnished where they might least be expected.

§ 630. After recognizing the fact that at the outset, active ecclesiastical headship is united with active military headship; and after recognizing the fact that throughout later stages these two headships remain nominally united with headship of the state; we may go on to observe that very soon, priests usually cease to be direct participators in war, and become indirect participators only.

During times when the characters of medicine-man and priest are vaguely represented in the person of one who is supposed to have power over, or influence with, supernatural beings, we see foreshadowed the advising and administrative functions of priests in war. The Dakotahs show this kind of action in its rudest form.

"The war chiefs often get some of the priests or jugglers to make war for them. In fact, any of the jugglers can make a war-party when they choose."

Then among the Abipones the medicine-man—

"teaches them the place, time, and manner proper for attacking wild beasts or the enemy. On an approaching combat, he rides round the ranks, striking the air with a palm bough, and with a fierce countenance, threatening eyes, and affected gesticulations, imprecates evil on their enemies."

And we are told that among the Khonds—

"The priest, who in no case bears arms, gives the signal to engage after the latter offering, by flourishing an axe in the air, and shouting encouragement to defiance."

To raise the courage of the soldiers by hopes of help from the gods, was in like manner a function of the priest among Spartans.

"Every expedition and every council of war was preceded by a sacrifice. A priest, called the fire-bearer ($\pi\upsilon\rho\phi\acute{\upsilon}\rho\sigma$), carried before the army a burning brand, which was kept always alight, taken from the altar in Sparta on which the king had offered sacrifice to Zeus Agetor."

And the Hebrews similarly availed themselves of the agency of the priest in promising supernatural aid; as witness Deuteronomy, xx, 1—4.

"And it shall be, when ye are come nigh unto the battle, that the priest shall approach and speak unto the people, And shall say unto them, O Israel, ye approach this day unto battle against your enemies: let not your hearts faint, fear not, and do not tremble, neither be ye terrified because of them; for the Lord your God is he that goeth with you to fight for you against your enemies, to save you."

In some cases of which I have notes, the functions of the priests who accompanied the armies, are not specified. On the Gold Coast, where "war is never undertaken by kings or states without consulting the national deities," the "fetishmen accompany the warriors to the field." And Herrera describes the armies of the Yucatanese as having "two Wings and a Center, where the Lord and the High Priest were." But the military functions of the priest during active war, are in other cases somewhat different. Among the primitive Germans—

"The maintenance of discipline in the field as in the council was left in great measure to the priests; they took the auguries and gave the signal for onset, they alone had power to visit with legal punishment, to bind or to beat."

In yet other cases the functions discharged are more exclusively of the kind called religious. The Samoans took a priest "to battle to pray for his people and curse the enemy." In New Caledonia, "the priests go to battle, but sit in the distance, *fasting* and praying for victory." Among the Comanches the supplicatory function was performed before going to war. "The priesthood," says Schoolcraft, " appear to exercise no influence in their general government, but, on war being declared, they exert their influence with the Deity." And in this conception of their office it seems that Christian priests agree with the priests of the Comanches; as witness the following prayer directed to be used by the Archbishop of Canterbury at the commencement of the late war in Egypt.

" O Almighty God, whose power no creature is able to resist, keep, we beseech Thee, our soldiers and sailors who have now gone forth to war, that they, being armed with Thy defence, may be preserved evermore from all perils, to glorify Thee, who art the only giver of all victory, through the merits of Thy only Son, Jesus Christ our Lord. Amen."

A noteworthy difference, however, being that whereas the priest among pagans in general, seeks some sign of divine approval as a first step, the Christian priest assumes that he has this approval; even though the case be that of attacking a people who are trying to throw off an intolerable tyranny.

Besides being direct or indirect aiders in battle, priests are in other cases relied on for military management, or appealed to for guidance. In Africa among the Eggarahs, a priest " officiates as minister of war." Of the ancient Mexicans we read—" The high-priests were the oracles whom the kings consulted in all the most important affairs of the state, and no war was ever undertaken without their approbation." Prescott speaks of the Peruvian priests as giving advice in matters of war; and Torquemada says that in Guatemala the priests had decisive authority on war questions. In San Salvador, too, the high-priest and his subordinates, after seeking supernatural knowledge, "called together the cazique and war chief, and advised them of the approach of their enemies, and whether they should go to meet them." And the like happened among the Hebrews. I Kings, xxii, tells us of consultations with the prophets concerning the propriety of a war, and especially with one of them:—

"So he [Micaiah] came to the king. And the king said unto him, Micaiah, shall we go against Ramoth-gilead to battle, or shall we forbear? And he answered him, Go, and prosper: for the Lord shall deliver it into the hand of the king."

§ 631. Anyone simple enough to suppose that men's professed creeds determine their courses of conduct, might infer that nations which adopted Christianity, if not deterred from war by their nominally-accepted beliefs, would at least limit the functions of their priests to those of a religious kind, or at any rate, a non-militant kind. He would be quite wrong however.

The fact is familiar that Christian Europe throughout many centuries, saw priests taking as active parts in war as do priests among some extant savages. In the seventh century in France, bishops went to battle; and "by the middle of the eighth century regular military service on the part of the clergy was already fully developed:" "under Charles Martel it was common to see bishops and clerks bearing arms." Says Guizot concerning the state of the church at this period, the bishops "took part in the national warfare; nay more, they undertook, from time to time, expeditions of violence and rapine against their neighbours on their own account." And in subsequent centuries Germany and France alike witnessed the union of military leadership with ecclesiastical leadership. In Germany the spiritual head "was now a feudal baron; he was the acknowledged leader of the military forces in his dioceses." Writing of events in France, Orderic describes the priests as leading their parishioners to battle, and the abbots their vassals, in 1094, and again in 1108; while in 1119 the bishops summoned the priests with their parishioners. Even after the middle of the fifteenth century the Cardinal de Balue mustered troops in Paris; and "the bishop, the heads of the university, the abbots, priors, and other churchmen," "appeared there with a certain number of men." Not until nearly the middle of the seventeenth century was there issued an edict which exempted the clergy from personal service in the armies. Even now, Christendom is not without an example of union between the man-slaying and soul-saving functions. It is remarked that the Montenegrins form "the only community now in Europe governed by a military bishop;" and the Rev. W. Denton says "the priests carry arms, and 'are generally good heroes,' the first at a gathering, the leaders of their flocks in war."

To a direct participation in war exhibited by actual service in the army, must be added an indirect participation implied by administrative control of the fighting organizations. Cardinal Richelieu was director of both navy and army. Moreover, his policy "was the opening of a new era for France, an era of great and systematized warfare;" and he, "in his *Testament politique*, recalls with pride the discipline he established in the army of Italy and among the troops which besieged La Rochelle. 'They obeyed like monks under arms.'"

Now-a-days people have become unaccustomed to these connexions, and forget that they ever existed. The military duties of priests among ourselves have dwindled down to the consecration of flags, the utterances by army-chaplains of injunctions of forgiveness to men who are going to execute vengeance, joined with occasional prayers to the God of love to bless aggressions, provoked or unprovoked.

§ 632. Thus, contemplation of facts supplied by all places and times, reverses that association of ideas which the facts immediately around us produce. Recognizing the truth that the gods of savages and partially-civilized peoples, were originally ferocious chiefs and kings whose ghosts were propitiated by carrying out their aggressive or revengeful projects; we see that their official propitiators, so far from being at first associated in doctrine and deed with the higher traits of human nature, were in both associated with the lower. Hence the naturalness of that militancy which characterizes them in early stages.

Under a more concrete form this union of the sacerdotal and belligerent characters, is shown by the fact that in the normal order of social evolution, the political head is at the same time the leader in war and the leader in worship. Evidently the implication is that these two functions, at first united, can acquire separate agencies but gradually; and that these separate agencies must long continue to show some community of character: a truth indicated by that nominal headship of the church and the army which the head of the state in many cases retains when actual headship has ceased.

That other priests besides that head priest who is also head warrior, should take active parts in war, is therefore to be expected. We need feel no surprise on finding that in various barbarous societies they share in battle—sometimes as actual soldiers, at other times as inspiring prompters, at other times as advisers divinely enlightened; while occasionally they act as war ministers.

Moreover this original relation is, as we see, not easily obliterated. The history of mediæval Europe proves undeniably that conditions which cause a great recrudescence of militancy, re-establish the primitive union of soldier and priest, notwithstanding a cult which forbids bloodshed— re-establish it just as completely as though the cult were of the most sanguinary kind. Only as war becomes less chronic, and the civilizing influences of peace begin to predominate, does the priest lose his semi-warlike character.

Lastly, let us note that the differentiation of these two functions of fighting enemies and propitiating deities, which were originally joined with headship of the State, has gone furthest in those religious organizations which are separate from the State. Unlike the ministers of the established church, who ordinarily belong to families which furnish military and naval officers, and who, though not actively militant, have their militant sympathies occasionally indicated by the votes of bishops in the House of Lords, dissenting ministers, derived from classes engaged in one or other form of industrial activity, are the least militant of religious functionaries.

CHAPTER XI.

THE CIVIL FUNCTIONS OF PRIESTS.

§ 633. OF course where the head of the State, himself regarded as god-descended, plays the part of priest in propitiating the ancestral gods, and, unlimited in his authority, carries his rule into all spheres, the union of civil functions with sacerdotal functions is complete. A good example of this condition in an early stage of social development, is furnished by the Polynesians.

"This system of civil polity, disjointed and ill adapted as it was to answer any valuable purpose, was closely interwoven with their sanguinary system of idolatry, and sanctioned by the authority of the gods. The king was not only raised to the head of this government, but he was considered as a sort of vicegerent to those supernatural powers presiding over the invisible world. Human sacrifices were offered at his inauguration; and whenever any one, under the influence of the loss he had sustained by plunder, or other injury, spoke disrespectfully of his person and administration, not only was his life in danger, but human victims must be offered, to cleanse the land from the pollution it was supposed to have contracted."

Various extinct societies presented kindred fusions of civil with sacerdotal headships. In Assyria, where the king "was either supposed to be invested with divine attributes, or was looked upon as a type of the Supreme Deity," and where "all his acts, whether in war or peace, appear to have been connected with the national religion, and were believed to be under the special protection and superintendence of the deity;" he, while civil head of the State, is represented in the sculptures as the chief sacrificer to the gods. The like connexion existed in ancient Egypt, in ancient Mexico, in ancient Peru; and in Japan, until recently, it continued to exist under a nominal form if not under a real form.

Obviously this is the normal connexion in those societies which have preserved that primitive structure in which, along with a general ancestor-worship there has arisen a special worship of the founder of the conquering tribe, whose descendant is at once head propitiator of him, and inheritor of his civil headship along with his military headship.

§ 634. This union, most conspicuous where the divine nature or divine descent of the king is an article of faith, continues also where he is believed to have divine sanction only. For habitually in such cases he is either nominal head or real head of the ecclesiastical organization; and while ordinarily occupied with civil functions, assumes on great occasions sacerdotal functions.

Where the religion is indigenous, this maintenance of the connexion is naturally to be expected; but we have proof that even where the religion is an invading one, which suppresses the indigenous one, there is apt to be a re-establishment of the connexion. This is shown by the growth of the ecclesiastical organization throughout Europe. At first diffused and local, it advanced towards a centralized union of religious with civil authority. According to Bedollierre, during the fourth and fifth centuries in France, senators, governors of provinces, great proprietors, imperial officers, were elected bishops; and Guizot writes that in the fifth century, "the bishops and the priests became the principal municipal magistrates." In the codes of Theodosius and Justinian are numerous regulations which remit municipal affairs to the clergy and the bishops. The jurisdiction of a bishop in Germany, beginning with his own clergy only, came to be by usage "extended to laymen, in cases where the duties of religion, the rights or discipline of the church, were concerned; and the execution of his decrees was confided to the care of the local courts." When, in the tenth century, by the growth of the feudal system, bishops had become " temporal barons themselves, and were liable like the merest laymen, to military service, to the *jurisdictio herilis*, and the other obligations of the dignity;" they became ministers of justice like secular barons, with the exception only that they could not pronounce or execute sentences of death. Similarly in the twelfth century in England.

"The prelates and abbots . . . were completely feudal nobles. They swore fealty for their lands to the king or other superior, received the homage of their vassals, enjoyed the same immunities, exercised the same jurisdiction, maintained the same authority as the lay lords among whom they dwelt."

To all which facts we must join the fact that with this acquisition of local civil authority by local ecclesiastics, there went the acquisition of a central civil authority, by the central ecclesiastic. The public and private actions of kings became in a measure subject to the control of the pope; so that in the thirteenth century there had taken place a "conversion of kingdoms into spiritual fiefs."

§ 635. We pass by a step, in many cases only nominal, from the civil functions of the priest as central or local ruler, to the civil function of the priest as judge only—as judge coexisting with, but separate from, the political head.

That devolution of the judicial function upon the priesthood which often takes place in early stages of social development, results from the idea that subordination to the deceased ruler who has become a god, is a higher obligation than subordination to the living ruler; and that those who, as priests, are in communication with the ghost of the deceased ruler, are channels for his commands and decisions, and are therefore the proper judges. Hence various facts which uncivilized and semi-civilized peoples present. Of the Coast Negroes we read that "in Badagry the fetish-priests are the sole judges of the people." In ancient Yucatan "the priests of the gods were so much venerated that they were the lords who inflicted punishments and assigned rewards." Already in § 525, when speaking of judicial systems, I have referred to the judicial functions of priests among the Gauls and Scandinavians. With more ancient peoples the like relation held for the like reason. Of the Egyptians we are told that—

"Besides their religious duties, the priests fulfilled the important offices of judges [Ælian, Hist. Var., lib. xiv, c. 34] and legislators, as well as counsellors of the monarch; and the laws as among many other nations of the East [the Jews, Moslems, and others], forming part of the sacred books, could only be administered by members of their order."

Unlike as was originally the relation of the priest to the ruler throughout Christendom, yet when the Christian priest came eventually to be regarded, like the priests of indigenous religions, as divinely inspired, there arose a tendency to recognize his judicial authority. In the old English period the bishop had "to assist in the administration of justice between man and man, to guard against perjury, and to superintend the administration of the ordeals." And this early participation with laymen in judicial functions afterwards became something like usurpation. Beginning as tribunals enforcing the discipline of superior priests over inferior priests, ecclesiastical courts, both here and abroad, extended their range of action to cases in which clerical and lay persons were simultaneously implicated, and eventually made the actions of laymen also, subject to their decisions. At first taking cognizance of offences distinguished as spiritual, these courts gradually extended the definition of such until in some places—

"All testamentary and matrimonial questions—all matters relating to bankers, usurers, Jews, Lombards—everything involving contracts and engagements upon oath—all cases arising out of the Crusades—the management of hospitals and other charitable institutions—all charges of sacrilege, perjury, incontinence," &c., fell under the "arbitration of the Church."

And at the same time there had been developed a body of canon law derived from papal judgments. These encroachments of ecclesiastical jurisdiction on the sphere of civil jurisdiction, led eventually to struggles for supremacy; until, in the thirteenth century, ecclesiastical jurisdiction began to be restricted, and has since become relatively small in range.

§ 636. Along with a large share in the administration of justice possessed by priests in countries where, or times when, they are supposed to be inspired with divine wisdom, or utterers of divine injunctions, priests also have in such places and times, a large share in the control of State-affairs as ministers or advisers.

In some cases the political ruler seeks their aid not because he believes they have supernatural wisdom but because they are useful controlling agents. Says Cruikshank, "many, also, among the higher and more intelligent ranks of the natives [of the Gold Coast], who have very little faith in the Fetish, [or fetish-man] acknowledge its value as an engine of civil government." The Fijian chiefs admitted "that they have little respect for the power of the priests, and use them merely to govern the people." Or, as Williams says, " a good understanding exists between the chief and the priests, and the latter take care to make the gods' utterances to agree with the wishes of the former." Probably a kindred relation exists in Abyssinia, where the king of Shoa rules his people " principally through the church."

In other and more numerous cases, however, the power of the priest (or the medicine-man, or the man uniting both characters,) as political counsellor, results from belief in his supernatural knowledge. Writing of the Marutse, Holub says that in King Sepopo's employment were "two old wizen-looking magicians or doctors, . . . who exercised almost a supreme control over state affairs." Similarly,

Boyle writes of the Dyaks that "next door to the Tuah [chief] lived the 'manang' or medicine man." And this reminds us of Huc's remark concerning the Tartar emperor, Mangou-khan, who "was given to a number of superstitious practices, and the principal soothsayer was lodged opposite his tent . . . having under his care the cars that bore the idols." So has it been where the sacerdotal character has become decided. We have seen that in Mexico "the high-priests were the oracles whom the kings consulted in all the most important affairs of the State." So was it among other ancient American peoples; as in primitive Michoacan, where the priests "had the greatest influence in secular as well as ecclesiastical affairs." In ancient Egypt it was the same. "Next to the king, the priests held the first rank, and from them were chosen his confidential and responsible advisers." And it is still so in Burmah, where, Sangermano says, "all is regulated by the opinions of the Brahmins, so that not even the king shall presume to take any step without their advice."

That this advising function in civil affairs should be joined with the sacerdotal function, in societies having cults originating from worship of dead rulers, is to be expected. We see, however, that even the priests of a conquering religion acquire in this, as in other respects, the same essential positions as the priests of an indigenous religion. The history of mediæval Europe shows how prelates became agents of civil rule; alike as ministers, as diplomatic agents, and as members of councils dealing with political affairs.

§ 637. But as with the military functions of priests so with their civil functions, social development, ever accompanied by specialization, more and more restricts them.

At the one extreme we have, in the primitive king, a complete fusion of the two sets of functions; while in the governments of advanced societies we see approach to an extreme in which priests, instead of taking prominent parts in civil affairs, are almost excluded from them. Among ourselves, save in the occasional instances of clerical magistrates, the judicial and executive powers once largely shared in by leading ecclesiastics, have lapsed out of their hands; while that remnant of legislative power still exercised by the bishops, appears not likely to be retained much longer. At the same time this differentiation has so established itself in the general mind, that it is commonly thought improper for clergymen to take active parts in politics.

Good reason exists for associating this change, or at any rate the completion of it, with development of the industrial type. Resistance to the irresponsible rule of priests, like resistance to other irresponsible rule, is ultimately traceable to that increasing assertion of personal freedom, with accompanying right of private judgment, which industrial life fosters by habituating each citizen to maintain his own claims while respecting the claims of others. But this connexion will be made more manifest as we proceed with the subject of the next chapter.

CHAPTER XII.

CHURCH AND STATE.

§ 638. In various ways it has been shown that originally Church and State are undistinguished. I do not refer only to the fact that in China and Japan the conceptions of this world and the other world have been so mingled that both worlds have had a living ruler in common. Nor am I recalling only the truth that the primitive ruler, vicegerent of his deceased ancestor, whom, as priest, he propitiates not only by sacrifices but by carrying out his dictates, thus becomes one in whose person are united government by the dead and government by the living. But I have in view the further fact that where the normal order has not been broken, the organizations for sacred rule and for secular rule remain practically blended, because the last remains in large measure the instrument of the first. Under a simple form this relation is well shown us in Mangaia, where—

"Kings were . . . 'the mouth-pieces, or priests, of Rongo.' As Rongo was the tutelar divinity and the source of all authority, they were invested with tremendous power—the temporal lord having to obey, like the multitude, through fear of Rongo's anger."

And this theocratic type of government has been fully developed in various places. Much more pronounced than among the Hebrews was it among some of the Egyptians.

"The influence of the priests at Meroë, through the belief that they spoke the commands of the Deity, is more fully shown by Strabo and Diodorus, who say it was their custom to send to the king, when it pleased them, and order him to put an end to himself, in obedience to the will of the oracle imparted to them ; and to such a degree had they contrived to enslave the understanding of those princes by superstitious fears, that they were obeyed without opposition."

Other cases of the subjection of the temporal power to the spiritual power, if less extreme than this, are still sufficiently marked.

"The Government of Bhutan, as of Tibet, and of Japan, is a theocracy, assigning the first place to the spiritual chief. That chief being by profession a recluse, the active duties are discharged ordinarily by a deputy."

But in these cases, or some of them, the supremacy of the spiritual head has practically given place to that of the temporal head: a differentiation of the two forms of rule which has arisen in Polynesia also, under kindred conditions.

Where Church and State are not so completely fused as by thus making the terrestrial ruler a mere deputy for the celestial ruler, there still continues a blending of the two where primitive beliefs survive in full strength, and where, consequently, the intercessors between gods and men continuing to be all-powerful merge civil rule in ecclesiastical rule. In Egypt for example—

"The priesthood took a prominent part in everything. . . . Nothing was beyond their jurisdiction : the king himself was subject to the laws established by them for his conduct, and even for his mode of living."

Along with religious beliefs equally intense with those in Egypt, there went in the ancient American societies a like unity of Church and State. The Peruvians exhibited a complete identity of the ecclesiastical government with the political ; in Yucatan the authority of priests rivalled that of kings; and in harmony with the tradition of the ancient Mexicans that the priests headed their immigration, there was such mingling of sacerdotal with civil rule as made the two in great measure one.

That this blending of Church and State is not limited to societies in which the gods are apotheosized rulers more or less ancient, but is found also in societies characterized by cults which are not indigenous, and that it continues as long as religious beliefs are accepted without criticism, we are shown by the history of mediæval Europe.

But in this case as in all cases, various causes subsequently conspire to produce differentiation and increasing separation. Co-operating efficiently though they at first do as having interests in large measure the same, yet the agencies for carrying on celestial rule and terrestrial rule eventually begin to compete for supremacy; and the competition joins with the growing unlikenesses of functions and structures in making the two organizations distinct.

§ 639. That we may understand the struggle for supremacy which eventually arises, and tends to mark off more and more the ecclesiastical structure from the political structure, we must glance at the sources of sacerdotal power.

First comes the claim of the priest, as representing the deity, to give a sanction to the authority of the civil ruler. At the present time among some of the uncivilized, as the Zulus, we find this claim recognized.

"As to the custom of a chief of a primitive stock of kings among black men, he calls to him celebrated diviners to place him in the chieftainship, that he may be really a chief."

In ancient Egypt the king, wholly in the hands of ecclesiastics, could be crowned only after having been made one of their body. Then among the Hebrews we have the familiar case of Saul who was anointed by Samuel in God's name. Passing without further cases to the acquired power of the popes, which became such that kings, receiving their crowns from them, swore obedience; we are shown that the consecration of rulers, continuing in form down to our own day, was, when a reality, an element of priestly power.

Next may be named the supposed influence of the priest with supernatural beings. Wherever faith is unqualified, dread of the evils which his invocations may bring, or trust in his ability to obtain blessings, gives him immense advantages. Even where each man could offer sacrifices, yet the professional priests profited by their supposed special knowledge. Instance the case of Rome, where their power was thus enhanced.

"Every suppliant and inquirer addressed himself directly to the divinity—the community of course by the king as its mouthpiece, just as the *curia* by the *curio*, and the *equites* by their colonels. . . . But . . . the god had his own way of speaking. . . . One who did rightly understand it knew not only how to ascertain, but also how to manage, the will of the god, and even in case of need to overreach or to constrain him. It was natural, therefore, that the worshipper of the god should regularly consult such men of skill and listen to their advice."

Of course where propitiation of a deity could be made only by sacerdotal agency—where, as among the Chibchas, "no sacrifice or offering, public or private, could be made but by the hands of the priest"—the ecclesiastical organization gained great strength.

To the influence possessed by priests as intercessors, may be added some allied influences similarly rooted in the accepted superstitions. One is the assumed power to grant or refuse forgiveness of sins. Then there is the supposed need for a passport to the other world; as shown us by usages in ancient Mexico, in Japan, and in Russia. Once more there is the dreaded excommunication, which, under the Christian system, as under the system of the druids, was visited especially on those who disregarded ecclesiastical authority.

To powers which priests acquire from their supposed relations with the gods, must be added powers of other kinds. In early societies they form the cultured class. Even the medicine-man of the savage is usually one who has some information not possessed by those around; and the developed priesthoods of established nations, as of the Egyptians and the Chaldeans, show us how knowledge of surrounding phenomena, accumulated and transmitted, enabling them to predict astronomical occurrences and do other astonishing things, greatly exalts them in the eyes of the uninitiated. With the further influence thus gained must be joined that gained by acquaintance with the art of writing. Beyond the wonder excited among the common people by the ability to convey ideas in hieroglyphics, ideographs, etc., there is the immense aid to co-operation throughout the ecclesiastical hierarchy which an exclusive means of communicating intelligence gives; and the history of mediæval Europe shows how power to read and write, possessed by priests but rarely by others, made their assistance indispensable in various civil transactions and secured great advantages to the Church. Nor must we forget the kindred enhancements of influence arising from the positions of prelates as the teachers of civil rulers. In mediæval Europe, bishops " were the usual preceptors of the princes;" and in Mandalay at the present time, the highest church dignitary, who stands next to the king in authority, " is generally made patriarch from having been the King's instructor during youth."

Lastly may be named the power resulting from accumulation of property. Beginning with payments to exorcisers and diviners among savages, progressing to fees in kind to sacrificing priests, and growing by-and-by into gifts made to temples and bribes to their officials, wealth everywhere tends to flow to the ecclesiastical organization. Speaking of ancient Mexico, Zurita says that "besides many towns, a great number of excellent estates were set apart for the maintenance of public worship." Among the Peruvians the share of the annual produce reserved for religious services was "from a third to a fourth." In ancient Egypt "the priests lived in abundance and luxury. The portion of the soil allotted to them, the largest in the threefold division, was [at one period] subject to no taxes." So again in Rome.

"The public service of the gods became not only more tedious, but above all more and more costly. . . . The custom of instituting endowments, and generally of undertaking permanent pecuniary obligations, for religious objects prevailed among the Romans in a manner similar to its prevalence in Roman Catholic countries at the present day."

And the analogy thus drawn introduces the familiar case of Europe during the middle ages; in which, besides offerings, tithes, etc., the Church had at one time acquired a third of the landed property.

§ 640. Holding in its hands powers, natural and supernatural, thus great and varied, an ecclesiastical organization seems likely to be irresistible, and in sundry places and times has proved irresistible. Where the original blending of Church with State has given place to that vague distinction inevitably resulting from partial specialization of functions accompanying social evolution, there are certain to arise differences of aim between the two; and a consequent question whether the living ruler, with his organization of civil and military subordinates, shall or shall not yield to the organization of those who represent dead rulers and profess to utter their commands. And if, throughout the society, faith is unqualified and terror of the supernatural extreme, the temporal power becomes subject to the spiritual power.

We may trace back this struggle to early stages. Respecting weather-doctors among the Zulus, and the popular valuation of them as compared with chiefs, we read:—

"The hail then has its doctors in all places; and though there is a chief in a certain nation, the people do not say, 'We have corn to eat through the power of the chief;' but they say, 'We have corn to eat through the son of So-and-so; for when the sky rolls cloud upon cloud, and we do not know that it will go back to another place, he can work diligently and do all that is necessary, and we have no more any fear.'"

To which it should be added that the chief among the Zulus, habitually jealous of the medicine-man, in some cases puts him to death. In another form, an example of the conflict comes to us from Samoa. At a council of war which the Samoans held to concert measures of vengeance on the Tongans, the high priest, "a bold, violent, unscrupulous man, who combined in his own person the threefold office of warrior, prophet, and priest," urged that the Tongan prisoners should be put to immediate death. The king opposed this proposal, and hence originated a feud between the priest and the king, which resulted in a civil war, the overthrow and exile of the king, and usurpation of his place by the priest. Though this contest between a merciful king and a merciless priest does not in all respects parallel that between Saul and Samuel, since Samuel, instead of usurping the kingship himself, merely anointed David; yet the two equally illustrate the struggle for authority which arises between the political head and the supposed mouthpiece of divine commands. Similarly among the Greeks. Curtius, speaking of the time when the *Iliad* took form, says:—

"The priests, especially the soothsayers, also oppose themselves to the royal power; themselves constituting another authority by the grace of God, which is proportionately more obstinate and dangerous.'

And we find traces of resistance to civil power among the Romans.

"The priests even in times of grave embarrassment claimed the right of exemption from public burdens, and only after very troublesome controversy submitted to make payment of the taxes in arrear."

In various ways among various peoples this conflict is shown. Of the Japanese priests in the sixteenth century, Dickson writes:—

"By their wealth, and from among their vassals, they were able to keep up a respectable army; and not by their vassals alone—the priests themselves filled the ranks.'

Among the Nahuan nations of ancient America, the priests "possessed great power, secular as well as sacerdotal. Yopaa, one of their principal cities, was ruled absolutely by a pontiff, in whom the Zapotec monarchs had a powerful rival." And the relation between spiritual and temporal rulers here indicated, recalling that between spiritual and temporal rulers in Christendom, reminds us of the long fights for supremacy which Europe witnessed between political heads wielding natural forces and the ecclesiastical head claiming supernatural origin and authority.

§ 641. There are reasons for thinking that the change from an original predominance of the spiritual power over the temporal power to ultimate subjugation of it, is mainly due to that cause which we have found in other cases chiefly operative in determining the higher types of social organization—the development of industrialism.

Already in § 618 we have noted that while their extreme servility of nature made the peoples of ancient America yield unresistingly to an unqualified political despotism appropriate to the militant type of society, it also made them submit humbly to the enormously developed priesthoods of their bloody deities; and we have seen that kindred connexions of traits were shown by various races of the old world in past times. The contrast with other ancient peoples presented by the Greeks, who, as before pointed out, (§§ 484-5, 498) were enabled by favouring conditions to resist consolidation under a despot, at the same time that, especially in Athens, industrialism and its arrangements made considerable progress among them, must here be joined with the fact that there did not arise among the Greeks a priestly hierarchy. And the connexion thus exemplified in classic times between the relatively free institutions proper to industrialism, and a smaller development of the sacerdotal organization, is illustrated throughout European history, alike in place and in time.

The common cause for these simultaneous changes is, as above implied, the modification of nature caused by substitution of a life carried on under voluntary co-operation for a life carried on under compulsory co-operation—the transition from a social state in which obedience to authority is the supreme virtue, to a social state in which it is a virtue to resist authority when it transgresses prescribed limits. This modification of nature proceeds from that daily habit of insisting on self-claims while respecting the claims of others, which the system of contract involves. The attitude of mind fostered by this discipline does not favour unqualified submission, either to the political head and his laws or to the ecclesiastical head and his dogmas. While it tends ever to limit the coercive action of the civil ruler, it tends ever to challenge the authority of the priest; and the questioning habit having once commenced, sacerdotal inspiration comes to be doubted, and the power flowing from belief in it begins to wane.

With this moral change has to be joined an intellectual change, also indirectly resulting from development of industrial life. That spreading knowledge of natural causation which conflicts with, and gradually weakens, belief in supernatural causation, is consequent on development of the industrial arts. This gives men wider experiences of uniformities of relation among phenomena; and makes possible the progress of science. Doubtless in early stages, that knowledge of Nature which is at variance with the teachings of priests, is accumulated exclusively by priests; but, as we see in the Chaldean astronomy, the natural order is not at first considered inconsistent with supernatural agency; and then, knowledge of the natural order, so long as it is exclusively possessed by priests, cannot be used to disprove their pretensions. Only as fast as knowledge of the natural order becomes so familiar and so generally diffused as insensibly to change men's habits of thought, is sacerdotal authority and power diminished by it; and general diffusion of such knowledge is, as we see, a concomitant of industrialism.

CHAPTER XIII.

NONCONFORMITY.

§ 642. NOTHING like that which we now call Nonconformity can be traced in societies of simple types. Devoid of the knowledge and the mental tendencies which lead to criticism and scepticism, the savage passively accepts whatever his seniors assert. Custom in the form of established belief, as well as in the form of established usage, is sacred with him: dissent from it is unheard of. And throughout long early stages of social evolution there continues, among results of this trait, the adhesion to inherited religions. It is true that during these stages numerous cults co-exist side by side; but, products as these are of the prevailing ancestor-worship, the resulting polytheism does not show us what we now understand as Nonconformity; since the devotees at the various shrines neither deny one another's gods, nor call in question in pronounced ways the current ideas concerning them. Only in cases like that of Socrates, who enunciated a conception of supernatural agents diverging widely from the popular conception of them, do we see in early societies Nonconformity properly so-called.

What we have here to deal with under this name occurs chiefly in societies which are substantially, if not literally, monotheistic; and in which there exists nominally, if not really, a tolerably uniform creed administered by a consolidated hierarchy.

Even as thus restricted, Nonconformity comprehends phenomena widely unlike in their natures; and that we may understand it, we must exclude much that is allied with it only by outward form and circumstance. Though in most cases a separating sect espouses some unauthorized version of the accepted creed; and though the nature of the espoused version is occasionally not without its significance; yet the thing specially to be noted is the attitude assumed towards ecclesiastical government. Though there is always some exercise of individual judgment; yet in early stages this is shown merely in the choice of one authority as superior to another. Only in late stages does there come an exercise of individual judgment which goes to the extent of denying ecclesiastical authority in general.

The growth of this later attitude we shall see on comparing some of the successive stages.

§ 643. Ancient forms of dissent habitually stand for the authority of the past over the present; and since tradition usually brings from more barbarous ages, accounts of more barbarous modes of propitiation, ancient forms of dissent are habitually revivals of practices more ascetic than those of the current religion. It was shown in § 620, that the primitive monachism originated in this way; and as Christianity, with the higher moral precepts on which it insisted, joined renunciation of ordinary life and its aims (said to be derived from the Essenes), there tended to be thereafter a continual re-genesis of dissenting sects characterized in common by austerities.

Kinds of dissent differing from these and differing from modern kinds of dissent, arose during those times in which the early church was spreading and becoming organized. For before ecclesiastical government had established itself and acquired sacredness, resistance to each new encroachment made by it, naturally led to divisions. Between the time when the authority dwelt in the Christian congregations themselves, and the time when the authority was centred in the pope, there necessarily went successive usurpations of authority, each of which gave occasion for protest. Hence such sects, arising in the third century and onward to the seventh century, as the Noetians, Novatians, Meletians, Aerians, Donatists, Joannites, Haesitantes, Timotheans, and Athingani.

Passing over that period during which ecclesiastical power throughout Europe was rising to its climax, we come, in the twelfth century, to dissenters of more advanced types; who, with or without differences of doctrine, rebelled against the then-existing church government. Such sects as the Arnoldists in Italy, the Petrobrusians, Caputiati and Waldenses in France, and afterwards the Stedingers in Germany and the Apostolicals in Italy, are examples; severally characterized by assertion of individual freedom, alike in judgment and action. Ordinarily holding doctrines called heretical, the promulgation of which was itself a tacit denial of ecclesiastical authority (though a denial habitually based on submission to an alleged higher authority) sects of this kind went on increasing in the fourteenth and fifteenth centuries. There were the Lollards in England; the Fraticelli in Italy; the Taborites, Bohemian Brethren, Moravians and Hussites, in Bohemia: all setting themselves against church-discipline. And then the rebellious movement of the reformation, as carried forward by the Lutherans in Germany, the Zwinglians and Calvinists in Switzerland, the Huguenots in France, the Anabaptists and Presbyterians in England, exhibited, along with repudiation of various established doctrines, ceremonies and usages, a more pronounced anti-sacerdotalism. Characterized in common by opposition to Episcopacy, protestant or catholic, we see first of all in the government by presbyters, adopted by sundry of these dissenting bodies, a step towards freedom of judgment and practice in religious matters, accompanied by denial of priestly inspiration. And then in the subsequent rise of the Independents, taking for their distinctive principle the right of each congregation to govern itself, we see a further advance in that anti-sacerdotal movement which reached its extreme in the next century with the Quakers; who, going directly to the fountain head of the creed, and carrying out more consistently than usual the professed right of private judgment, repudiated the entire paraphernalia of ecclesiasticism

It is true that the histories of these various non-conforming bodies, not excluding even the Society of Friends, show us the re-growth of a coercive rule, allied to that against which there had been rebellion. Of religious revolutions as of political revolutions, it is true that in the absence of differences of character and culture greater than can be expected in the same society at the same time, they are followed by gradually established forms of rule only in some degree better than those diverged from. In his assumption of infallibility, and his measures for enforcing conformity, Calvin was a pope comparable with any who issued bulls from the Vatican. The discipline of the Scottish Presbyterians was as despotic, as rigorous, and as relentless, as any which Catholicism had enforced. The Puritans of New England were as positive in their dogmas, and as severe in their persecutions, as were the ecclesiastics of the church they left behind. Some of these dissenting bodies, indeed, as the Wesleyans, have developed organizations scarcely less priestly, and in some respects more coercive, than the organization of the church from which they diverged. Even among the Quakers, notwithstanding the pronounced individuality implied by their theory, there has grown up a definite creed and a body exercising control.

§ 644. Modern Nonconformity in England has much more decidedly exhibited the essential trait of anti-sacerdotalism. It has done this in various minor ways as well as in a major way.

There is the multiplication of sects, with which by foreign observers England is reproached, but which, philosophically considered, is one of her superior traits. For the rise of every new sect, implying a re-assertion of the right of private judgment, is a collateral result of the nature which makes free institutions possible.

Still more significant do we see this multiplication of sects to be if we consider the assigned causes of division. Take for instance the case of the Wesleyans. In 1797 the Methodist New Connexion organized itself on the principle of lay participation in church government. In 1810 the Primitive Methodists left the original body : the cause being a desire to have " lay representatives to the Conference." Again, in 1834, prompted by opposition to priestly power, the Wesleyan Methodist Association was formed : its members claiming more influence for the laity, and resisting central interference with local government. And then in 1849, there was yet another secession from the Methodist body, similarly characterized by resistance to ministerial authority.

Of course in sects less coercively governed, there have been fewer occasions for rebellions against priestly control; but there are not wanting illustrations, some of them supplied even by the small and free bodies of the Unitarians, of this tendency to divide in pursuance of the right of private judgment. Moreover, in the absence of a dissidence sufficiently great to produce secession, there is everywhere a large amount of expressed disagreement on minor points, among those holding what is supposed to be the same body of beliefs. Perhaps the most curious instance of this is furnished by the established Church. I do not refer simply to its divisions into high, and low, and broad ; all implying more or less of the nonconforming spirit within it. I refer more especially to the strange anomaly that the ritualists are men who, while asserting priestly authority, are themselves rebels against priestly authority—defy their ecclesiastical superiors in their determination to assert ecclesiastical supremacy.

But the universally admitted claim to religious freedom shown in these various ways, is shown still more by the growing movement for disestablishment of the Church. This movement which, besides tacitly denying all sacerdotal authority, denies the power of a government, even though elected by a majority of votes, to prescribe religious belief or practice, is the logical outcome of the Protestant theory. Liberty of thought, long asserted and more and more displayed, is about to be carried to the extent that no man shall be constrained to support another man's creed.

Evidently the arrival at this state completes that social differentiation which began when the primitive chief first deputed his priestly function.

§ 645. As implied in the last sentence, the changes above sketched out are concomitants of the changes sketched out in the last chapter. The prolonged conflict between Church and State accompanying their differentiation, and ending in the subordination of the Church, has been accompanied by these collateral minor conflicts between the Church and recalcitrant portions of its members, ending in separation of them.

There is a further implication. In common with the subjection of the Church to the State, the spread of Nonconformity is an indirect result of growing industrialism. The moral nature proper to a social organization based on contract instead of *status*—the moral nature fostered by a social life carried on under voluntary co-operation instead of compulsory co-operation, is one which works out religious independence as it works out political freedom. And this conclusion, manifest *a priori*, is verified *a posteriori* in sundry ways. We see that Nonconformity, increasing as industrialism has developed, now characterizes in the greatest degree these nations which are most characterized by development of the industrial type—America and England. And we also see that in England itself, the contrast between urban and rural populations, as well as the contrast between populations in different parts of the kingdom, show that where the industrial type of life and organization predominates, Nonconformity is the most pronounced.

CHAPTER XIV.

THE MORAL INFLUENCES OF PRIESTHOODS.

§ 646. As was said when treating of "The Military Functions of Priests," there exists in most minds an erroneous association between religious ministrations and moral teachings. Though priests habitually enforce conduct which in one way or other furthers preservation of the society; yet preservation of the society is so often furthered by conduct entirely unlike that which we now call moral, that priestly influence serves in many cases rather to degrade than to elevate.

Reading as we do of the Tahitian god Oro, that when war "proceeded in its bloodiest forms, it was supposed to afford him the highest satisfaction"—reading again of the Mexican king Montezuma, that he avoided subduing the neighbouring Tlascalans "that he might have Men to sacrifice" (thus making Tlascala a preserve of victims for the gods)—reading once more of the Chibchas that "the sacrifices which they believed to be most welcome to their gods were those of human blood;" we are reminded that priests who carry on propitiations of cannibal deities and deities otherwise atrocious (deities almost everywhere worshipped in early days) have done anything but foster high forms of conduct. Robbery as well as murder has had, and has still in some places, a religious sanctification. Says Burton of the Beloochis, "these pious thieves never rob, save in the name of Allah." Of a robber-tribe among the Chibchas, Piedrahita writes, "they regard as the most acceptable sacrifice that which they offer up out of the robbery to certain idols of gold, clay, and wood, whom they worship." And at the present time in India, we have freebooters like the Domras, among whom "a successful theft is always celebrated by a sacrifice" to their chief god Gandak. Nor is it only by encouraging disregard for life and property, that various cults, and by implication their priests, have aided in demoralizing men rather than in moralizing them. On finding that "among the Friendly Islanders the chief priest was considered too holy to be married, but he had the right to take as many concubines as he pleased"—that among the Caribs, "the bride was obliged to pass the first night with the priest, as a form essentially necessary to constitute the legality of the marriage"—that among some Brazilian tribes, "the Pajé [priest], like the feudal lord of former times in some parts of England, enjoys the *jus primæ noctis;*" or again on being reminded of the extent to which prostitution in temples was a religious observance among Eastern peoples; we are shown in yet another way that there is no necessary connexion between priestly guidance and right action: using the word right in the sense at present given to it.

But now carrying with us the implied qualifications, let us ask in what ways Ecclesiastical Institutions have affected men's natures. We shall find that they have been instrumental in producing, or furthering, certain all-important modifications.

§ 647. When describing the action of "An Ecclesiastical System as a Social Bond," it was pointed out that a common worship tends to unify the various groups which carry it on; and that, by implication, the priests of such worship usually act as pacificators. While often instigating wars with societies of other blood, worshipping other gods, they, on the average of cases, check hostilities between groups of the same blood worshipping the same gods. In this way they aid social co-operation and development.

This function, however, is but a collateral display of their fundamental function—the maintenance of subordination: primarily to the deified progenitor, or the adopted god, and secondarily to his living descendant or appointed vice-gerent. It is scarcely possible to emphasize enough the truth that, from the earliest stages down to existing stages, the one uniform and essential action of priesthoods, irrespective of time, place, or creed, has been that of insisting on obedience. That primitive men may be moulded into fitness for social life, they must be held together; and that they may be held together, they must be made subject to authority. Only by restraints of the most powerful kinds can the unregulated explosive savage be made to co-operate permanently with his fellows; and of such restraints the strongest, and apparently the indispensable one, is fear of vengeance from the god of the tribe, if his commands, repeated by his successor, are disobeyed. How important is the agency of Ecclesiastical Institutions as thus re-inforcing Political Institutions, is well seen in the following description Ellis gives of the effects produced by undermining local religions in Polynesia.

"The sacrificing of human victims to the idols had been one of the most powerful engines in the hands of the government, the requisition for them being always made by the ruler, to whom the priests applied when the gods required them. The king, therefore, sent his herald to the petty chieftain, who selected the victims. An individual who had shewn any marked disaffection towards the government, or incurred the displeasure of the king and chiefs, was usually chosen. The people knew this, and therefore rendered the most unhesitating obedience. Since the subversion of idolatry, this motive has ceased to operate; and many, free from the restraint it had imposed, seemed to refuse all lawful obedience and rightful support."

The result, as described by Ellis, being that social order was in a considerable degree disturbed.

This maintenance of subordination, to which an ecclesiastical system has been instrumental, has indirectly subserved other disciplines of an indispensable kind. No developed social life would have been possible in the absence of the capacity for continuous labour; and out of the idle improvident savage there could not have been evolved the industrious citizen, without a long-continued and rigorous coercion. The religious sanction habitually given in early societies to rigid class-distinctions and the concomitant slavery, must be regarded as having conduced to a modification of nature which furthered civilization.

A discipline allied and yet different, to which superior as well as inferior classes have been subjected by Ecclesiastical Institutions, has been the discipline of asceticism. Considered in the abstract asceticism is indefensible. As already shown (§§ 140 and 620) it grew out of the desire to propitiate malicious ghosts and diabolical deities; and even as displayed among ourselves at present, we may trace in it the latent belief that God is pleased by voluntarily-borne mortifications and displeased by pursuit of gratifications. But if instead of regarding self-infliction of suffering, bodily or mental, from the stand-point of absolute ethics, we regard it from the stand-point of relative ethics, as an educational regimen, we shall see that it has had a use, and perhaps a great use. The common trait of all ascetic acts is submission to a pain to avoid some future greater pain, or relinquishment of a pleasure to obtain some greater pleasure hereafter. In either case there is sacrifice of the immediate to the remote. This is a sacrifice which the uncivilized man cannot make; which the inferior among the civilized can make only to a small extent; and which only the better among the civilized can make in due degree. Hence we may infer that the discipline which, beginning with the surrendering of food, clothing, etc., to the ancestral ghost, and growing into the voluntary bearing of hunger, cold, or pain, to propitiate deities, has greatly aided in developing the ability to postpone present to future. Possibly only a motive so powerful as that of terror of the supernatural, could have strengthened the habit of self-denial in the requisite degree—a habit which, we must remember, is an essential factor in right conduct towards others, as well as in the proper regulation of conduct for self-benefit.

Irrespective, then, of the particular traits of their cults, Ecclesiastical Institutions have, in these ways, played an important part in moulding human nature into fitness for the social state.

§ 648. Among more special moral effects wrought by them, may be named one which, like those just specified, has been wrought incidentally rather than intentionally. I refer to the respect for rights of property, curiously fostered by certain forms of propitiation. Whether or not Mariner was right in saying that the word *taboo*, as used in the Tonga Islands, literally meant "sacred or consecrated to a god," the fact is that things tabooed, there and elsewhere, were at first things thus consecrated: the result being that disregard of the taboo became robbery of the god. Hence such facts as that throughout Polynesia, "the prohibitions and requisitions of the tabu were strictly enforced, and every breach of them punished with death" (the delinquent being sacrificed to the god whose tabu he had broken); and that in New Zealand "violators of the tapu were punished by the gods and also by men. The former sent sickness and death; the latter inflicted death, loss of property, and expulsion from society. It was a dread of the gods, more than of men, which upheld the tapu."

Obviously a sacredness thus given to anything bearing a sign that it belongs to a god, may easily be simulated. Though the mark on an animal or a fruit implies that an offering to a god will eventually be made of it; yet, since the time of sacrifice is unspecified, there results the possibility of indefinite postponement, and this gradually opens the door to pretended dedication of things which never are sacrificed —things which nevertheless, bearing the sign of dedication, no one dares meddle with. Thus we read that in the New Hebrides "the tapu is employed in all the islands to preserve persons and objects;" that in New Zealand, tapu, from being originally a thing made sacred, has come to mean a thing forbidden. Fiji, Tonga, and Samoa furnish kindred facts: the last place being one in which the name of the tabu indicates the sort of curse which the owner of a tabued thing hopes may fall on the thief. In Timor, "a few palm leaves stuck outside a garden as a sign of the 'pomali' [tabu] will preserve its produce from thieves as effectually as the threatening notice of man-traps, spring guns, or a savage dog, would do with us." Bastian tells us that the Congoese make use of the fetich to protect their houses from thieves; and he makes a like statement respecting the negroes of the Gaboon. Livingstone, too, describes the Balonda as having this usage; and evidence of kindred nature is furnished by the Malagasy and by the Santals.

As, originally, this dedication of anything to a god is made either by a priest or by a chief in his priestly capacity, we must class it as an Ecclesiastical Institution; and the fostering of respect for proprietary rights which grows out of it, must be counted among the beneficial disciplines which Ecclesiastical Institutions give.

§ 649. Respecting the relation which exists between alleged supernatural commands and the right ruling of conduct at large, it is difficult to generalize. Many facts given in foregoing chapters unite to show that everything depends on the supposed character of the supernatural being to be propitiated. Schoolcraft says of the Dakotahs—

"They stand in great awe of the spirits of the dead, because they think it is in the power of the departed spirits to injure them in any way they please; this superstition has, in some measure, a salutary effect. It operates on them just as strong as our laws of hanging for murder."

But if, as happens in many cases, a dying man's peremptory injunction to his son (like that of David to Solomon) is to wreak vengeance on those who have injured him, fear of his ghost becomes not a moralizing but a demoralizing influence; using these words in their modern acceptations. When, concerning the deities of Mangaia, we read that "the cruel Kereteki, twice a fratricide, and his brother Utāhea, were worshipped as gods in the next generation;" we are shown that divine example, if not precept, is in some cases a prompter to crime rather than otherwise. But on the average an opposite effect may be inferred. As the deified chief must be supposed to have had at heart the survival and spread of his tribe, sundry of his injunctions are likely to have had in view that maintenance of order conducing to tribal success. Hence rules traditionally derived from him are likely to be restraints on internal aggressions. Ferocious as were the Mexicans, and bloody as were their religious rites, they nevertheless had, as given by Zurita, a moral code which did not suffer by comparison with that of Christians: the one like the other claiming divine authority. Concerning the Peruvians, who like various of these semi-civilized American peoples had confessors, the account runs that—

"The sin of which they mostly accuse themselves was—to have killed somebody in time of peace, to have robbed, to have taken the wife of another, to have given herbs or charms to do harm. The most notable sin was neglect in the service of the huacas [gods] . . . abuse of, and disobedience towards, the Ynca."

And in this case, as in many other cases, we see that after the first and greatest sin of insubordination to the deity, come sins constituted by breaches of those laws of conduct needful for social concord.

Evidently through long stages of individual and social evolution, belief in the alleged divine origin of such laws is beneficial. The expected supernatural punishments for breaches of them, usefully re-inforce the threats of natural punishments. And various cases might be given showing that the moral code required for each higher stage, gaining alleged divine authority through some intermediating priest or inspired man, thus becomes more effective for the time being than it would otherwise be: the cases of Moses and of the later Hebrew prophets serving as examples.

§ 650. Multitudinous anomalies occur, however—anomalies which seem unaccountable till we recognize the truth that in all cases the one thing which precedes in importance the special injunctions of a cult, is the preservation of the cult itself and the institutions embodying it. Hence the fact that everywhere the duty which stands higher than duties properly called moral, is the duty of obedience to an alleged divine will, whatever it may be. Hence the fact that to uphold the authority of a sacerdotal hierarchy, by which the divine will is supposed to be uttered, is regarded by its members and adherents as an end yielding in importance only to recognition of the divine will itself. And hence the fact that the histories of Ecclesiastical Institutions show us how small is the regard paid to moral precepts when they stand in the way of ecclesiastical supremacy.

Of course the atrocities perpetrated in inquisitions and the crimes committed by popes will come into all minds as illustrations. But there are more remarkable illustrations even than these. The bitterest animosity shown by established churches against dissenting sects, has been shown against those which were distinguished by endeavours to fulfil the precepts of Christianity completely. The Waldenses, who "adopted, as the model of their moral discipline, the Sermon of Christ on the Mount," but who at the same time rebelled against ecclesiastical rule, suffered a bloody persecution for three centuries. The Quakers, who alone among protestants sought to obey the commands of the Christian creed not in some ways only but in all, were so persecuted that before the accession of James II. more than 1500 out of their comparatively small number were in prison. Evidently, then, the distinctive ethics of a creed, restrain but little its official administrators when their authority is called in question.

Not only in such cases, however, are we shown that the

chief concern of a sacerdotal system is to maintain formal subordination to a deity, as well as to itself as his agency, and that the ordering of life according to the precepts of the professed religion is quite a secondary matter; but we are shown that such a right ordering of life is little insisted on even where insistence does not conflict with ecclesiastical supremacy. Through all these centuries Christian priests have so little emphasized the virtue of forgiveness, that alike in wars and in duels, revenge has continued to be thought an imperative duty. The clergy were not the men who urged the abolition of slavery, nor the men who condemned regulations which raised the price of bread to maintain rents. Ministers of religion do not as a body denounce the unjust aggressions we continually commit on weak societies; nor do they make their voices loudly heard in reprobating such atrocities as those of the labour-traffic in the Pacific, recently disclosed by a Royal Commission (see *Times*, June 18th, 1885). Even where they are solely in charge, we see not a higher, but rather a lower, standard of justice and mercy than in the community at large. Under clerical management, public schools have in past times been the scenes of atrocities not tolerated in the world outside of them; and if we ask for a recent instance of juvenile savagery, we find it at King's College School, where the death of a small boy was caused by the unprovoked blows given in sheer brutality by cowardly bigger boys: King's College being an institution established by churchmen, and clerically governed, in opposition to University College, which is non-clerical in its government and secular in its teaching.

§ 651. Contemplating Ecclesiastical Institutions at large, apart from the particular cults associated with them, we have, then, to recognize the fact that their presence in all societies which have made considerable progress, and their immense predominance in those early societies which reached relatively high stages of civilization, verify inductively the deductive conclusion, that they have been indispensable components of social structures from the beginning down to the present time: groups in which they did not arise having failed to develop.

As furnishing a principle of cohesion by maintaining a common propitiation of a deceased ruler's spirit, and by implication checking the tendencies to internal warfare, priesthoods have furthered social growth and development. They have simultaneously done this in sundry other ways: by fostering that spirit of conservatism which maintains continuity in social arrangements; by forming a supplementary regulative system which co-operates with the political one; by insisting on obedience, primarily to gods and secondarily to kings; by countenancing the coercion under which has been cultivated the power of application; and by strengthening the habit of self-restraint.

Whether the modifications of nature produced by this discipline, common to all creeds, are accompanied by modifications of higher kinds, depends partly on the traditional accounts of the gods worshipped, and partly on the social conditions. Religious obedience is the primary duty; and this, in early stages, often furthers increase of ferocity. With the change from a more militant to a more industrial state, comes a reformed ethical creed, which increases or decreases in its influence according as the social activities continue peaceful or again become warlike. Little as such reformed ethical creed (presently accepted as of divine origin) operates during periods when war fosters sentiments of enmity instead of sentiments of amity, advantage is gained by having it in reserve for enunciation whenever conditions favour.

But clerical enunciation of it habitually continues subject to the apparent needs of the time. To the last as at first, subordination, religious and civil, is uniformly insisted on—"fear God, honour the king;" and providing subordination is manifested with sufficient emphasis, moral shortcomings may be forgiven.

CHAPTER XV.

ECCLESIASTICAL RETROSPECT AND PROSPECT.

§ 652. Among social phenomena, those presented by Ecclesiastical Institutions illustrate very clearly the general law of evolution.

Subjection to the family-head during his life, continues to be shown after his death by offering to his double the things he liked, and doing the things he wished; and when the family multiplies into a tribe, presents to the chief, accompanied by compliments and petitions, are continued after his death in the shape of oblations, praises, and prayers to his ghost. That is to say, domestic, civil, and religious subordination have a common root; and are at first carried on in like ways by the same agencies.

Differentiation early begins, however. First some contrast arises between the private cult proper to each family, and the public cult proper to the chief's family; and the chief, as propitiator of his dead ancestor on behalf of the tribe, as well as on his own behalf, unites the functions of civil head and spiritual head. Development of the tribe, bringing increased political and military functions, obliges the chief more and more to depute, usually to a relative, his priestly function; and thus, in course of time, this acquires a separate agency.

From integration of societies effected by conquest, there results the coexistence of different cults in different parts of the same society; and there arise also deputed priests, carrying on the more important of these cults in the different localities. Hence polytheistic priesthoods; which are made heterogeneous by the greater increase of some than of others. And eventually, in some cases, one so immensely enlarges that it almost or quite excludes the rest.

While, with the union of simple societies into compound ones, and of these again into doubly compound ones, there go on the growths of priesthoods, each priesthood, differentiating from others, also differentiates within itself. It develops into an organized whole subordinate to an arch priest, and formed of members graduated in their ranks and specialized in their functions.

At the same time that an ecclesiastical hierarchy is becoming within itself more closely integrated and clearly differentiated, it is slowly losing that community of structure and function which it originally had with other parts of the body politic. For a long time after he is distinguishable as such, the priest takes an active part, direct or indirect, in war; but where social development becomes high, what military character he had is almost or quite lost. Similarly with his civil functions. Though during early stages he exercises power as ruler, minister, counsellor, judge, he loses this power by degrees; until at length there are but traces of it left.

This development of Ecclesiastical Institutions, which, while it makes the society at large more definitely heterogeneous, shows us increase of heterogeneity within the ecclesiastical organization itself, is further complicated by successive additions of sects. These, severally growing and organizing themselves, make more multiform the agencies for carrying on religious ministrations and exercising religious control.

Of course the perpetual conflicts among societies, ending now in unions and now in dislocations, here breaking up old institutions and there superposing new ones, has made the progress of Ecclesiastical Institutions irregular. But amid all the perturbations, a course essentially of the kind above indicated may be traced.

§ 653. With structural differentiations must here be joined a functional differentiation of deep significance. Two sacerdotal duties which were at first parts of the same, have been slowly separating; and the one which was originally unobtrusive but is now conspicuous, has become in large measure independent. The original duty is the carrying on of worship; the derived duty is the insistence on rules of conduct.

Beginning as the entire series of phenomena does with propitiation of the dead parent or dead chief, and dependent as the propitiatory acts are on the desires of the ghost, which are supposed to be like those of the man when alive; worship in its primitive form, aiming to obtain the goodwill of beings in many cases atrocious, is often characterized by atrocious observances. Originally, there is no moral element in it; and hence the fact that extreme attention to religious rites characterizes the lower types, rather than the higher types, of men and of societies. Renouf remarks that "the Egyptians were among the most religious of the ancient nations. Religion in some form or other was dominant in every relation of their lives;" or, as M. Maury has it, "l'Égyptien ne vivait en réalité que pour pratiquer son culte." This last statement reminds us of the ancient Peruvians. So onerous were their sacrifices to ancestors, and deities derived from ancestors, that it might truly be said of them that the living were the slaves of the dead. So, too, of the sanguinary Mexicans, whose civilization was, in a measure, founded on cannibalism, it is remarked that "of all nations which God has created, these people are the strictest observers of their religion." Associated with their early stages and arrested stages, we find the same trait in Aryan peoples.

"The Vedas represent the ancient Indo-Aryans to have been eminently religious in all their actions. According to them, every act of life had to be accompanied by one or more mantras, and no one could rise from his bed, or wash his face, or brush his teeth, or drink a glass of water, without going through a regular system of purifications, salutations and prayers."

Similarly with the Romans. "Religion everywhere met the public life of the Roman by its festivals, and laid an equal yoke on his private life by its requisition of sacrifices, prayers, and auguries." And speaking of the existing Hindu, the Rev. M. A. Sherring says—

"He is a religious being of wonderful earnestness and persistency. His love of worship is a passion, is a frenzy, is a consuming fire. It absorbs his thoughts; it influences and sways his mind on every subject."

Everywhere we find kindred connexions; be it in the ancient Thracian who with great cruelty of character joined "ecstatic and maddening religious rites," or in the existing Mahometan with his repeated daily prayers and ablutions. Even if we compare modern Europeans with Europeans in mediæval times, when fasts were habitual and penances common, when anchorites were numerous and self-torturings frequent, when men made pilgrimages, built shrines, and counted their numerous prayers by beads, we see that with social progress has gone a marked diminution of religious observances. Evidence furnished by many peoples and times thus shows us that the propitiatory element, which is the primary element, diminishes with the advance of civilization, and becomes qualified by the growing ethical element.

This ethical element, like all other elements in the religion, is propitiatory in origin and nature. It begins with fulfilment of the wishes or commands of the dead parent, or departed chief, or traditional god. There is at first included in the ethical element no other duty than that of obedience. Display of subordination is in this, as in all other religious acts, the primary thing; and the natures of the particular commands obeyed the secondary things: their obligations being regarded not as intrinsic, but as extrinsically derived from their alleged origin. But slowly, experience establishes ethical conceptions, round which there gather private sentiments and public opinions, giving them some independent authority. More especially when a society becomes less occupied in warlike activities, and more occupied in quietly carrying on production and distribution, do there grow clear in the general consciousness those rules of conduct which must be observed to make industrial co-operation harmonious.

For these there is eventually obtained a supernatural authority through some alleged communication of them to an inspired man; and for long periods, conformity to them is insisted on for the reason that they are God's commands. The emphasizing of moral precepts which are said to be thus derived, comes, however, to occupy a larger space in religious services. With offerings, praises, and prayers, forming the directly propitiatory part, come to be joined homilies and sermons, forming the indirectly propitiatory part: largely composed of ethical injunctions and exhortations. And the modified human nature produced by prolonged social discipline, evolves at length the conception of an independent ethics—an ethics so far independent that it comes to have a foundation of its own, apart from the previously-alleged theological foundation. Nay, more than this happens. The authority of the ethical consciousness becomes so high that theological dogmas are submitted to its judgments, and in many cases rejected because of its disapproval. Among the Greeks, Socrates exemplified the way in which a developed moral sentiment led to a denial of the accepted beliefs concerning the gods and their deeds; and in our own days we often see current religious doctrines brought to the bar of conscience, and condemned as untrue because they ascribe to a deity who claims worship, certain characters which are the reverse of worshipful. Moreover, while we see this—while we see, too, that in daily life, criticisms passed on conduct approve or condemn it as intrinsically good or bad, irrespective of alleged commands; we also see that modern preaching tends more and more to assume an ethical character.

Dogmatic theology, with its promises of rewards and threats of damnation, bears a diminishing ratio to the insistences on justice, honesty, kindness, sincerity, etc.

§ 654. Assuming, as we must, that evolution will continue along the same general lines, let us now, after this retrospect, ask—What is the prospect? Though Ecclesiastical Institutions hold less important places in higher societies than in lower societies, we must not infer that they will hereafter wholly disappear. If in times to come there remain functions to be fulfilled in any way analogous to their present functions, we must conclude that they will survive under some form or other. The first question is—Under what form?

That separation of Ecclesiastical Institutions from Political Institutions, foreshadowed in simple societies when the civil ruler begins to depute occasionally his priestly function, and which, in many ways with many modifications according to their types, societies have increasingly displayed as they have developed, may be expected to become complete. Now-a-days, indeed, apart from any such reasons as are above assigned, the completing of it, already effected in some cases, is recognized as but a question of time in other cases. All which it concerns us here to observe is that separation is the ending of a process of evolution, partially carried out in societies of the more militant type, characterized by the predominance of structures which maintain subordination, and carried out in greater degrees in societies that have become more industrial in their type, and less coercive in their regulative appliances.

The same emotional and intellectual modifications which, while causing the diminished power of State-churches, has caused the multiplication of churches independent of the State, may be expected to continue hereafter doing the like. We may look for increased numbers of religious bodies having their respective differences of belief and practice. Though along with intellectual advance there may probably go, in the majority of sects thus arising, approximation to a unity of creed in essentials; yet analogy suggests that shades of difference, instead of disappearing, will become more numerous. Divergences of opinion like those which, within our generation, have been taking place in the established church, may be expected to arise in all existing religious bodies, and in others hereafter formed.

Simultaneously there will probably continue, in the same direction as heretofore, changes in church government. That fostering of individuality which accompanies development of the industrial type of society, must cause increase of local independence in all religious organizations. And along with the acquirement of complete autonomy by each religious body, there is likely to be a complete loss of the sacerdotal character by any one who plays the part of minister. That relinquishment of priestly authority which has already gone far among Dissenters, will become entire.

These conclusions, however, proceed on the assumption that development of the industrial type will advance as it has advanced during recent times; and it is quite possible, or even probable, that this condition will not be fulfilled during an epoch on which we are entering. The recrudescence of militancy, if it goes on as it has been lately going on, will bring back ideas, sentiments, and institutions appropriate to it; involving reversal of the changes above described. Or if, instead of further progress under that system of voluntary co-operation which constitutes Industrialism properly so called, there should be carried far the system of production and distribution under State-control, constituting a new form of compulsory co-operation, and ending in a new type of coercive government, the changes above indicated, determined as they are by individuality of character, will probably be arrested and opposite changes initiated.

§ 655. Leaving structures and turning to functions, it remains to ask—What are likely to be the surviving func tions, supposing the evolution which has thus far gone on is not reversed ? Each of the two functions above described, may be expected to continue under a changed form.

Though with the transition from dogmatic theism to agnosticism, all observances implying the thought of propitiation may be expected to lapse ; yet it does not follow that there will lapse all observances tending to keep alive a consciousness of the relation in which we stand to the Unknown Cause, and tending to give expression to the sentiment accompanying that consciousness. There will remain a need for qualifying that too prosaic and material form of life which tends to result from absorption in daily work, and there will ever be a sphere for those who are able to impress their hearers with a due sense of the Mystery in which the origin and meaning of the Universe are shrouded. It may be anticipated, too, that musical expression to the sentiment accompanying this sense will not only survive but undergo further development. Already protestant cathedral music, more impersonal than any other, serves not unfitly to express feelings suggested by the thought of a transitory life, alike of the individual and of the race—a life which is but an infinitesimal product of a Power without any bounds we can find or imagine ; and hereafter such music may still better express these feelings.

At the same time, that insistence on duty which has formed an increasing element in religious ministration, may be expected to assume a marked predominance and a wider range. The conduct of life, parts of which are already the subject-matters of sermons, may hereafter probably be taken as subject-matter throughout its entire range. The ideas of right and wrong, now regarded as applying only to actions of certain kinds, will be regarded as having applications coextensive with actions of every kind. All matters concerning individual and social welfare will come to be dealt with ; and a chief function of one who stands in the place of a minister, will be not so much that of emphasizing precepts already accepted, as that of developing men's judgments and sentiments in relation to those more difficult questions of conduct arising from the ever-increasing complexity of social life.

In brief, we may say that as there must ever continue our relations to the unseen and our relations to one another, it appears not improbable that there will survive certain representatives of those who in the past were occupied with observances and teachings concerning these two relations; however unlike their sacerdotal prototypes such representatives may become.

CHAPTER XVI.*

RELIGIOUS RETROSPECT AND PROSPECT.

§ 656. As, before describing the origin and development of Ecclesiastical Institutions, it was needful to describe the origin and development of Religion; so the probable future of Ecclesiastical Institutions could not be forecast without indicating the probable future of Religion. Unavoidably therefore, the close of the last chapter has partially forestalled the contents of this. Here, after briefly recapitulating the leading traits of religious evolution, I propose to give reasons for the conclusions just indicated respecting the ultimate form of religion.

Unlike the ordinary consciousness, the religious consciousness is concerned with that which lies beyond the sphere of sense. A brute thinks only of things which can be touched, seen, heard, tasted, etc.; and the like is true of the young child, the untaught deaf-mute, and the lowest savage. But the developing man has thoughts about existences which he regards as usually intangible, inaudible, invisible; and yet which he regards as operative upon him. What suggests this notion of agencies transcending perception? How do these ideas concerning the supernatural evolve out of ideas concerning the natural? The transition cannot be sudden; and an

---

* With the exception of its introductory paragraph and an added sentence in its last paragraph, this Chapter stands as it did when first published in *The Nineteenth Century* for January 1884: a few verbal improvements being the only other changes.

account of the genesis of religion must begin by describing the steps through which the transition takes place.

The ghost-theory exhibits these steps quite clearly. We are shown by it that the mental differentiation of invisible and intangible beings from visible and tangible beings progresses slowly and unobtrusively. In the fact that the other-self, supposed to wander in dreams, is believed to have actually done and seen whatever was dreamed—in the fact that the other-self when going away at death, but expected presently to return, is conceived as a double equally material with the original; we see that the supernatural agent in its primitive form, diverges very little from the natural agent—is simply the original man with some added powers of going about secretly and doing good or evil. And the fact that when the double of the dead man ceases to be dreamed about by those who knew him, his non-appearance in dreams is held to imply that he is finally dead, shows that these earliest supernatural agents are conceived as having but temporary existences: the first tendencies to a permanent consciousness of the supernatural, prove abortive.

In many cases no higher degree of differentiation is reached The ghost-population, recruited by deaths on the one side but on the other side losing its members as they cease to be recollected and dreamed about, does not increase; and no individuals included in it come to be recognized through successive generations as established supernatural powers. Thus the Unkulunkulu, or old-old one, of the Zulus, the father of the race, is regarded as finally or completely dead; and there is propitiation only of ghosts of more recent date. But where circumstances favour the continuance of sacrifices at graves, witnessed by members of each new generation who are told about the dead and transmit the tradition, there eventually arises the conception of a permanently-existing ghost or spirit. A more marked contrast in thought between supernatural beings and natural beings is thus established. There simultaneously results an increase in the number of these supposed supernatural beings, since the aggregate of them is now continually added to; and there is a strengthening tendency to think of them as everywhere around, and as causing all unusual occurrences.

Differencies among the ascribed powers of ghosts soon arise. They naturally follow from observed differences among the powers of living individuals. Hence it results that while the propitiations of ordinary ghosts are made only by their descendants, it comes occasionally to be thought prudent to propitiate also the ghosts of the more dreaded individuals, even though they have no claims of blood. Quite early there thus begin those grades of supernatural beings which eventually become so strongly marked.

Habitual wars, which more than all other causes initiate these first differentiations, go on to initiate further and more decided ones. For with those compoundings of small societies into greater ones, and re-compounding of these into still greater, which war effects, there, of course, with the multiplying gradations of power among living men, arises the idea of multiplying gradations of power among their ghosts. Thus in course of time are formed the conceptions of the great ghosts or gods, the more numerous secondary ghosts or demi-gods, and so on downwards—a pantheon: there being still, however, no essential distinction of kind; as we see in the calling of ordinary ghosts *manes*-gods by the Romans and *elohim* by the Hebrews. Moreover, repeating as the other life in the other world does, the life in this world, in its needs, occupations, and social organization, there arises not only a differentiation of grades among supernatural beings in respect of their powers, but also in respect of their characters and kinds of activity. There come to be local gods, and gods reigning over this or that order of phenomena; there come to be good and evil spirits of various qualities; and where there has been by conquest a posing of one society upon another, each having its own system of ghost-derived beliefs, there result an involved combination of such beliefs, constituting a mythology.

Of course primitive ghosts being doubles like their originals in all things; and gods (when not the living members of a conquering race) being doubles of the more powerful men; it results that they are primarily conceived as no less human than other ghosts in their physical characters, their passions, and their intelligences. Like the doubles of the ordinary dead, they are supposed to consume the flesh, blood, bread, wine, given to them; at first literally, and later in a more spiritual way by consuming the essences of them. They not only appear as visible and tangible persons, but they enter into conflicts with men, are wounded, suffer pain: the sole distinction being that they have miraculous powers of healing and consequent immortality. Here, indeed, there needs a qualification; for not only do various peoples hold that gods die a first death (as naturally happens where they are members of a conquering race, called gods because of their superiority), but, as in the case of Pan, it is supposed, even among the cultured, that there is a second and final death of a god, like that second and final death of a man supposed among existing savages. With advancing civilization the divergence of the supernatural being from the natural being becomes more decided. There is nothing to check the gradual de-materialization of the ghost and of the god; and this de-materialization is insensibly furthered in the effort to reach consistent ideas of supernatural action: the god ceases to be tangible, and later he ceases to be visible or audible. Along with this differentiation of physical attributes from those of humanity, there goes on more slowly a differentiation of mental attributes. The god of the savage, represented as having intelligence scarcely if at all greater than that of the living man, is deluded with ease. Even the gods of the semi-civilized are deceived, make mistakes, repent of their plans; and only in course of time does there arise the conception of unlimited vision and universal knowledge. The emotional nature simultaneously undergoes a parallel transformation. The grosser passions, originally conspicuous and carefully ministered to by devotees, gradually fade, leaving only the passions less related to corporeal satisfactions; and eventually these, too, become partially de-humanized.

Ascribed characters of deities are continually adapted and re-adapted to the needs of the social state. During the militant phase of activity, the chief god is conceived as holding insubordination the greatest crime, as implacable in anger, as merciless in punishment; and any alleged attributes of milder kinds occupy but small space in the social consciousness. But where militancy declines and the harsh despotic form of government appropriate to it is gradually qualified by the form appropriate to industrialism, the foreground of the religious consciousness is increasingly filled with those ascribed traits of the divine nature which are congruous with the ethics of peace: divine love, divine forgiveness, divine mercy, are now the characteristics enlarged upon.

To perceive clearly the effects of mental progress and changing social life, thus stated in the abstract, we must glance at them in the concrete. If, without foregone conclusions, we contemplate the traditions, records, and monuments, of the Egyptians, we see that out of their primitive ideas of gods, brute or human, there were evolved spiritualized ideas of gods, and finally of a god; until the priesthoods of later times, repudiating the earlier ideas, described them as corruptions: being swayed by the universal tendency to regard the first state as the highest—a tendency traceable down to the theories of existing theologians and mythologists. Again, if, putting aside speculations, and not asking what historical value the *Iliad* may have, we take it simply as indicating the early Greek notion of Zeus, and compare this with the notion contained in the Platonic dialogues; we see that Greek civilization had greatly modified (in the better minds, at least) the purely anthropomorphic conception of him: the lower human attributes being dropped and the higher ones transfigured. Similarly, if we contrast the Hebrew God described in early traditions, man-like in appearance, appetites, and emotions, with the Hebrew God as characterized by the prophets, there is shown a widening range of power along with a nature increasingly remote from that of man. And on passing to the conceptions of him which are now entertained, we are made aware of an extreme transfiguration. By a convenient obliviousness, a deity who in early times is represented as hardening men's hearts so that they may commit punishable acts, and as employing a lying spirit to deceive them, comes to be mostly thought of as an embodiment of virtues transcending the highest we can imagine.

Thus, recognizing the fact that in the primitive human mind there exists neither religious idea nor religious sentiment, we find that in the course of social evolution and the evolution of intelligence accompanying it, there are generated both the ideas and sentiments which we distinguish as religious; and that through a process of causation clearly traceable, they traverse those stages which have brought them, among civilized races, to their present forms.

§ 657. And now what may we infer will be the evolution of religious ideas and sentiments throughout the future? On the one hand, it is irrational to suppose that the changes which have brought the religious consciousness to its present form will suddenly cease. On the other hand, it is irrational to suppose that the religious consciousness, naturally generated as we have seen, will disappear and leave an unfilled gap. Manifestly it must undergo further changes; and however much changed it must continue to exist. What, then, are the transformations to be expected? If we reduce the process above delineated to its lowest terms, we shall see our way to an answer.

As pointed out in *First Principles*, § 96, Evolution is

throughout its course habitually modified by that Dissolution which eventually undoes it: the changes which become manifest being usually but the differential results of opposing tendencies towards integration and disintegration. Rightly to understand the genesis and decay of religious systems, and the probable future of those now existing, we must take this truth into account. During those earlier changes by which there is created a hierarchy of gods, demi-gods, manes-gods, and spirits of various kinds and ranks, Evolution goes on with but little qualification. The consolidated mythology produced, while growing in the mass of supernatural beings composing it, assumes increased heterogeneity along with increased definiteness in the arrangement of its parts and the attributes of its members. But the antagonist Dissolution eventually gains predominance. The spreading recognition of natural causation conflicts with this mythological evolution; and insensibly weakens those of its beliefs which are most at variance with advancing knowledge. Demons and the secondary divinities presiding over divisions of Nature, become less thought of as the phenomena ascribed to them are more commonly observed to follow a constant order; and hence these minor components of the mythology slowly dissolve away. At the same time, with growing supremacy of the great god heading the hierarchy, there goes increasing ascription to him of actions which were before distributed among numerous supernatural beings: there is integration of power. While in proportion as there arises the consequent conception of an omnipotent and omnipresent deity, there is a gradual fading of his alleged human attributes: dissolution begins to affect the supreme personality in respect of ascribed form and nature.

Already, as we have seen, this process has in the more advanced societies, and especially among their higher members, gone to the extent of merging all minor supernatural powers in one supernatural power; and already this one supernatural power has, by what Mr. Fiske aptly calls deanthropomorphization, lost the grosser attributes of humanity. If things hereafter are to follow the same general course as heretofore, we must infer that this dropping of human attributes will continue. Let us ask what positive changes are hence to be expected.

Two factors must unite in producing them. There is the development of those higher sentiments which no longer tolerate the ascription of inferior sentiments to a divinity; and there is the intellectual development which causes dissatisfaction with the crude interpretations previously accepted. Of course in pointing out the effects of these factors, I must name some which are familiar; but it is needful to glance at them along with others.

§ 658. The cruelty of a Fijian god who, represented as devouring the souls of the dead, may be supposed to inflict torture during the process, is small compared with the cruelty of a god who condemns men to tortures which are eternal; and the ascription of this cruelty, though habitual in ecclesiastical formulas, occasionally occurring in sermons, and still sometimes pictorially illustrated, is becoming so intolerable to the better-natured, that while some theologians distinctly deny it, others quietly drop it out of their teachings. Clearly, this change cannot cease until the beliefs in hell and damnation disappear.* Disappearance of them will be aided by an increasing repugnance to injustice. The visiting on Adam's descendants through hundreds of generations, dreadful penalties for a small transgression which they did not commit; the damning of all men who do not avail themselves of an alleged mode of obtaining forgiveness, which most men have never heard of; and the effecting a reconciliation by sacrificing a son who was perfectly innocent, to satisfy the assumed necessity for a propitiatory victim; are modes of action

---

* To meet a possible criticism, it may be well to remark that, whatever force they have against deists (and they have very little), Butler's arguments concerning these and allied beliefs do not tell at all against agnostics.

which, ascribed to a human ruler, would call forth expressions of abhorrence; and the ascription of them to the Ultimate Cause of things, even now felt to be full of difficulties, must become impossible. So, too, must die out the belief that a Power present in innumerable worlds throughout infinite space, and who during millions of years of the Earth's earlier existence needed no honouring by its inhabitants, should be seized with a craving for praise; and having created mankind, should be angry with them if they do not perpetually tell him how great he is. As fast as men escape from that glamour of early impressions which prevents them from thinking, they will refuse to imply a trait of character which is the reverse of worshipful.

Similarly with the logical incongruities more and more conspicuous to growing intelligence. Passing over the familiar difficulties that sundry of the implied divine traits are in contradiction with the divine attributes otherwise ascribed—that a god who repents of what he has done must be lacking either in power or in foresight; that his anger presupposes an occurrence which has been contrary to intention, and so indicates defect of means; we come to the deeper difficulty that such emotions, in common with all emotions, can exist only in a consciousness which is limited. Every emotion has its antecedent ideas, and antecedent ideas are habitually supposed to occur in God: he is represented as seeing and hearing this or the other, and as being emotionally affected thereby. That is to say, the conception of a divinity possessing these traits of character, necessarily continues anthropomorphic; not only in the sense that the emotions ascribed are like those of human beings, but also in the sense that they form parts of a consciousness which, like the human consciousness, is formed of successive states. And such a conception of the divine consciousness is irreconcilable both with the unchangeableness otherwise alleged, and with the omniscience otherwise alleged. For a consciousness constituted of ideas and feelings caused by objects and occurrences, cannot be simultaneously occupied with all objects and all occurrences throughout the universe. To believe in a divine consciousness, men must refrain from thinking what is meant by consciousness—must stop short with verbal propositions; and propositions which they are debarred from rendering into thoughts will more and more fail to satisfy them. Of course like difficulties present themselves when the will of God is spoken of. So long as we refrain from giving a definite meaning to the word will, we may say that it is possessed by the Cause of All Things, as readily as we may say that love of approbation is possessed by a circle; but when from the words we pass to the thoughts they stand for, we find that we can no more unite in consciousness the terms of the one proposition than we can those of the other. Whoever conceives any other will than his own, must do so in terms of his own will, which is the sole will directly known to him: all other wills being only inferred. But will, as each is conscious of it, presupposes a motive—a prompting desire of some kind. Absolute indifference excludes the conception of will. Moreover will, as implying a prompting desire, connotes some end contemplated as one to be achieved, and ceases with the achievement of it: some other will, referring to some other end, taking its place. That is to say, will, like emotion, necessarily supposes a series of states of consciousness. The conception of a divine will, derived from that of the human will, involves like it, localization in space and time. The willing of each end, excludes from consciousness for an interval the willing of other ends; and therefore is inconsistent with that omnipresent activity which simultaneously works out an infinity of ends. It is the same with the ascription of intelligence. Not to dwell on the seriality and limitation implied as before, we may note that intelligence, as alone conceivable by us, presupposes existences independent of it and objective to it. It is carried on in terms of changes primarily wrought by alien activities—the impressions generated by things beyond consciousness, and the ideas derived from such impressions. To speak of an intelligence which exists in the absence of all such alien activities, is to use a meaningless word. If to the corollary that the First Cause, considered as intelligent, must be continually affected by independent objective activities, it is replied that these have become such by act of creation, and were previously included in the First Cause; then the reply is that in such case the First Cause could, before this creation, have had nothing to generate in it such changes as those constituting what we call intelligence, and must therefore have been unintelligent at the time when intelligence was most called for. Hence it is clear that the intelligence ascribed, answers in no respect to that which we know by the name. It is intelligence out of which all the characters constituting it have vanished.

These and other difficulties, some of which are often discussed but never disposed of, must force men hereafter to drop the higher anthropomorphic characters given to the First Cause, as they have long since dropped the lower. The conception which has been enlarging from the beginning must go on enlarging, until, by disappearance of its limits, it becomes a consciousness which transcends the forms of distinct thought, though it for ever remains a consciousness.

§ 659. "But how can such a final consciousness of the Unknowable, thus tacitly alleged to be true, be reached by successive modifications of a conception which was utterly untrue? The ghost-theory of the savage is baseless. The material double of a dead man in which he believes, never had any existence. And if by gradual de-materialization of this double was produced the conception of the supernatural agent in general—if the conception of a deity, formed by the dropping of some human attributes and transfiguration of others, resulted from continuance of this process; is not the developed and purified conception reached by pushing the process to its limit, a fiction also? Surely if the primitive belief was absolutely false, all derived beliefs must be absolutely false."

This objection looks fatal; and it would be fatal were its premiss valid. Unexpected as it will be to most readers, the answer here to be made is that at the outset a germ of truth was contained in the primitive conception—the truth, namely, that the power which manifests itself in consciousness is but a differently-conditioned form of the power which manifests itself beyond consciousness.

Every voluntary act yields to the primitive man, proof of a source of energy within him. Not that he thinks about his internal experiences; but in these experiences this notion lies latent. When producing motion in his limbs, and through them motion in other things, he is aware of the accompanying feeling of effort. And this sense of effort which is the perceived antecedent of changes produced by him, becomes the conceived antecedent of changes not produced by him—furnishes him with a term of thought by which to represent the genesis of these objective changes. At first this idea of muscular forces as anteceding unusual events around him, carries with it the whole assemblage of associated ideas. He thinks of the implied efforts as efforts exercised by beings like himself. In course of time these doubles of the dead, supposed to be workers of all but the most familiar changes, are modified in conception. Besides becoming less grossly material, some of them are developed into larger personalities presiding over classes of phenomena which, being comparatively regular in their order, suggest a belief in beings who, while far more powerful than men, are less variable in their modes of action. So that the idea of force as exercised by such beings, comes to be less associated with the idea of a human ghost. Further advances, by which minor supernatural agents are merged in one general agent, and by which the personality of this general agent is rendered vague while becoming widely extended, tend still further to dissociate the notion of objective force from the force known as such in consciousness; and the dissociation reaches its extreme in the thoughts of the man of science, who interprets in terms of force not only the visible changes of sensible bodies, but all physical changes whatever, even up to the undulations of the ethereal medium. Nevertheless, this force (be it force under that statical form by which matter resists, or under that dynamical form distinguished as energy) is to the last thought of in terms of that internal energy which he is conscious of as muscular effort. He is compelled to symbolize objective force in terms of subjective force from lack of any other symbol.

See now the implications. That internal energy which in the experiences of the primitive man was always the immediate antecedent of changes wrought by him—that energy which, when interpreting external changes, he thought of along with those attributes of a human personality connected with it in himself; is the same energy which, freed from anthropomorphic accompaniments, is now figured as the cause of all external phenomena. The last stage reached is recognition of the truth that force as it exists beyond consciousness, cannot be like what we know as force within consciousness; and that yet, as either is capable of generating the other, they must be different modes of the same. Consequently, the final outcome of that speculation commenced by the primitive man, is that the Power manifested throughout the Universe distinguished as material, is the same Power which in ourselves wells up under the form of consciousness.

It is untrue, then, that the foregoing argument proposes to evolve a true belief from a belief which was wholly false. Contrariwise, the ultimate form of the religious consciousness, is the final development of a consciousness which at the outset contained a germ of truth obscured by multitudinous errors.

§ 660. Those who think that science is dissipating religious beliefs and sentiments, seem unaware that whatever of mystery is taken from the old interpretation is added to the new. Or rather, we may say that transference from the one to the other is accompanied by increase; since, for an explanation which has a seeming feasibility, science substitutes an explanation which, carrying us back only a certain distance, there leaves us in presence of the avowedly inexplicable.

Under one of its aspects scientific progress is a gradual transfiguration of Nature. Where ordinary perception saw perfect simplicity it reveals great complexity; where there seemed absolute inertness it discloses intense activity; and in what appears mere vacancy it finds a marvellous play of forces. Each generation of physicists discovers in so-called "brute matter," powers which but a few years before, the most instructed physicists would have thought incredible; as instance the ability of a mere iron plate to take up the complicated aerial vibrations produced by articulate speech, which, translated into multitudinous and varied electric pulses, are re-translated a thousand miles off by another iron plate and again heard as articulate speech. When the explorer of Nature sees that quiescent as they appear, surrounding solid bodies are thus sensitive to forces which are infinitesimal in their amounts—when the spectroscope proves to him that molecules on the Earth pulsate in harmony with molecules in the stars—when there is forced on him the inference that every point in space thrills with an infinity of vibrations passing through it in all directions; the conception to which he tends is much less that of a Universe of dead matter than that of a Universe everywhere alive: alive if not in the restricted sense, still in a general sense.

This transfiguration which the inquiries of physicists continually increase, is aided by that other transfiguration resulting from metaphysical inquiries. Subjective analysis compels us to admit that our scientific interpretations of the phenomena which objects present, are expressed in terms of our own variously-combined sensations and ideas—are expressed, that is, in elements belonging to consciousness, which are but symbols of the something beyond consciousness. Though analysis afterwards reinstates our primitive beliefs, to the extent of showing that behind every group of phenomenal manifestations there is always a *nexus*, which is the reality that remains fixed amid appearances which are variable; yet we are shown that this *nexus* of reality is for ever inaccessible to consciousness. And when, once more, we remember that the activities constituting consciousness, being rigorously bounded, cannot bring in among themselves the activities beyond the bounds, which therefore seem unconscious, though production of either by the other seems to imply that they are of the same essential nature; this necessity we are under to think of the external energy in terms of the internal energy, gives rather a spiritualistic than a materialistic aspect to the Universe: further thought, however, obliging us to recognize the truth that a conception given in phenomenal manifestations of this ultimate energy can in no wise show us what it is.

While the beliefs to which analytic science thus leads, are such as do not destroy the object-matter of religion, but simply transfigure it, science under its concrete forms enlarges the sphere for religious sentiment. From the very beginning the progress of knowledge has been accompanied by an increasing capacity for wonder. Among savages, the lowest are the least surprised when shown remarkable products of civilized art: astonishing the traveller by their indifference. And so little of the marvellous do they perceive in the grandest phenomena of Nature, that any inquiries concerning them they regard as childish trifling. This contrast in mental attitude between the lowest human beings and the higher human beings around us, is paralleled by contrasts among the grades of these higher human beings themselves. It is not the rustic, nor the artizan, nor the trader, who sees something more than a mere matter of course in the hatching of a chick; but it is the biologist, who, pushing to the uttermost his analysis of vital phenomena, reaches his greatest perplexity when a speck of protoplasm under the microscope shows him life in its simplest form, and makes him feel that however he formulates its processes the actual play of forces remains unimaginable. Neither in the ordinary tourist nor in the deer-stalker climbing the mountains above him, does a highland glen rouse ideas beyond those of sport or of the picturesque; but it may, and often does, in the geologist. He, observing that the glacier-rounded rock he sits on has lost by weathering but half an inch of its surface since a time far more remote than the beginnings of human civilization, and then trying to conceive the slow denudation which has cut out the whole valley, has thoughts of time and of power to which they are strangers—thoughts which, already utterly inadequate to their objects, he feels to be still more futile on noting the contorted beds of gneiss around, which tell him of a time, immeasurably more remote, when far beneath the Earth's surface they were in a half-melted state, and again tell him of a time, immensely exceeding this in remoteness, when their components were sand and mud on the shores of an ancient sea. Nor is it in the primitive peoples who supposed that the heavens rested on the mountain tops, any more than in the modern inheritors of their cosmogony who repeat that "the heavens declare the glory of God," that we find the largest conceptions of the Universe or the greatest amount of wonder excited by contemplation of it. Rather, it is in the astronomer, who sees in the Sun a mass so vast that even into one of his spots our Earth might be plunged without touching its edges; and who by every finer telescope is shown an increased multitude of such suns, many of them far larger.

Hereafter as heretofore, higher faculty and deeper insight will raise rather than lower this sentiment. At present the most powerful and most instructed mind has neither the knowledge nor the capacity required for symbolizing in thought the totality of things. Occupied with one or other division of Nature, the man of science usually does not know enough of the other divisions even rudely to conceive the extent and complexity of their phenomena; and supposing him to have adequate knowledge of each, yet he is unable to think of them as a whole. Wider and stronger intellect may hereafter help him to form a vague consciousness of them in their totality. We may say that just as an undeveloped musical faculty, able only to appreciate a simple melody, cannot grasp the variously-entangled passages and harmonies of a symphony, which in the minds of composer and conductor are unified into involved musical effects awakening far greater feeling than is possible to the musically uncultured; so, by future more evolved intelligences, the course of things now apprehensible only in parts may be apprehensible all together, with an accompanying feeling as much beyond that of the present cultured man, as his feeling is beyond that of the savage.

And this feeling is not likely to be decreased but to be increased by that analysis of knowledge which, while forcing him to agnosticism, yet continually prompts him to imagine some solution of the Great Enigma which he knows cannot be solved. Especially must this be so when he remembers that the very notions, origin, cause and purpose, are relative notions belonging to human thought, which are probably irrelevant to the Ultimate Reality transcending human thought; and when, though suspecting that explanation is a word without meaning when applied to this Ultimate Reality, he yet feels compelled to think there must be an explanation.

But one truth must grow ever clearer—the truth that there is an Inscrutable Existence everywhere manifested, to which he can neither find nor conceive either beginning or end. Amid the mysteries which become the more mysterious the more they are thought about, there will remain the one absolute certainty, that he is ever in presence of an Infinite and Eternal Energy, from which all things proceed.

REFERENCES.

To find the authority for any statement in the text, the reader is to proceed as follows:—Observing the number of the section in which the statement occurs, he will first look out, in the following pages, the corresponding number, which is printed in conspicuous type. Among the references succeeding this number, he will then look for the name of the tribe, people, or nation concerning which the statement is made (the names in the references standing in the same order as that which they have in the text); and that it may more readily catch the eye, each such name is printed in Italics. In the parenthesis following the name, will be found the volume and page of the work referred to, preceded by the first three or four letters of the author's name; and where more than one of his works has been used, the first three or four letters of the title of the one containing the particular statement. The meanings of these abbreviations, employed to save the space that would be occupied by frequent repetitions of full titles, is shown at the end of the references; where will be found arranged in alphabetical order, these initial syllables of authors' names, &c., and opposite to them the full titles of the works referred to.

CEREMONIAL INSTITUTIONS.

§ 343. *Australians* (Mitch. ii, 68; i, 87; Ang. i, 59)—*Tasmanians* (Bon. pp. 3, 37, 226)—*Esquimaux* (ref. lost)—*Comanches* (Banc. i, 519)—*Araucanians* (Smith, 196)—*Bedouins* (Bur. —)—*Arabs* (Lyon, 53)—*Balonda* (Liv. 296)—*Malagasy* (Ell. "Hist." i, 258)—*Samoans* (Tur. 289). § 344. *Chinese* (Will. ii, 69)—*Tahitians* (Ell. "Pol. Res." i, —; ii, 369)—*Tongans* (Mar. ii, 78, 100)—*Ancient Mexicans* (Dur. i, ch. 26)—*Peru* (Gar. bk. ii, ch. 12)—*Japanese* (Alc. i, 63)—*England* (Whar. 469)—*Tahitians* (Ell. "Pol. Res." ii, 216)—*Sandwich Islanders* (Ell. "Hawaii," 393-4)—*Nicaraguans* (Ovi. bk. xlii, ch. 2 & 3)—*Peruvians* (Acos. bk. v, ch. 25)—*Hebrews* (Kue. i, 292-3)—*Mediæval Europe* (ref. lost). § 345. *Tongans* (Mar. i, 146, note)—*Fijians* (Wil. i, 233)—*Siamese* (La Loub. i, 353)—*Chinese* (Will. i, 313)—*Japanese* (Stein. —). § 346. *Mongol* (Timk. i, 196) — *Philippines* (Jag. 161) — *Chittagong Hill Tribes* (Lew. 118) — *Burmese* (Fyt. ii, 69) — *Samoans* (Tur. 346) — *Esquimaux* (Beech. i, 242)—*New Zealanders* (Cook, "Last Voy." 49)—*Snake Indians* (Lew. & Cl. 266)—*Comanches* (Marcy, 29)—*Fuegians* (Eth. S. "Trans." i, 263)—*Loango* (Pink. Voy. xvi, 331)—*Batoka* (Liv. 551)—*Balonda* (Liv. 276)—*Loango* (Ast. iii, 228) — *Fuegians* (U. S. Ex. i, 127) — *Fiji* (Wil. i, 37) — *Australians* (Mitch. i, 87) — *New Zealanders* (Ang. ii, 32-75) — *Central*

South Africa (Liv. —)—*Shoshones* (Banc. i, 438)—*Australians* (Ang. i, 59)—*Vaté* (Ersk. 334)—*Samoan* (Tur. 194)—*Africa* (Liv. —)—*Peruvians* (Cie. 168) — *Egyptians* (Wilk. plates) — *Moslem* (Klun. 106) — *Tahitians* (Hawk. i. 417)— *Kaffirs* (Bar. i, 175)—*Tasmanians* (West, ii, 7)—*Arabs* (Bnk. 86)—*Kamschadales* (Krash. 212-3). § 347. *Patagonians* (Falk. 121)—*Madagascar* (Ell. "Hist." ii, 258)—*Samoans* (Tur. 348) — *Fijians* (Ersk. 254) — *Ashantees* (Dup. 43) — *Yorubas* (Lan. i, 125) — *Madagascar* (ref. lost)—*China* (Staun. 345) — *Chibchas* (Sim. 267) — *Samoa* (Tur. 314) — *Madagascar* (Ell. " Visits," 127) — *Japanese* (Stein. —) — *Chinese* (Mil. 94)—*Rome* (Beck. 213)—*Assyrians* (Raw. i, 503-4)—*Mexico* (Her. iii, 203; Torq. bk. ix, ch. 20)—*Nicaragua* (Squ. ii, 346)—*Peru* (Piz. 225; Xer. 48)— *Chibchas* (Pied. bk. i, ch. 5)—*Uganda* (Speke, 294—*Dahomey* (Bur. i, 244) — *Abyssinians* (Duf. 71 ; Bru. iv, 454, 417) — *New Zealand* (Thom. i, 111) —*Egypt* (Eb. i, 352)—*China* (Huc, "Trav." ii, 261; Gutz. ii, 311; Will. i, 331-2 ; ii, 68-9)—*Japanese* (Dick. 79 ; Mit. ii, 43)—*Chivalry* (Scott, 3-4) —*France* (Leb. vol. xiii, passim ; Cher. 536-7) — *England* (Nob. passim)— *Peru* (Acos. bk. v, ch. 6)—*Madagascar* (Ell. " Hist." i, 356)—*England* (Nob. 46 & passim)—*France* (Leb. vol. xiii, passim) — *England* (Nob. 315-6). § 349. *Vaté* ('Tur. 393)—*Shoshones* (Banc. i, 438)—*Mishmis* (Coop. 190)— *Santals* (As. S. B. xx, 582)—*Koossas* (Lich. i, 288)—*Ashantee* (Beech. 211) —*Ceris and Opatas* (Banc. i, 581)—*Chichimecs* (Banc. i, 629). § 350. *Hebrews* (Judges vii, 25 ; 1 Samuel xvii, 54) — *Chichimecs* (Banc. i, 629)— *Abipones* (Dob. ii, 408)—*Mundrucus* (Hen. 475)—*New Zealanders* (Thom. i, 130) — *Congo* (Tuck. 101)—*Ashantee* (Dup. 227)—*Persia* (Mor. 186)—*Timour* (Gib. ch. lxv)—*Dahomey* (Bur. i, 218 ; Dal. 76)—*Northern Celebes* (ref. lost) —*Dyaks* (Boyle, 170-1)—*Kukis* (As. S. B. ix, 837)—*Borneo* (St. John, ii, 27). § 351. *Ashantee* (Ram. 130)—*Tahitians* (Hawk. ii, 161)—*Vaté* (Tur. 393) — *Boigu* (Roy. G. S. xx, 96) — *Tupis* (South. i, 222) — *Caribs* (Ed. i, 35) — *Moxos* (Hutch. 31)—*Central Americans* (Fan. 315)—*Poland* (Gib. ch. lxiv) — *Constantine* (Gib. ch. xlviii) — *Montenegro* (*The Times*, Dec. 14, 1876). § 352. *Mexicans* (Nouv. xcix, 134 ; Saha. bk. ix, c. 15)—*Yucatan* (Her. iv, 174) — *Abipones* (Dob. ii, 408) — *Shoshones* (Lew. & Cl. 309) — *Nagas* (As. S. B. ix, 959)—*Mandans* (Cat. i, 136)—*Cochimis* (Banc. i, 567). § 353. *Mexicans* (Banc. i, 581)—*Californians* (Banc. i, 380) — *Khonds* (Macph. 57) —*Egyptians* (Dun. i, 131)—*Abyssinians* (Bru. vi, 116-17; Heri. 188-9)— *Hebrews* (1 Sam. xviii, 25, 27). § 354. *Osages* (Tylor, " Prim. Cult." i, 416)—*Ojibways* (Hind, ii, 123). § 355. *Gauls* (Lehuërou, 371; Par. 320,658)—*Jews* (2 Maccabees xv, 30 ; 2 Sam. iv, 12). § 356. *Gauls* (Diod. i, 315) — *Timour* (Gib. ch. lxv) — *Khonds* (Macph. 57) — *Tahitians* (Ell. " Pol. Res." i, 488)—*Philistines* (1 Sam. xxxi, 10)—*Greeks* (Pot. ii, 109-10) —*Fijians* (U.S. Ex. iii, ch. lxxx)—*Flemish* (Chór. 358)—*French* (Leb. vi, 127). § 357. *Scotland* (Burt. i, 398)—*Khonds* (Macph. 46)—*Athenians* (Grote, iii, 382)—*Fiji* (Wil. i, 31)—*Panthay* (Baber)—*Fiji* (Ersk. 454)— *Shoshones* (Banc. i, 433) — *Chichimecs* (Banc. i, 629) — *Hebrews* (1 Sam. xi, 1-2)—*Bulgarians* (Gib. ch. lv). § 358. *Araucanians* (Thomp. G. i, 406)— *Bactrians* (Dun. i, 174)—*Hebrews* (Judges i, 6-7) — *Fiji* (Wil. i, 30, 198, 177) — *Charruas* (Hutch. 48 *et seq.*) — *Mandans* (ref. lost) — *Tonga* (Mar. ii, 210-11)—*Australians* (Mitch. ii, 345)—*Hottentot* (Pink. Voy. xvi, 141) — *Egypt* (Wilk. i, 307) — *Japanese* (Busk, 241). § 359. *Central Americans* (Her. iv, 136) —*Ashantees* (Ram. 216)—*Anc. Mexico* (Clav. bk. vii, c. 17)—*Hondouras* (Her. iv, 140)—*Miztecs* (Her. iii, 262-3)—*Zapotecas* (Her. iii, 269) — *Hebrews* (Knobel, 226-7) — *Burmese* (Sang. 124) — *Gond* (Fors. 164)—*Astrachan* (Bell, i, 43)—*Hebrews* (2 Kings, xix). § 360. *Sandwich Islands* (Ell. " Hawaii," 165-6 ; Ell. W. ii, 69)—*Australians* (Ang. ii, 217 ; Hay. 103-4)—*Anc. Peruvians* (Cie. 177, 181). § 361. *Britain* (Cox and Jones, 88)—*Kalmucks* (Pal. —)—*Chinese* (Will. ii, 224)—*Greeks and Romans* (Smith, W. *s.v.* " Coma ") — *Nootkas* (Banc. i, 195) — *Caribs* (Ed. i, 4.)—*Nicaragua* (Her. iii, 298)—*Central Americans* (Cog. bk. iv, ch. 4)—

REFERENCES.

Ancient Mexicans (Zur. 111) — *Chibchas* (Pied. bk. i, ch. 2) — *Itzaex* (Fan. 313)—*Ottomans* (Pax. iv, 87)—*Greeks* (Beck. 453-55)—*Franks* (Guer. "Polyp." i, 300; Bouq. ii, 49; Greg. bk. iii, ch. 18) —*Japanese* (Busk, 144) — *Samoans* (Tur. 205-6) — *New Caledonians* (Eth. S. "Jour." iii, 56)— *Europe* (Duc. 379) — *Cloris and Alaric* (Duc. 383) — *Dacotahs* (Lew. & Cl. 64) — *Caribs* (Ed. i, 42) — *Hebrews* (Leviticus xxi, 5; Jer. xvi, 6) — *Greeks and Romans* (Smith, W. *s.v.* "Coma") — *Greeks* (Pot. ii, 198-9; Soph. 47; Beck. 398; Smith, W. *s.v.* "Coma") — *Romans* (ref. lost). — *Hebrews* (Jer. xli, 5) —*Arabians* (Krehl, 32-3) — *Ancient Peru* (Acosta, bk. v, ch. 5)—*Tahitians* (Hawk. i, 468)—*France* (Guizot "Col." —).
§ 362. *Spoleto* (Gib. —) — *Phrygian* (Dun. i, 531) — *Mexicans* (Brin. 147)—*Hottentots* (Kol. i, 112)—*Phœnicians* (Mov. i, 362)—*San Salvador* (Squ. "Coll." 87)—*Moses* (Exod. iv, 24-26)—*Antiochus* (1 Macc. i, 48-60)— *Mattathias* (1 Macc. ii, 45-6)—*Hyrcanus* (Jos. i, 525)—*Aristobulus* (Jos. i, 532)—*Tongans* (Mar. ii, 79)—*Berbers* (Rohlfs, 45). **§ 363.** *Kaffirs* (Gard. 264)—*Jews* (Jerem. xli, 5)—*Samoans* (Tur. 187)—*Central Americans* (Mart. 338). **§ 364.** *Huns* (Jor. 215)—*Turks* (Pell. i, 158, note)— *Lacedæmonians* (Pot. ii, 204) — *Hebrews* (Levit. xix, 28 — *Scandinavians* (Heim. i, 224, 225) —*Andamans* (Eth. S. "Trans." ii, 36) —*Abeokuta* (Bur. i, 104) — *Cuebas* (Banc. i, 753) — *Peruvians* (Cic. 311) — *Sandwich Islanders* (Ell. W. ii, 152) — *Darian Indians* (Banc. i, 771) — *Sandwich Islanders* (Ell. "Hawaii," 166)—*Eastern* (reference lost)—*Hebrews* (Deut. xxxii, 5; Rev. vii, 2-3; xiv, 1, 9, 10)—*Arabs* (Thomson, i, 91)—*Christians* (Kal. ii, 429-30) — *Mexico* (Torq. bk. ix, ch. 31) — *Angola* (Bast. 76) — *Tongans* (Mar. ii, 268). **§ 365.** *Bechuanas* (Lich. ii, 331) — *Damaras* (And. 224)—*Congo* (Tuck. 80)—*Itzaex* (Fan. 313)—*Abipones* (Dob. ii, 35). **§ 368.** *Ancient Peruvians* (Gar. bk. ii, ch. 4). **§ 369.** *Mexico* (Torq. bk. xiv, ch. 9) — *Chibchas* (Sim. 251) — *Yucatan* (Landa, § xx) — *Tahitians* (Forst. 370) — *Fiji* (Wil. i, 28) — *Tahiti* (Ell. "Pol. Res." i, 319) — *Fiji* (Ersk. —)—*Malagasy* (Drur. 220). **§ 370.** *Timbuctoo* (Cail. ii, 53)— *Kaffirs* (Lich. i, 287, 271) — *Vera Paz* (Torq. bk. xi, ch. 19) — *Chibchas* (Pied. bk. i, ch. 5) — *Mexicans* (Tern. x, 404) — *Peru* (Cuz. 91) — *Hebrews* (2 Chron. ix, 23-4; 1 Sam. x, 27) — *Japan* (Dick. 325; Kæm. 49) — *China* (Chin. Rep. iii, 110-11)—*Burmah* (Yule, 76)—*Merovingians* (Bouq. ii, 617) —*England* (Rob. 20). **§ 371.** *Persia* (Mal. ii, 477-8)—*Tonga* (Mar. i, 232, note)—*Mexicans* (Dur. i, ch. 25; Tern. xvi, 288-9.—*Montezuma* (Gal. 117; Tern. x, 405) — *Merovingians and Carolingians* (Wai. ii, 557; iv, 91-5-8; Guer. "St. Père," introd.; Leber, vii, —; Guer. "St. Père," introd.)—*English* (Stubbs, i, 278). **§ 372.** *Chibchas* (Pied. bk. ii, ch. 4)— *Sumatra* (Mars. 211)—*Jummoo* (Drew "Jum." 15)—*Anglo-Saxons* (Broom, 27)—*Normans* (Moz. *s.v.* "Orig. Writ.;" Black iii, 279)—*Kirghis* (ref. lost)—*France* (Guizot, "Hist." iii, 260; Cher. *s.v.* "Epices")—*English* (Rob. 1; Stubbs, i, 384) - *Spain* (Rose, i, 79)—*Bechuanas* (Burch. i, 544)— *Dahomey* (For. i, 34)—*East* (Van Len. ii, 592). **§ 373.** *Congo* (Tuck. 116) — *Tonquin* (Tav. description of plates)—*New Caledonians* (Tur. 88) — *Veddah* (Eth. S. "Trans." ii, 301)—*Dyaks* (Brooke, ii, 73)—*Greeks* (Guhl, 283) — *Zulu* (Gard. 96) — *Hebrews* (Levit. i) — *Greeks* (Pot. i, 239) — *Hebrews* (1 Sam. xxi, 6) — *England* (Hook, 541). **§ 374.** *Ancient Mexico* (Saha. bk. iii, ch. 1, § 3-4) — *Kukis* (As. S. B. xxiv, 630) — *Battas* (Mars. 386)—*Bustars* (His. 17) - *Dahomey* (Bur. ii, 153; For. i, 174)— *Ashantees* (Beech. 189) — *Tahitians* (Ell. "Pol. Res." ii, 271) — *Central America* (Ovi. bk. xlii, ch. 2 and 3) — *Greeks* (Pot. i, 172, 247) — *Early Christians* (Hook, 540-1) — *Mediæval* (Guer. "N. Dame," i, p. xiv). **§ 375.** *China* (Staun. 351)—*Kukis* (But. 94)—*Dahomey* (For. ii, 243)—*Germans* (Tac. xiv)—*French* (Duc. 96; Mons. bk. i, ch. 59). **§ 376.** *Australians* (Hawk. iii, 634)—*Ostyaks* (Bell, ii, 189)—*Julifunda* (Park, —)—*North American Indians* (Cat. i, 223, note)—*Yucatanese* (Landa, § xxiii)—*Japanese* (Mit. i, 112, 142)—*Himalayas* (Mark. 108)—*Bootan* (Turn. 223, 72)—*Rome*

CEREMONIAL INSTITUTIONS.

(Cor. 14–15)—*France* (Du M. 115). § 379. *Joloffs* (Mol. 31)—*Kaffirs* (Shoot. 99)—*Ancient Peruvians* (Cic. 262; X r. 68)—*Mexico* (Tern. xvi, 333–4)—*Ashantee* (Beech. 94–6)—*Dahomey* (Bur. i, 296))—*Madagascar* (Ell. "Visits," 127)—*Siam* (Bowr. ii, 108)—*Mogul* (Tav. ii, 67)—*Jummoo* (Drew, "North. Bar." 47)—*Japan* (Kæm. 49, 66, 11)—*France* (Tocq. 225). § 380. *Spain* (Rose i, 119)—*Japan* (Kæm. 51; 46). § 381. *Wahhabees* (Pal. ii, 110) — *Persia* (Tav. bk. v, ch. xiv, 235) — *Africa* (Grant, 48) — *French* (Rules, 150). § 383. *Shoshones* (Lew. & Cl. 265) — *Batoka* (Liv. 551)—*Tonga* (Forst. 361) — *Africa* (Laird i, 192)—*Peru* (Gar. bk. iii, ch. 2; Markham 94). § 384. *Chibcha* (Sim. 264)—*Borghoo* (Lan. ii, 183) —*Asia* (Camp. 147; Bowr. ii, 270)—*Polynesia* (Cook, "Last Voy." 304)— *Jews* (2 Sam. ix, 6) — *Bithynia* (Mon. —) — *Bootan* (Turn. 80) — *Coast Negroes* (Bos. 317)—*Brass* (Laird i, 97)— *Congo* (Tuck. 125) — *Niger* (All. & T. i, 392) — *Russia* (ref. lost) — *China* (Will. ii, 68-9) — *Hebrews* (Gen. xxxiii, 3; xvii. 17; Dan. ii, 46; iii, 6)—*Mongols* (Pall. —)—*Japanese* (Kæm. 50). § 385. *Dahomey* (Bur. i, 261) — *Mexicans* (Dur. i, 207) - - *New Caledonians* (Ersk. 356)—*Dahomey* (Bur. i, 262)—*Siam* (Bowr. i, 128)—*Cambodia* (Bowr. ii, 31)—*Zulu* (Gard. 203)—*Loango* (Ast. iii, 221) — *Dahomey* (Bur. i, 250; ii, 45) — *Japan* (Dick. 30) — *China* (Pink. vii, 238) — *Europe* (Ste. Pal. ii, 197-8) — *Japanese* (Chin. Rep. iii, 200) — *China* (Will. ii, 68)—*Soosoos* (Wint. i, 123) — *Samoa* (Tur. 332) — *Ancient Mexicans* (Nouv. xcviii, 20) — *Chinese* (Will. ii, 68) — *Congoese* (Bast. 143). § 386. *Loango* (Ast. iii, 228) — *Uganda* (Speke. 331)—*Balonda* (Liv. 296) —*Karague* (Grant, 140)—*Fiji* (Wil. i, 35-6)—*Eboe* (Laird i, 388)—*Ancient Mexicans* (Diaz, ch. 71) —*Abyssinians* (Har. iii, 170) — *Malagasy* (Drur. 67-8)—*Ancient Peru* (Xer. 68) — *Persia* (Por. i, 464) — *Tonga* (Mar. i, 227 note)—*Arabian* (Pax. iv,43)—*Orientals* (ref. lost)—*Mexico* (Clav. bk. vi, ch. 8) —*Peru* (Acos. bk. v, ch. 4; Gar. bk. ii, ch. 8)—*Greeks* (Smith, W. *s.v.* "Saltatio")—*Pepin* (Bouq. v. 433). § 387. *Africa* (Bur. "Dah." i, 259-60; All. & T. i, 345; Liv. 276, 296; All. & T. i, 392)—*Jews* (Jos. ii, 287)—*Turkey* (White ii, 239; i,232)—*Jews* (1 Kings xx, 32; Josh. vii,6). § 388. *Uganda* (Grant, 228)—*Chinese* (Doo. i, 121)—*Mongol* (Huc, "Chin. Emp." i, 54)— *Malagasy* (Drur. 78)—*Siamese* (La Loub. i, 179) — *Unyanyembe* (Grant, 52)—*Sumatra* (Mars. 281)—*Greeks* (ref. lost)—*Siamese* (Bowr. i, 128)— *China* (Will. ii, 68). § 389. *Fijians* (Ersk. 297)—*Otaheitans* (Hawk. ii, 84)—*Soudan* (Tylor, "Early Hist." 50) — *Uganda* (Speke, 374)—*Abyssinia* (Har. iii, 171) — *Tahitians* (Ell. "Pol. Res." ii, 352; Forst. 361) — *Gold Coast* (Cruic. ii, 282; ref. lost)—*Spain* (Ford, "Gatherings," 249)—*Dahomey* (Bur. i, 49) — *Gold Coast* (Cruic. ii, 282) — *Ancient America* (Anda. 58; Tern. —)—*Burmah* (Yule, 79)—*Persia* (Mor. 241)—*Ancient Mexico* (Diaz, ch. 91) — *Peru* (Anda. 58) — *Dahomey* (Dal. p. vii) — *France* (Com. bk. ii, ch. 3; St. Sim. xi, 378) — *Hebrews* (Isa. xxxii, 11) — *East* (Pax. iv, 136) — *Peru* (Gar. bk. vi, ch. 21)—*Damaras* (And. 231) — *Turks* (White ii, 96). § 390. *Toorkee* (Grant, 333) — *Slave Coast* (Bos. 318) — *China* (Gray, i, 211)—*Mosquitos* (Banc. i, 741)—*Arabs* (Mal. —; Nieb. ii, 247). § 391. *Kamschadales* (Krash. 177)—*Uganda* (Grant, 228). § 392. *Poles* (Spen. i, 156-7)—*Turkish* (White ii, 303)—*Siam* (Bowr. i, 127; La Loub. ii, 178)—*Russia* (ref. lost). § 393. *Tupis* (Stade, 151, 59)—*Africa* (Mol. 288)—*Sandwich Is.* (Ell. "Hawaii," 385)—*France* (La Sale, 196)— *Spain* (Ford, "Handbook," p. lxi). § 394. *France* (Cher. ii, 1131)— *Hebrews* (2 Sam. xiv, 22; Isaiah xlviii, 20; 2 Kings xvi, 7) — *Europe* (Duc. 393) — *Samoan* (Tur. 348). § 395. *Egypt* (ref. lost) — *Siam* (Bowr. i, 127) — *Turkey* (White ii, 52) —*Bulgarians* (*Times*, 12 Dec. 1876) —*French* (Sully —) — *Delhi* (Tav. ii, 84-5) — *Russia* ref. lost) — *France* (ref. lost)—*Chinese* (Gray i, 211) — *India* (Pax. ii, 74) — *Persians* (Tav. bk. v, ch. iii, 205). § 396. *Snakes* (Lew. & Cl. 266) — *Araucanians* (Smith, 195-6)—*Arabs* (Lyon, 53)—*Chinese* (Du H. ii, 185)—*France* (Mon. —). § 397. *Abipones* (Dob. ii, 204)—*Samoa* (Ersk. 107)—*Javans* (Raf. i, 366)—

Mexican (**Gal.** 28)—*Kaffirs* (Shoot. 221)—*Samoa* (Ersk. 44)—*Siam* (Bowr. i, 276) — *China* (Chin. Rep. iv, 157) — *Siam* (Bowr. i, 127-9) — *Chinese* (Du H. ii, 177)—*Siamese* (La Loub. i, 166-7)—*Japanese* (Stein. 299-300)— *Germany* (Ger. 124; May. i, 395) — *France* (Chal. ii. 31) — *Samoa* (Tur. 310). § 398. *Dacotahs* (ref. lost)—*Veddahs* (Eth. S. "Trans." ii, 298) —*China* (Chin. Rep. iv, 157). § 400. *Tupis* (South. i, 222; Stade, 145)— *Creeks* (ref. lost) — *Nicaragua* (Ovi. bk. xlii, ch. 1) — *Fiji* (Wil. i, 55) — *Mexico* (Dur. i, 102-3) — *Fiji* (ref. lost). § 401. *Tupis* (South. i, 239) — *Guatemala* (Xim. 163, etc.) — *Dahomey* (Bur. ii, 407)— *Usambara* (Krapf, 395)—*Zulu* (Gard. 91; Shoot. 290)— *Kaffir* (Shoot. 99) — *Samoa* (Ersk. 44) — *Mexicans* (Her. iii, 201) — *Chibchas* (Her. v, 86)— *Peruvians* (Gar. bk. iii, ch. 8) — *Burmah* (Daily News, 24 Mar. 1879). § 402. *Todas* (ref. lost) — *Tartars* (Pink. vii, 591) — *Madagascar* (Ell. "Hist." i, 261)—*Dahomey* (Bur. i, 262) — *Ancient Mexicans* (Mot. 81) — *Kasias* (As. S. B. xiii, 620). § 403. *China and Japan* (Alc. ii, 343)—*Zulus* (ref. lost)—*Nicaraguans* (Squ. ii, 357-8)—*Dahomey* (Bur. i, 273) — *Asia* (Tav. ii, 24) — *Zulus* (Gard. 91) — *Japanese* (Mit. i, 202) — *Siam* (Bowr. i, 275) — *China* (Hue, i, 268) — *Siam* (Pink. ix, 86) — *Russia* (Wahl, 35) — *Dyaks* (St. John ii, 103)—*Kasias* (As. S. B. xiii, 620) — *Bechuana* (Thomp. i, 174). § 404. *Teutonic* (Mul. ii, 280). § 405. *King* (Mul. ii, 281) — *Abyssinia* (Bru. iv, 452) — *France* (Chér. 66-7)—*Merovingian* (Mich. i, 174, note). § 406. *Samoa* (Tur. 281) —*Siam* (Pink. ix, 584; La Loub. i, 237)—*Chinese* (Will. ii, 71; i, 521)— *Rome* (Mom. ii, 368-9)—*Mecklenburgh* (Spen. i, 44)—*Spain* (Ford "Handbook," p. lxi). § 407. *Dahomey* (Bur. i, 52)—*Burman* (Yule, 194)— *China* (Will. i, 317)—*Europe* (Ger. 91)—*Russia* (Salu, 252). § 408. *Ukumi* (Grant, 92) — *Zulus* (ref. lost) — *Uganda* (Speke, 290) — *Chichimecs* (Church. iv, 513) — *Yucatanese* (Landa, § xxix). § 409. *Japan* (Busk, 21)—*Madagascar* (Ell. "Visits," —)—*Uganda* (Speke, 375)—*Japan* (Dick. 49) — *Hebrews* (Ew. iii, 73) — *Zeus* (Pau. bk. ix, c. 40) — *Franks* (Wai. ii, 130; Greg. bk. vii, ch. 33; Leb. xiii, 259-65) — *Araucanians* (ref. lost) — *Uganda* (Speke, 429) — *France* (ref. lost). § 410. *Peruvians* (Gar. bk. vii, ch. 6; Markham, 54, note) — *Sandwich Is.* (Ell. "Hawaii," 142) — *Fijians* (U. S. Ex. iii, 79) — *Chibchas* (Sim. 269) — *Mexicans* (Clav. bk. vii, chs. 22 & 24). § 411. *Thlinkeets* (Banc. i, 109)—*China* (Du H. i, 278). § 412. *Africa* (ref. lost; Heug. 92-3) — *Greeks* (Guhl, 232)—*Sandwich Is.* (Hawk. ii, 192)—*Tonga* (Hawk. —) —*Fundah* (Laird i, 202)—*Arabs* (Pal. —)—*Gaul* (Quich. 25-31; 57-66)— *Rome* (Guhl, 485)—*Madagascar* (Ell. "Hist." i, 279)—*Siam* (La Loub. i, 75)—*Mongol* (Bell i, 344)—*France* (Le Grand, ii, 184; — ref. lost)— *China* (Staun. 244)—*Japan* (Krœm. 43). § 413. *Guatemala* (Ath. p. 1537)—*Chibchas* (Ur. 24-5)— *Cimbri* (Tac. 15)—*Ashantee* (Dup. 71)— *Malagasy* (Ell. "Hist." i, 284)—*Dakotas* (Lew. & Cl. 44)—*Kukis* (As. S. B. xxiv, 646)—*Dyaks* (Boyle, 95)—*New Zealand* (Thom. i, 164)—*Mandans* (Cat. i, 101)—*Nagas* (As. S. B. viii, 464)—*Hottentots* (Kol. i, 198)—*Snakes* (Lew. & Cl. 315)—*Congo* (Tuck. 362)—*Chibchas* (Acos. 219; Sim. 253)— *Peru* (Gar. bk. iv, ch. 11)—*France* (ref. lost)—*New Zealanders* (Hawk. iii, 457)—*Astrachan* (Bell. i, 43). § 414. *Rome* (Mom. ii, 335, n.; Guhl, 497-8)— *France* (ref. lost). § 415. *Tahitians* (Ell. "Pol. Res." ii, 354)—*Rome* (Mom. i, 72)—*Mexicans* (Torq. bk. xiv, ch. 4)—*Peru* (Gar. bk. i, ch. 213)—*Rome* (Guhl, 479)—*Russia* (Cust. —; Wag. ii, 21)—*Germany* (Spen. ii, 176). § 416. *Lombock* (Wal. i, 344)—*Burma* (Yule, 163)— *Siam* (Bowr. i, 125)—*Davotahs* (School. iv, 69)—*Abipones* (Dob. ii, 106)— *Mishmis* (As. S. B. v, 195-6)—*Bambaras* (Cail. i, 377)— *Gold Coast* (Bos. 112). § 417. *Guatemala* (Juar. 194-5)—*Tanna* (Tur. 77)—*Mexicans* (Dur. i, 55; Her. iii, 198)—*Hottentot* (Kol. i, 50-51)—*Egyptians* (Wilk. iii, 360-3). § 418. *Mexico* (Clav. —)—*Dahomey* (Dal. 98; Bur. i, 217)— *Japan* (Stein. —) — *Burmah* (Yule, 139; Sang. 127; Symes —, 185-6).

§ **419.** *Chibchas* (Sim. 253)—*Madagascar* (Ell. "Hist." i, 283)—*Romans* (Guhl, 513)—*Japan* (Kœm. 70)—*China* (Will. i, 404)—*Turkey* (White, i, 43)—*Siam* (Bowr., i, 117)—*Congo* (Bast. 57)—*Assyrians* (Raw. i, 495)—*India* (ref. lost)—*Siam* (Bowr. 1, 425)—*China* (Gutz. ii, 278)—*Java* (Raf. i, 312)—*Utlatlan* (Torq. bk. xi, ch. 18)—*Dahomey* (Waitz, ii, 87)—*Siamese* (Bowr. i, 116)—*Joloffs* (Bast. 57).   § **420.** *Tasmanians* (Bon. "Daily Life," 64)—*Australia* (Sturt, ii, 54)—*Khond* (Macph. 56)—*Tahiti* (Ell. "Pol. Res." i, 222)—*Fijians* (U. S. Ex. iii, 332; See. 179)—*Chibchas* (Sim. 253)—*San Salvador* (Her. iv, 149)—*Peru* (Acos. bk. iv, ch. 22).   § **421.** *Society Islands* (Forst. 271)—*Fijian* (Ersk. 430)—*Sumatra* (Mars. 47)—*Indians* (ref. lost)—*Tahiti* (Ell. "Pol. Res." i, 173)—*Karague* (Speke, 210 & 231)—*Tahiti* (Cham. s.v. "Ava")—*Guatemala* (Xim. 157).   § **424.** *Fiji* (—; Wil. i, 39)—*Darfur* (ref. lost)—*Burgundy* (Quich. 299)—*France* (Ste. Beuve, ref. lost).   § **425.** *New Zealand* (Ang. i, 319; Thom. i, 190). § **428.** *Abyssinia* (Bru. vi, 16) — *Mexicans* (Clav. bk. vi, ch. 20). § **429.** *Fiji* (Ersk. 462; Wil. i, 39; i, 37)—*Uganda* (Speke, 298; Stan. i, 369; Speke, 256 & 258)—*Siamese* (Bowr. i, 434)—*Fiji* (U. S. Ex. iii, 326)—*Loango* (Ast. iii, 226)—*Ashantee* (Cruic. i, 109)—*Siamese* (La Loub. i, 186 & 172)—*China* (Pink. vii, 265; Huc, "Chin. Empire," i, 212)—*Japan* (Dick. 45)—*Russia* (Cust. —)—*Siamese* (La Loub. i, 172; Bowr. i, 435)—*Burma* (Symes, 244) — *China* (Will. i, 509; Huc, "Chin." ii, 289). § **431.** *Japan* (ref. lost)—*Russia* (Cust. —)—*Spain* (ref. lost).   § **432.** *China* (Will. i, 509).

TITLES OF WORKS REFERRED TO.

(If not otherwise specified, London is to be understood as the place of publication).

Acos.—Acosta (Jos. de) *Historia natural y moral de las Indias.* Sevilla, 1590.
Alc.—Alcock (Sir R.). *The Capital of the Tycoon.* 1863.
All. & T.—Allen (W.) and Thomson (T. R. H.) *Expedition to River Niger in* 1841. 1848.
Anda.—Andagoya (P. de) *Proceedings of P. Davila.* (Hakluyt Society.) 1865.
And.—Andersson (C. J.) *Lake Ngami.* 1856.
Ang.—Angas (G. F.) *Savage life and scenes in Australia and New Zealand.* 1847.
As. S. B.—Asiatic Society of Bengal. *Journal.* Calcutta, v.y.
Ast.—Astley (T.) *New general collection of voyages and travels.* [By J. Green.] 1745-7.
Ath.—*Athenæum* for 1856.
Baber.—Baber (E. C.) *Notes of a journey through Western Yunnan.* (Foreign Office Papers.) 1877.
Bak.—Baker (Sir S. W.) *The Nile tributaries of Abyssinia.* 1871.
Banc.—Bancroft (H. H.) *The native races of the Pacific States of North America.* 1875-6.
Bar.—Barrow (Sir J.) *Travels into the interior of Southern Africa.* 1801-4.
Bast.—Bastian (A.) *Africanische Reisen.* Bremen, 1859.
Beck.—Becker (W. A.) *Gallus;* or *Roman scenes of the time of Augustus.* Trans. 1844.
„   ——— *Charicles; illustrations of the private life of the ancient Greeks.* Trans. 1854.
Beech.—Beecham (John) *Ashantee and the Gold Coast.* 1841.
Beechey—Beechey (F. W.) *Voyage to the Pacific and Behring's Strait.* 1831.
Bell—Bell (John) *Travels from St. Petersburgh to various parts of Asia.* Edin. 1788.

REFERENCES.

Black.—Blackstone (Sir W.) *Commentaries of the laws of England.* Ed. by R. M. Kerr. 1857.
Bon.—Bonwick (J.) *Last of the Tasmanians.* 1870.
" ———— *Daily life and origin of the Tasmanians.* 1870.
Bos.—Bosman (W.) *Description of the coast of Guinea.* Trans. 1721.
Bouq.—Bouquet (Dom. M.) *Recueil des historiens des Gaules et de la France.* Paris, 1738-1855.
Bowr.—Bowring (Sir John) *Kingdom and people of Siam.* 1857.
Boyle—Boyle (F.) *Adventures among the Dyaks of Borneo.* 1865.
Brin.—Brinton (D. G.) *Myths of the New World.* New York, 1868.
Brooke—Brooke (C.) *Ten years in Saráwak.* 1866.
Broom—Broom (H.) *Commentaries on the common law.* 1880.
Bru.—Bruce (James) *Travels to discover the source of the Nile.* Edin. 1804.
Burch.—Burchell (W. J.) *Travels into the interior of South Africa.* 1822-4.
Burt.—Burton (J. H.) *History of Scotland.* Edin. 1867-70.
Bur.—Burton (Sir R. F.) *Mission to Gelele, King of Dahomé.* 1864.
" ———— *Abeokuta and the Cameroon Mountains.* 1863.
" ———— *Pilgrimage to El Medineh and Mecca.* 1855-6.
" ———— *Lake regions of Central Africa.* 1860.
Busk—Busk (Mrs.) *Manners and customs of the Japanese.* 1841.
But.—Butler (Maj. J.) *Travels and adventures in Assam.* 1855.
Cail.—Caillié (R.) *Travels to Timbuctoo.* Trans. 1830.
Camp.—Campbell (Gen. John) *The wild tribes of Khondistan.* 1864.
Cat.—Catlin (G.) *Letters, &c., on North American Indians.* 1841.
Chal.—Challamel (Aug.) *Mémoires du peuple français.* Paris, 1866-73.
Cham.—*Chamber's Encyclopædia.* Edin. 1874. Vol. I.
Chér.—Chéruel (A.) *Dictionnaire historique de la France.* Paris, 1855.
Chin. Rep.—*Chinese Repository.* Canton, 1832-44.
Church.—*Churchill's Collection of Voyages.* 1744-46.
Cie.—Cieza de Leon (P de) *Travels, A.D.* 1532-50. (Hakluyt Soc.) 1864.
Clav.—Clavigero (Fr. S.) *History of Mexico.* Trans. 1787.
Cog.—Cogolludo (D. L.) *Historia de Yucatan.* Merida, 1867-8.
Com.—Comines (P. de) *Historie of Louis XI.* Trans. 1614.
Cook—Cook (Capt. J.) *A narrative of second voyage.* 1777-8.
" ———— *Journal of last royage.* 1781.
Coop.—Cooper (T. T.) *The Mishmee Hills.* 1873.
Cor.—Cortet (E.) *Essai sur les fêtes religieuses.* Paris, 1867.
Cox & Jones—Cox (Rev. Sir G. W.) and Jones (E. H.) *Popular romances of the Middle Ages.* 1871.
Cruic.—Cruickshank (B.) *Eighteen years on the Gold Coast of Africa.* 1853.
Cust.—Custine (Marq. de) *Russia.* Trans. 1844.
Dal.—Dalzel (A.) *History of Dahomey.* 1793.
Diaz—Diaz de Castillo (B.) *Memoirs* [1598]. Trans. 1844.
Dick.—Dickson (W.) *Japan.* 1879.
Diod.—Diodorus Siculus. *Historical Library.* Trans. Booth. 1814.
Dob.—Dobrizhoffer (M.) *Account of Abipones of Paraguay.* Trans. 1822.
Doo.—Doolittle (Rev. J.) *Social life of the Chinese.* New York, 1867.
Drew—Drew (F.) *The Jummoo and Kashmir territories.* 1875.
" ———— *The northern barrier of India.* 1877.
Drur.—Drury (R.) *Madagascar : fifteen years' captivity on that island.* 1731.
Duc.—Duange (Ch. Dufresne, sr.) *Dissertations sur l'histoire de S. Louys.* (In *Petitot, Collection de Mémoires,* tome iii. Paris, 1819.)
Duf.—Dufton (H.) *Narrative of a journey through Abyssinia.* 1867.
Du H.—Du Halde (J. B.) *General description of China.* Trans. 1736.
Du M.—Du Méril (Edél.) *Études sur quelques points d'archéologie.* Paris, 1862.
Dun.—Duncker (Max.) *History of antiquity.* Trans. 1877-82.
Dup.—Dupuis (Jas.) *Journal of a residence in Ashantee.* 1824.

Dur.—Duran (Fr. D.) *Historia de las Indias de Nueva España.* Mexico, 1867.
Eb.—Ebers (G.) *Ægypten und die Bücher Mose's.* Leipzig, 1868.
Ed.—Edwards (B.) *History of the British Colonies in the West Indies.* 1793.
Ell.—Ellis (Rev. W.) *Polynesian Researches.* 1829.
„ ——————— *Tour through Hawaii.* 1827.
„ ——————— *History of Madagascar.* 1838.
„ ——————— *Three visits to Madagascar.* 1858.
Ell., W.—Ellis (W.) *Narrative of voyage of Capts. Cook and Clerke in search of a North-West Passage.* 1782.
Ersk.—Erskine (Capt. J. E.) *Journal of a cruise among the islands of the Western Pacific.* 1853.
Eth. S.—Ethnological Society. *Journal.* Vol. iii, 1854.
„ „ ——————— *Transactions.* N.S.
Ew.—Ewald (G. H. A.) *History of Israel.* Trans. Vol. iii, 1878.
Falk.—Falkner (T.) *Description of Patagonia.* Hereford, 1774.
Fan.—Fancourt (C. St. J.) *History of Yucatan.* 1854.
For.—Forbes (F. E.) *Dahomey and the Dahomans.* 1851.
Ford—Ford (R.) *Gatherings from Spain.* 1846.
„ ——————— *Handbook for travellers in Spain.* 1847.
Forst.—Forster (G.) *Observations during a voyage round the world.* 1778.
Fors.—Forsyth (Capt. J.) *Highlands of Central India.* 1871.
Fyt.—Fytche (Gen. A.) *Burma past and present.* 1878.
Gal.—Gallatin (A.) *Notes on the semi-civilized nations of Mexico.* (In *Transactions of the American Ethnological Soc.*, vol. i. New York, 1845.)
Galt.—Galton (F.) *Narrative of an explorer in tropical south Africa.* 1853.
Gar.—Garcilasso de la Vega. *First part of the Royal Commentaries of the Yncas.* Trans. (Hakluyt Soc.). 1869-71.
Gard.—Gardiner (A. F.) *Narrative of a journey to the Zoolu Country.* 1836.
Ger.—*German Home Life.* 1877.
Gib.—Gibbon (E.) *Decline and fall of the Roman Empire.* Edited by H. H. Milman. 1838.
Grant—Grant (J. A.) *A walk across Africa.* 1864.
Gray—Gray (Archdn. J. H.) *China, its laws, manners, and customs.* 1878.
Greg.—Gregory of Tours. *Historia ecclesiastica Francorum.* Paris, 1836-8.
Grote—Grote (G.) *History of Greece.* 1872.
Guér.—Guérard (B.) *La Polyptique de l'Abbé Irminon.* Paris, 1844.
„ ——————— *Cartulaire de l'Église de Nôtre-Dame de Paris.* Paris, 1850.
„ ——————— *Cartulaire de l'Abbaye de Saint-Père de Chartres.* Paris, 1840.
Guhl—Guhl (E.) and Koner (W.) *Life of the Greeks and Romans.* Trans. 1877.
Guizot—Guizot (F.) *The History of Civilization.* Trans. Bohn's Ed. 1856.
„ ——————— *Collection des mémoires relatifs à l'histoire de France.* Paris, 1823.
Gütz.—Gützlaff (Rev. K. F. A.) *China opened.* 1838.
Guz.—Guzman (A. E. de) *Life and Acts*, A.D. 1518 to 1543. (Hakluyt Soc.) 1862.
Har.—Harris (Sir W. C.) *Highlands of Æthiopia.* 1844.
Hawk.—Hawkesworth (J.) *Account of the voyages undertaken for making discoveries in the southern hemisphere.* 1773.
Hay.—Haygarth (H. W.) *Recollections of bush life in Australia.* 1848.
Heim.—*Heimskringla; or, Chronicle of the Kings of Norway.* Trans. from Snorro Sturleson by S. Laing. 1844.

Hen.—Henderson (J.) *History of the Brazil.* 1821.
Heri.—Hericourt (Rochet d') *Seconde voyage.* Paris, 1846.
Her.—Herrera (Ant. de) *The general history of the continent and islands of America.* Trans. 1725-6.
Heug.—Heuglin (Th. von) *Reise in das Gebiet des Weissen Nil.* Leipzig, 1869.
Hind—Hind (H. Y.) *Canadian Red River exploring expedition.* 1860.
His.—Hislop (Rev. S.) *Aboriginal tribes of the central provinces.* 1860.
Hook—Hook (Dean W. F.) *A church dictionary.* 1854.
Huc—Huc (L'Abbé) *Travels in Tartary, Thibet, and China.* (In National Illustrated Library.)
„ ——— *The Chinese Empire.* Trans. 1855.
Hutch.—Hutchinson (T. J.) *The Paraná.* 1868.
Jag.—Jagor (F.) *Travels in the Philippines.* Trans. 1875.
Jor.—Jornandes (Episc. Ravenn.) *De Getarum sive Gothorum origine et rebus gestis.* (In L. A. Muratori, *Rerum Ital. Script.* Mediol. 1723. Tom. i.)
Jos.—Josephus (Flavius) *Works.* Trans. Whiston. 1825.
Juar.—Juarros (Dom.) *Statistical and commercial history of Guatemala.* Trans. 1824.
Kæm.—Kæmpfer (E.) *Account of Japan.* (Universal Lib.) 1853.
Kal.—Kalisch (M.) *Commentary on the Old Testament—Leviticus.* 1867-72.
Klun.—Klunzinger (C. B.) *Upper Egypt.* 1878.
Knobel—Knobel (Aug.) *Die Bücher Exodus und Leviticus.* Leipzig, 1880.
Kol.—Kolben (P.) *Present state of the Cape of Good Hope.* Trans. 1731.
Krapf—Krapf (J. L.) *Travels, &c., in Eastern Africa.* 1860.
Krash.—Krasheninnikov (S. P.) *History of Kamschatka.* Trans. by J. Grieve. Glocester, 1764.
Krehl—Krehl (L.) *Ueber die Religion der Vorislamischen Araber.* Leipzig, 1863.
Kue.—Kuenen (A.) *The Religion of Israel.* Trans. 1874-5.
Laird—Laird (M.) and Oldfield (R. A. K.) *Expedition into the interior of Africa, by the Niger.* 1837.
La Loub.—La Loubère (M. de) *Du royaume de Siam en 1687-8.* Amst. 1691.
La Sale—La Sale (A. de) *The history of little Jehan de Saintré.* Trans. 1862.
Landa—Landa (Diego de) *Relation des choses de Yucatan.* (In *Collection de documents; par Brasseur de Bourbourg,* vol. iii. Paris, 1864).
Lan.—Lander (Richard) *Records of Capt. Clapperton's last expedition.* 1830.
Leb.—Leber (C.) *Collection des meilleures dissertations relatives à l'histoire de France.* Paris, 1826-38.
Le Grand—Le Grand d'Aussy (P. J. B.) *Fabliaux ou contes du XIIe et du XIIIe siècle.* Paris, 1779-81.
Lehuërou—Lehuërou (J. M.) *Histoire des institutions Carolingiennes.* Paris, 1843.
Lew.—Lewin (T. H.) *Wild races of south-eastern India.* 1870.
Lew. & Cl.—Lewis (M.) and Clarke (W.) *Travels to the source of the Missouri.* 1817.
Lich.—Lichtenstein (H.) *Travels in southern Africa.* Trans. 1812-15.
Liv.—Livingstone (D.) *Missionary travels and researches in south Africa.* 1857.
Lyon—Lyon (Capt. G. F.) *Travels in northern Africa.* 1821.
Macph.—Macpherson (Lieut.) *Report upon the Khonds of Ganjam and Cuttack.* Calcutta, 1842.
Mal.—Malcolm (Sir J.) *History of Persia.* 1815.
Marcy—Marcy (Col. R. B.) *Thirty years of army life on the border.* New York, 1866.

Mar.—Mariner (W.) *Account of the natives of the Tonga islands.* 1818.
Markham—Markham (C. R.) *Reports on the discovery of Peru.* (Hakluyt Soc.) 1872.
Mark.—Markham (Col. F.) *Shooting in the Himalayas.* 1854.
Mars.—Marsden (W.) *History of Sumatra.* 1811.
Mart.—Martyr ab Angleria (Petrus) *De rebus oceanicis Decades tres.* Coloniæ, 1574.
May.—Mayhew (H.) *German life and manners.* 1864.
Mich.—Michelet (J.) *History of France* Trans. 1844-6.
Mil.—Milne (Rev. W. C.) *Life in China.* 1858.
Mitch.—Mitchell (Sir T. L.) *Three expeditions into the interior of Eastern Australia.* 1839.
Mit.—Mitford (A. B.) *Tales of old Japan.* 1871.
Mol.—Mollien (G. T.) *Travels in the interior of Africa to the sources of the Senegal and Gambia.* Trans. 1820.
Mom.—Mommsen (Th.) *History of Rome.* Trans. 1868.
Mons.—Monstrelet (E. de) *Chronicles.* Trans. 1840.
Mor.—Morier (J.) *Second journey through Persia.* 1818.
Mot.—Motolinia (Fr. T. Benavente) *Historia de los Indios de Nueva España.* (In *Coleccion de documentos para la historia de Mexico.* Mexico, 1858.)
Mov.—Movers (F. C.) *Die Phönizier.* Bonn, 1841-56.
Moz.—Mozley (H. N.) and Whiteley (G. S.) *Concise law dictionary.* 1876.
Mul.—Müller (F. Max) *Lectures on the science of language.* 1873.
Nieb.—Niebuhr (M.) *Travels through Arabia.* Trans. Edinb. 1792.
Nob.—Noble (Rev. M.) *History of the College of Arms.* 1804.
Nouv.—*Nouvelles annales des voyages.* Tomes 98, 99. Paris, 1843.
Ovi.—Oviedo y Valdés (G. F. de) *Historia general y natural de las Indias.* Madrid, 1851-55.
Pal.—Palgrave (W. G.) *Narrative of a year's journey through central and eastern Arabia.* 1865.
Pall.—Pallas (P. S.) *Voyages dans les gouvernements méridionaux de la Russie.* Trad. Paris, 1805.
Par.—Pardessus (J. M.) *Loi salique.* Paris, 1843.
Park—Park (Mungo) *Travels in Africa.* (Pinkerton's Voyages, vol. xvi.)
Pau.—Pausanias. *Description of Greece.* Trans. 1824.
Pax.—Paxton (G.) *Illustrations of Scripture.* Edinb. 1843.
Pell.—Pelloutier (S.) *Histoire des Celtes.* Paris, 1770-71.
Pied.—Piedrahita (L. Fernandez de) *Historia del nuevo reyno de Granada.* Amberes (1688).
Pink—Pinkerton (J.) *General collection of voyages.* 1808-14.
Piz.—Pizarro (P.) *Relacion del descubrimiento y conquista de los reinos de Perú, Año 1571.* (In F. Navarrete, Salvá y Baranda, *Coleccion de documentos inéditos para la historia de España.* Madrid, 1844.)
Por.—Porter (Sir R. K.) *Travels in Georgia, Persia, Armenia, ancient Babylonia.* 1821-2.
Pot.—Potter (J.) *Archæologia Græca.* Edinb. 1827.
Quich.—Quicherat (J.) *Histoire du costume en France.* Paris, 1875.
Raf.—Raffles (Sir T. S.) *History of Java.* 1817.
Ram.—Ramseyer (F. A.) and Kühne (J.) *Four years in Ashantee.* Trans. 1875.
Raw.—Rawlinson (G.) *The five great monarchies of the ancient eastern world.* 1871.
Rob.—Roberts (George) *Social history of the southern counties of England.* 1856.
Rohlfs—Rohlfs (G.) *Adventures in Morocco.* 1874.
Rose—Rose (Rev. H. J.) *Untrodden Spain.* 1874.
Roy G. S.—Royal Geographical Society. *Proceedings,* vol. xx. 1876.

Rules—*Rules (The) of civility.* Trans. 1685.
Saha.—Sahagun (Bernardino de) *Historia general de las cosas de nueva España.* Mexico, 1829-30.
St. John—St. John (Sir Spencer) *Life in the forests of the far east.* 1862.
St. Sim.—Saint Simon (Duc de) *Mémoires.* Paris, 1839-41.
Ste Beuve—Sainte-Beuve (C. A.) *Nouveaux Lundis.* Paris, 1863-72.
Ste. Pal.—Ste. Palaye (La Curne de) *Mémoires sur l'ancienne chevalerie.* Paris, 1781.
Sala—Sala (G. A.) *Journey due north.* 1858.
Sang.—Sangermano (Father) *Description of the Burmese empire.* Trans. Rome, 1833.
Schom.—Schomburgk (Sir R. H.) *Reisen in Britisch-Guiana.* Leipzig, 1847-49.
School.—Schoolcraft (H. R.) *Information respecting the Indian tribes of the U.S.* 1853-56.
Scott—Scott (Sir W.) *Chivalry, romance, and the drama.* (In *Miscellaneous Prose Works.* Edinb. 1841.)
See.—Seemann (B.) *Viti; a mission to the Vitian or Fijian islands.* Camb. 1862.
Sel.—*Selections from the Records of Government of India.* (Foreign Depart.)
Shoot.—Shooter (Rev. J.) *The Kafirs of Natal and the Zulu country.* 1857.
Sim.—Simon (P.) *Tercera (y cuarta) noticia.* (In Lord Kingsborough's *Antiquities of Mexico*, vol. viii, 1830.)
Smith—Smith (E. R.) *The Araucanians.* 1855.
Smith, W.—Smith (Dr. W.) *Dictionary of Greek and Roman antiquities.* 1849.
Soph —Sophocles. *The Electra.* Ed. by R. C. Jebb. 1880.
South.—Southey (R.) *History of Brazil.* 1810-19.
Speke—Speke (J. H.) *Journal of the discovery of the source of the Nile.* 1863.
Spen.—Spencer (Capt. E.) *Germany and the Germans.* 1836.
Squ.—Squier (E. G.) *Nicaragua.* 1852.
„ ———— *Collection of documents concerning the discovery and conquest of America.* New York, 1860.
Stade—Stade (Hans) *Captivity in Brazil.* Trans. (Hakluyt Soc.) 1874.
Stan.—Stanley (H. M.) *How I found Livingstone.* 1872.
Staun.—Staunton (Sir G.) *Account of embassy to China.* 1797.
Stein.—Steinmetz (A.) *Japan and her people.* 1859.
Stubbs—Stubbs (Bp. W.) *Constitutional history of England.* Oxford, 1874.
Sturt—Sturt (Capt. Chas.) *Two expeditions into the interior of Australia.* 1833.
Sully—Sully (Max. Duc de) *Memoirs.* Trans. 1774.
Symes—Symes (M.) *Account of embassy to Ava.* 1800.
Tac.—Tacitus (C. C.) *Germania.* Trans. by John Aikin. 1823.
Tav.—Tavernier (J. B.) *Six voyages through Turkey into Persia and the East Indies.* Trans. 1678.
Tern.—Ternaux-Compans (H.) *Recueil de pièces relatives à la conquête du Mexique.* (In *Voyages, Relations, &c.*, vols. x, and xvi. Paris, 1837-41.)
Thomp.—Thompson (Geo.) *Travels and adventures in Southern Africa.* 1827.
Thomp., G.—Thompson (Col. Geo.) *The war in Paraguay.* 1869.
Thom.—Thomson (A. S.) *The story of New Zealand.* 1859.
Thomson—Thomson (W. M.) *The Land and the Book.* 1859.
Timk.—Timkowski (G.) *Travels through Mongolia.* Trans. 1827.
Tocq.—Tocqueville (A. de) *State of society in France before 1789.* Trans. 1856.
Torq.—Torquemada (J. de) *Monarquia Indiana.* Madrid, 1723.
Tuck.—Tuckey (Capt. J. K.) *Narrative of an expedition to the river Zaire.* 1818.
Tur.—Turner (Rev. G.) *Nineteen years in Polynesia.* 1861.

Turn.—Turner (Capt. S.) *Embassy to the court of the Teshoo Lama in Thibet.* 1800.
Tyl.—Tylor (E. B.). *Researches into the early history of mankind.* 1878.
—————— *Primitive culture.* 1871.
U. S. Ex.—*United States Exploring Expedition.* (Comm. C. Wilkes.) Phil. 1845.
Ur.—Uricoechea (E.) *Memoria sobre las antiguedades Neo-Granadinas.* Berlin, 1854.
Van Len.—Van Lennep (H. J.) *Bible lands, their modern customs and manners.* 1875.
Wag.—Wagner (M.) *Travels in Persia, Georgia, and Koordistan.* Trans. 1856.
Wahl—Wahl (O. W.) *The Land of the Czar.* 1875.
Wai.—Waitz (Geo.) *Deutsche Verfassungsgeschichte.* Kiel. Vols. i and ii (2nd ed.), 1865-70; vols. iii and iv, 1860-1.
Waitz—Waitz (T.) *Anthropologie der Naturvölker.* Leipzig, 1859-72.
Wal.—Wallace (A. R.) *The Malay Archipelago.* 1869.
West—West (J.) *History of Tasmania.* Launceston, Tasmania, 1852.
Whar.—Wharton (J. S.) *Law Lexicon.* 1876.
White—White (C.) *Three years in Constantinople.* 1845.
Wilk.—Wilkinson (Sir. J. G.) *Manners and customs of the ancient Egyptians.* Ed. by S. Birch, 1878.
Will.—Williams (S. W.) *The middle kingdom; geography, &c., of the Chinese empire.* 1848.
Wil.—Williams (Rev. T.) and Calvert (J.) *Fiji and the Fijians.* 1860.
Wint.—Winterbottom (T.) *Account of the native Africans in the neighbourhood of Sierra Leone.* 1803.
Xer.—Xeres (F. de) *Account of Cuzco.* (In *Reports on the discovery of Peru.* Trans. (Hakluyt Soc.) 1872.)
Xim.—Ximenes (F.) *Las historias del origen de los Indios de Guatemala.* Viena, 1857.
Yule—Yule (Col. H.) *Narrative of mission to Ava.* 1858.
Zur.—Zurita (Al. de) *Rapports sur les différentes classes de chefs de la Nouvelle-Espagne.* (In *Voyages,* &c., par H. Ternaux-Compans. Vol. xi. Paris, 1840.)

REFERENCES.

(For explanation see the first page of References.)

POLITICAL INSTITUTIONS.

§ **437.** *Santals* (Hunt. "Ann." i, 248)—*Sowrahs* (Shortt Pt. iii, 38)—*Todas* (Hark. 18; Metz, 13; Hark. 17)—*Tipperahs* (Hunt. "Stat." vi, 53)—*Marius [Gonds]* (Glas. No. xxxix, 41)—*Khonds* (Macph. vii, 196)—*Santals* (Hunt. "Ann." i, 215-6)—*Lepchas* (Eth. Soc. "Jour." N. S. i, 150)—*Bodo & Dhimals* (As. S.B. xviii, 745)—*Carnatics* (Hunt. "Dic." 10)—*Chakmás* (Hunt. "Stat." vi, 48)—*Santals* (Hunt. "Ann." i, 215-6; Dalt. 217)—*Bodo & Dhimals* (As. S.B. xviii, 745)—*Lepchas* (Hook. i, 175; Eth. Soc. "Jour." N.S. i, 154)—*New Guinea* (D'Alb. 45, 48, 58-9)—*Fijians* (ref. lost)—*Dahomey* (Bur. i, 195, note; ii, 190, note)—*Mexicans* (Torn. x. 212; Clav. bk. vi, ch. 18; Diaz, ch. 208; Her. iii, 208-9)—*Cent. Americans* (Landa § xxiv; Gall. i, 104; Her. iii, 223; Pres. bk. i, ch. iv; Her. iv, 174)—*Veddahs* (Bail. ii, 228; Ten. ii, 445; Prid. i, 461). § **442.** *Digger Indians* (Kel. i, 252-3)—*Chaco Indians* (Hutch. 280)—*Unyoro* (Eth. Soc. "Trans." 1867, 234-5)—*New Zealand* (Hawk. iii, 470)—*Beluchees* (Eth. Soc. "Jour." i, 109)—*Greeks* (Cur. i, 115-6)—*Carolingians* (Dun. i, 101). § **443.** *Egyptians* (Wilk. i, 330-336)—*Roman* (Lact. cc. 7, 23, Salv. bk. v)—*France* (Guiz. iii, 251-2; Clam. i, 355-438, ii, 160-230, i, pp. xxv-vi)—*Gwalior* ("The Statesman," Aug. 1880, 218-19)—*Japan* (ref. lost)—*Byzantium* (Gib. iii, 303, ch. liii). § **446.** *Rome* (Duruy iii, 126-7). § **448.** *Bechuanas* (Burch. ii, 532)—*Greeks* (Hom. "Iliad," bk. i)—*Khonds* (Macph. 43). § **449.** *Seminoles* and *Snakes* (School. "I.T." v. 260)—*Peruvians* (Squi. "Peru," 19; Cic. ch. xiii)—*Equatorial Africa* (Grant—)—*Abors* (As. S.B. xiv, 426)—*Damaras* (ref. lost)—*Kookies* (As. S. B. xxiv, 633)—*Mishmees* (Coop. 228)—*Bachapins* (Burch. ii, 512). § **450.** *Bushmen* (Lich. ii, 194)—*Rock Veddahs* (Ten. ii, 440)—*New Zealand* (ref. lost)—*S. Americans* (Humb. ii, 412)—*Athenians* (Gro. iii, 88)—*Romans* (Mom. i, 65)—*Greeks* (Gro. iii, 77)—*Rome* (Coul. "C. Ant." 146; Mom. i, 67)—*India* (Maine, "E. H." 107)—*Greeks* (Gro. ii, 312-3). § **451.** *Karens* (As. S. B. xxxvii, 152)—*Hottentots* (Kol. i, 287)—*New Cal.* (Tur. 85-6)—*Samoa* (Tur. 291)—*Greece* (Gro. iv, 430; ii, 350)—*Fúlbe* (Bar. ii, 510)—*Damaras* (Roy. G. S., 1852, 159)—*Peru* (Onde. 152-3). § **452.** *Patagonians* (Falk. 123)—*Chinooks* (Kane, 215)—*Abipones* (Dob. ii, 105)—*Balonda* (Liv. 208)—*Kukis*

(M'Cull. xxvii, 58)—*American Indians* (Morg. 341)—*Britain* (Burt. ii, 72; Mart. "Hist." i, 343)—*Mexicans* (Zur. —)—*Peru* (Garc. bk. iv, ch. 8, and bk. v, ch. 9)—*Japanese* (Dick. 305). § 454. *Fuegians* ([Hawk.] "Hawkesworth's Voyages," ii, p. 58)—*Coroados* (Spix. ii, 244). § 455. *Bodo and Dhimals* (Hodg. 158)—*Lepchas* (Eth. Soc. "Jour." N. S. i, 147)—*Arafuras* (ref. lost). § 456. *N. A. Indians* (Kane, 214-5—*Nootkas* (Banc. i, 195)—*Vera Pax* (Xim. 202-3)—*Honduras* (Her. iv, 136)—*Dyaks* (St. John —). § 457. *New Zealanders* (Thom. i, 148) —*Sandwich Islands* (Ell. "Tour" 397)—*Fiji* (Ersk. --)—*Scot.* (Maine, "E. I.," 133)—*British* (Pear. i, 12)—*English* (Stubbs, ii, 493)—*Scotland* (Innes, "Mid. Ages," 141-2). § 458. *Egypt* (Shar. i, 189; Ken. ii, 42)—*Rome* (Mom. i, 95)—*Germans* (Stubbs, i, 34)—*English* (Kem. i, 69; Hall. "M. A." ii, 295)—*Egyptians* (Wilk. i, 150, note)—*Roman* (Coul., Revue, xcix, 246)—*England* (Hall. "M. A." ch. ii, pt. 1 ; Ree. i, 34-6). § 459. *Danish* (Maine, "E. I." 84-5)—*Med. Eur.* (Free. "N. C." i, 96-7). § 460. *Fijians* (Sec. 179; Wilkes, iii, 73-4) —*Sandwich Islanders* (Ell. "Tour" 7-8)—*Tahitians* (Ell. "Pol. Res." ii, 16)—*Africa* (Rea. 241). § 461. *Sandwich Islanders* (Ell. "Tour." 392-3). § 462. *China* (Gutz. ii, 305-6)—*France* (ref. lost; Warn. i, 549-50—*Hottentots* (Thomp. ii, 30)—*Bechuanas* (Burch. ii, 347)— *Chinooks* (Wai. iii, 338)—*Albania* (Boué, iii, 254)—*Birth*, &c. (Maine, "E. H." 134)—*France* (A. L. F. ii, 645). § 464. *Australians* (Sm. i, 103)—*Chippewas*, &c. (School. "Travels," 340-1)—*Cent. Amer.* (Banc. i, 702)—*Khonds* (Macph. 32 and 27)—*New Zea.* (Thom. i, 95) —*Tahitians* (Ell. "P. R." ii, 363)—*Madag.* (Ell. "M." i, 378)— *Phœnicians* (Mov. ii, pt. i, 541)—*Greeks* (Gro. ii, 92)—*Pr. Ger.* (Tac. in Free. "Eng. Const." 17)—*Iceland* (Mall. 201-3)—*Swiss* (Free. "E. C." pp. 1-7)—*Old Eng.* (Free. "E. C." 60). § 466. *Greenlanders* (Crantz, i, 164-5)—*Australians* (Sturt, —)—*Salish* (ref. lost ; Dom. ii, 343-4)— *Bodo and Dhimals* (Hodg. 159)—*Australians* (Grey, ii, 240)—*Snakes* (L. and C. 306) *Chinooks* (L. and C. 443—*Dakotas* (School. "I. T." ii, 182)—*Creeks* (School. "I. T." i, 275)—*Khirgiz* (Wood, 338)—*Ostyaks* ("Rev. Sib." ii, 269) -*Nagas* (But. 146)-*Kor. Hottentots* (Thomp. ii, 30)—*Kaffirs* (Lich. i, 286-7). § 467. *Tupis* (Sou. i, 250)—*Juangs* (Dalt. 156)—*Kor. Hottentots* (Thomp. ii, 30)—*Kaffirs* (Shoo. 102)—*Damaras* (ref. lost—*Araucanians* (Smith, 243)—*Dyaks* (Broo. i, 129)—*Malagasy* (Ell. "H. M." i, 146)—*Savages* (Lubb. 445). § 468. *Arafuras* (Kolff, 161)—*Khirgiz* (Mich. —)—*Sumatrans* (Mars. 217)—*Madag.* (Ell. "Hist. Madag." i, 377 —*East Africans* (Bur. "C. A." ii. 361)—*Javans* (Raff. i, 274)—*Sumatra* (Mars. 217)--*Ashantee* (Beech. 90-1). § 469. *Congo* (Pink. xvi, 577)—*Dahomans* (Bur. i, 263). § 471. *Nicobarians* (Bast. iii, 384)—*Haidahs* (Banc. i, 168)—*Californians* (Banc. i, 348)—*Navajos* (Banc. i. 508)—*Angamies* (As. S. B. xxiv, 650—*Lower Californians* (Banc. i, 565)—*Flatheads* (Banc. i, 275)—*Sound Indians* (Banc. i, 217)—*Lower Californians* (Banc. i, 565)—*Chippewayans* (Frank. 159)—*Abipones* (Dob. ii, 102—*Bedouins* (Ram. 9). § 472. *Khonds* (Camp. 50)—*Cent. India* (Fors. 9)—*Esquimaux* (ref. lost)—*Fuegians* (Fitz. ii, 179)—*Rock Veddahs* (Ten. ii, 440)—*Dyaks* (ref. lost)—*Caribs* (Edw. i, 49)—*Bushmen* (Lich. ii, 194)—*Tasmanians* (Lloyd, 56 ; Dove, i, 253—*Tapajos* (Bates 222-3)—*Bedouins* (Bur. "El Med." iii, 44)—*Greece* (Gro. ii, 87)—*Scot.* (Martin, M. 101) *Snake Indians* (L. and C. 306)—*Creeks* (School. "I. T." v, 279)—*Comanches* (School. "I. T." ii, 130)—*Coroados* (Spix, ii, 234)—*Ostyaks* ("Rev. Sib." ii, 269)—*Tacullies* (Banc. i, 123)—*Tolewas* (Banc. i, 348)—*Spokanes* (ref. lost) —6)— *Navajos* (Banc. i, 508)— *Dōrs* (Heug. 195)—*Arabs* (Burck. i, 300)—*Sumatra* (Mars. 211). § 473. *Australians* (Eth. Soc. Trans., N. S., iii, 256)—*Comanches*

(School. "I. T." i, 231)—*Flatheads* (Banc. i, 275)—*Dyaks* (Low, 209; St. John —)—*Caribs* (Edw. i, 49)—*Abipones* (Dob. ii, 103)—*Egypt* (Tay. 16)—*Rome* (Mom. i, 79)—*Germans* (Sohm i, 9)—*French* (Ranke, i, 75). § 474. *Thlinkeets* (Banc. iii, 148)—*Fuegians* (Fitz. ii, 178)—*Tasmanians* (Bon. 175)—*Haidahs* (Banc. iii, 150)—*Dakotas* (School, "I. T." iv, 495)—*Amazulu* (Call. 340, note 86)—*Obbo* (Bak. i, 318-9)—*Mexicans* (Banc. iii, 295 ; Clav. bk. vii, ch. 7)—*Chibchas* (Pied. bk. ii, ch. 7)—*Egypt* (Brug. i, 406 ;—*Jews* (Sup. Rel. i, 117-18). §475. *Egypt* (Shar. ii, 2)—*Coroados* (Spix, ii, 244-5)—*Santals* (Hunt. "Ann." i, 216-7) —*Khonds* (Macph. 47). § 476. *Haidahs* (Banc. i, 167)—*Fiji* (See. 232) —*Tahitians* (Ell. "P. R." ii, 346 ; Hawk. ii, 121)—*Madagascar* (Ell. "H. M." i, 342-3)—*Congoese* (ref. lost)—*Coast Negroes* (ref. lost)—*Inland Negroes* (ref. lost)—*Peru* (Gom. ch. 124 ; Garc. bk. iv, ch. 9)—*Egypt* (Wilk. i, 161 note ; 162 note)—*Ceylon* (Ten. i, 497 ; ii, 459)—*New Caledonia* (ref. lost)—*Madagascar* (Ell. "H. M." i, 342)—*Abyssinia* (Bru. iv, 488)—*Timmanees* (Wint. i, 124)—*Kaffir* (Arb. 149)—*Aragon* (Hall. ii, 43-4). § 477. *Amazulu* (Call. 208 ; 390)—*Kukis* (As. S.B. xxiv, 625)—*Tahitians* (Ell. "P.R." ii, 341)—*Tonga* (Mar. ii, 76)—*Peru* (Garc. bk. i, ch. 23)—*Egyptians* (Wilk. i, 321-2 and note ; Brug. ii, 35-36) —*Aryans* (Gro. i, 618)—*Chibchas* (Sim. 261-2). § 478. *Chinooks* (L. and C. 443 ; Wai. iii, 338)—*Patagonians* (Falk. 121)—*Orinoco Indians* (ref. lost)—*Borneo* (Low, 183)—*Sabines* (ref. lost —*Germans* (Dunh. i, 17)—*Dyaks* (Boy. 183)—*Kalmucks* (Pall. i, 527)—*Araucanians* (Thomps. i, 405)—*Kaffirs* (Lich. i, 286)—*Greeks* (Glad. iii, 10-11)—*Karens* (As. S.B. xxxvii, 131)—*Congo* (Bast. "Af. R." 58)—*Yariba* (Lan. ii, 223)—*Ibu* (All. and T. i, 234)—*Kukis* (But. 91)—*Greeks* (Glad. iii, 51-2)—*Rome* (ref. lost)—*Europe* (ref. lost)—*French* (Hall. ch. i)—*Merovingians* (Wai. ii, 45-6, —)—*France* (Méray, 45 ; Boss. ii, 56 ; St. Sim. iii, 69). § 479. *Zulus* (Eth. Soc. "Trans." N.S., v, 291)—*Bheels* (Mal. "C. I." i, 551)—*Loango* (Ast. iii, 223 ; Pink. xvi, 577)—*East Africa* (Bur. "C. A."ii, 361)—*Msambara* (Krapf, 384 note)—*Dahome* (Bur. i, 226)—*Malagasy* (Ell. "H. M." i, 341)—*Sandwich Islands* (Ell. "Tour," 401)—*Siam* (Bowr. i, 422-3)—*Burmah* (Saug. 58)—*China* (Gutz. ii, 251)—*Japan* (Ad. i, 11). § 480. *Tonga* Ersk. 126)—*Gondar* (Har. iii, 10, 34)—*Bhotan* (Ren. 16-17)—*Japan* (Ad. i, 74, 17 ; Tits. 223 ; Ad. i. 11, 70)—*Merovingian* (Egin. 123-4). § 483. *Arafuras* (Kolff, 161)—*Todas* (Eth. Soc. "Trans." N. S., vii, 241)—*Bodo* and *Dhimáls* (As. S.B. xviii, 708)—*Papuans* (Kolff. 6 Earl —)—*Bodo* and *D.* (ref. lost)—*Lepchas* (Eth. Soc. "Jour," July, 1869)—*Nagas* (As. S. B. xxiv, 608-9 ; ix, 950)—*N. A. Indians* (School. "I. T." ii, 183)—*Comanches* (School. "I.T." ii, 130 ; Banc. i, 509)—*Central America* (Squi. "Nic." ii, 340-1)—*Nagas* (As. S. B. xxiv, 607)—*Africa* (Bur. "Abeo." i, 276). § 485. *Greece* (Toz. 284-5 ; Herm. 14 ; Gro. ii, 103)—*Scotland* (Ske. iii, 323-4)—*Crete* (Cur. i, 182 ; 178-9)—*Corinth* (Gro. iii, 2)—*Sparta* (Gro. ii, *passim*)—*Latins* (Mom. i, 30 ; 80 ; 87 ; 84). § 486. *Venice* (Sis. i, 300-313)—*Netherlands* (Gra. 10, 11, 20 ; Mot. i, 38 ;—*Switz.* (Vieus. 39)—*Grisons* (May, i, 355)—*San. Mar.* (Bent. 808 15). §487. *Ital. Repub.* (Sis. [Lard.] 21 ; Sis. i. 371 ; Sis. [Lard.] 22 ; 83). § 488. *Sparta* (ref. lost ; Gro. ii, 90) —*Rome* (Mom. ii, 326)—*Ital. Repub.* (Hall. i, 368 ; Sis. [Lard.] 280)—*Holland* (May, ii, 17-18)—*Berne* (May i, 373)—*Venice* (Sis. [Lard.] 121)—*Greece* (Gro. iii, 25 ; Cur. i, 250)—*Romans* (Macch. iii, 429)—*Ital. Repub.* (Sis. [Lard.] 80)—*Athens* (Gro. iii, 181-5)—*Rome* (Mom. bk. i., ch. 4, *passim*)—*Italian Reput.* (May, i, 281-2). § 490. *Samoa* (Tur. 284)—*Fulahs* (L. and O. ii, 85 —*Mandingo* (Park i, 15). § 491. *Italian Rep.* (Sis. [Lard.] 21-2)—*Poles* (Dunh. 278 ; 285)—*Hungarians* (Lévy, 165)—*Germans* (Stubbs, i, 63)—*Merov.* (Rich. 119-20)—*Appenzal* (Lav. 65)—*Uri* (Free.

16   POLITICAL INSTITUTIONS.

"E. C." 7)—*Scandinavia* (C. and W., i, 157-8; ref. lost)—*Tatars* (Gib. ii, 16)—*Sparta* (Gro.—).   § 492. *Kaffirs* (Lich. i, 286)—*Bechuanas* (Moff. 66)—*Wanyamwezi* (Bur. "C. A." ii, 362)—*Ashantee* (Beech. 91) —*Mexico* (Zur. 106; Clav. bk. vii, ch. 13)—*Vera Paz* (Tor. bk. xi, ch. 20) —*Poland* (Dunh. 278, 279-80)—*Germans* (Hall. ii, 93)—*France* (ref. lost)—*Madag.* (Ell. "H. M." ii, 252)—*Hebrews* (1 Samuel, ch. xv)— *Tahitians* (Ell. "P. R." ii, 489)—*Mexicans* (Saha. bk. viii, ch. 24)—*Egypt* (Wilk. i, 159)—*France* (Roth, 317-20).   § 493. *Denmark* (C. and W. i, 262-3)—*France* (Rich. 119-20)—*Madag.* (Ell. "H. M.," i, 378)—*England* (Free. "E.C." 60).   § 494. *Egypt* (Wilk. i, 160 note)—*Persia* (Raw. iii, 223)—*China* (Will. i, 324)—*France* (Boss. ii, 56, 113, v, 4; Pul. i, 8-9; St. Sim. iii, 69)—*Rome* (Mom. i, 71-2; iii, 361)—*Poland* (Dunh. 282).   § 496. *Scandinavia* (C. and W., i, 158)—*Hungary* (Patt. i, 66; 253)—*Rome* (Dur. iii, 376-8).   § 498. *Greece* (Gro. iii, 124-5; iv, 169)—*Italy* (Sis. [L.] 23; 291)—*Spain* (Dunh. iv, 158)—*England* (Hume, ii, 54).   § 499. *Spain* (Hall. ii, 7-8)—*France* (ref. lost)—*Scotland* (Burt. ii, 85).   § 500. *Scandinavia* (Mall. 291-5)—*France* (Mor. 379-80)—*England* (Stubbs, i, 448-9)—*Holland* (Mot. i, 35)—*Anglo-Sax.* (Stubbs, i, 192)—*Spain* (Dunh. iv, 158)—*England* (Stubbs, i, 450).   § 501. *England* (Hume, i, 466-7; Stubbs, i, 137)—*France* (Hall. i, 230)—*Spain* (Hall. ii, 25, 29)—*France* (Dar. "Ad." ii, 57-8; Clam. ii. 3-4; Dar. "Ad." i, 78)—*Scotland* (Innes, "Leg. An.," 116).   § 502. *France* (Ord. ii, 201)—*Hungary* (Lévy, 165) —*Scotland* (Innes, "Leg. An.," 119)—*England* (Hume, —).   § 504. *Egypt* (Wilk. iii, 371)—*Persia* (Raw. iii, 221)—*England* (Kem. ii, 105-11) —*Hebrews* (Ew. iii, 266-7)—*Rome* (Dur. iii, 175)—*France* (Gon. —)— *Eggarahs* (All. and T. i, 327)—*Mizteca* (Her. iii, 265).   § 505. *Normans and Old English* (Stubbs, i, 390)—*Scot.* (Innes, "Mid. Ages," 120-1)— *Russia* (Fowl. i, 379)—*France* (Jer. ii, 158-9; Kit. iii, 210)—*England* (Turn. vi, 132).   § 508. *Tahiti* (Ell. "P. R." ii, —)—*England* (Kem. ii, 142)—*France* (Gui. iii, 233-4)—*Mexico, &c.* (Zur. 66-7)—*Chibchas* (Acos. 188-90)—*Med. Europe* (Maine, "V. C." 235-6).   § 509. *England* (Free. "N. C." i, 80; Fis. 301; Hall. "M. A." ch. viii).   § 510. *Feudal* (Maine, "E. I." 77)—*France* (Man. cvii, 584)—*Persians* (Raw. iii, 418; 426)— *Rome* (Dur. v, 83 4)—*France* (Thie. i, 365-6; Cher. "Hist." ii, 138-9)— *England* (Hall. "C. H." ch. xii).   § 511. *Bedouins* (Burck. "Notes" 5; Pal. "Ency. Brit." ii, 249)—*Irish* (Maine, "E. I." 105-6)—*Albania* (Boué, ii, 86; iii, 359)—*England* (You. 147).   § 512. *Mexico* (Zur. 50 62) —*Russia* (Rav. i, 8, 9)—*Teutons* (Stubbs, i, 56; Cæs. vi, 22; Kem. i, 56-7) —*Bukwains* (Liv. 14)—*Japan* (Alc. ii, 241)—*Franks* (Kem. i, 238)— *England* (Thor. i, 274; 386; 450)—*Russia* (Kou. 229).   § 513. *England* (Kem. i, 240-3; Stubbs,—)—*Peru* (Pres. 72)—*Mexico* (Clav. bk. vii, ch. 5; Gom. —)—*Egypt* (Heer. ii, 139)—*Greece* (Herm. 10)—*China* (Will. i, 388)—*India* (Gho. *passim*)—*Scandinavia* (ref. lost; Bren. lxviii)— *England* (Bren. lxix-lxx.)   § 516. *Siam.* (Loub. i. 237)—*Ashante* (Beech. 129)—*Fulahs* (L. and O. ii, 87)—*Rome* Mom. i. 99-100).   § 517. *Suevi* Stubbs, i, 15).   § 518. *Guaranis* (Waitz, iii, 422)—*Nicaragua* (Squi. "Nic." ii, 342)—*New Zealand* (ref. lost)—*Bedouins* (Burck. —)— *Tahiti* (Forst. 377)—*Hebrews* (2 Sam. xxi, 17)—*Carolingian* (Wai. iv, 522) —*Japan* (Ad. i, 15)—*Peru* (Pres. 35).   § 519. *Hottentots* (Kol. i, 85)— *Malagasy* (Ell. "H. M." ii, 253)—*Chibchas* (Sim. 269)—*Rome* (Coul. "C. A." 158)—*Germans* (Stubbs, i, 34)—*Old England* (Kem. i. 69)—*France* (Kit. i, 399; Froiss. i, 168)—*Sparta* (Gro. —)—*Rome* (Mom. i, 98-9).   § 520. *France* (Ranke, i, 83).   § 522. *Chinooks* (Waitz, iii, 338)—*Arabs* (Bur. "El Med." iii, 47)—*Italy* (Sis. [L.] 90)—*France* (Maine, *Fort. Rev.* 614) —*England* (Rec, i, 153-4)—*France* (Gui. —).   § 523. *Hottentots* (Kol. i, 294-6)—*Greece* (Gro. ii, 99-100)—*Rome* (Mom. i, 159)—*Germans*

(Tac cap. xi, xii)—*Danes* (C. and W. i, 263)—*Irish* (Les. xvii, 312)
§ 524. *Hebrews* (Deut. xxi, 19)—*Romans* (Mom. i, 158)—*France* (Join. 10-11)—*Carolingian* (Mor. 379-80; Sohm, i, § 16)—*Frieslanders* (ref. lost)—Holland (Lav. 282-3). § 525. *Zulus* (Arb. 140)—*Eggarahs* (All. and T. i, 326)—*Germans* (Tac. c. 7)—*Scandinavia* (Grimm, i, 93) § 526. *Peru* (Her. iv, 337)—*Germany* (Dunh. 1, 120)—*France* (Bay. i, 70-1)—*Scotland* (Innes, "L. A." 221)—*England* (Stubbs, i, 443, 673)—*France* (Hal', i, 239). § 527. *Bedouins* ("Ram. in Syria," 9)—*Mexicans* (Dur. i, 216)—*Athens* (Cur. ii, 450)—*France and Germany* (Black. iii, 41)—*France* (Duc. 11-12; A. L. F., v, 346-7; Dar. "Ad." —)—*England* (Fis. 238; Stubbs, ii, 292). § 528. *Court, &c.* (Maine, "E. I." 289). § 529. *Sandwich I.* (Ell. 399)—*Bechuanas* (ref. lost)—*Karens* (As. S. B. xxxvii, 131)—*France* (Kœnigs. 186). § 530. *Scandinavia* (Mall. 117)—*Egypt* (Rec. ii, 11; xii, 48)—*Peru* (Santa C. 107; Gar. bk. i, ch. 23)—*Tahitians* (Ell. "P.R." ii, 235)—*Todas* (Metz, 17-18)—*Hebrews* (2 Sam. v. 22-25)—*India* (Maine, "A. L." 18)—*Greece* (Gro. ii, 111-2; Herm. 48)—*France* (Hinc. ii, 201). § 531. *Assyrians* (Lay. ii, 473-4)—*Greeks* (Tie. 217; Coul. 221)—*Egypt* (Wilk. i, 164). § 532. *Zulus* (Arb. 161 note)—*Peru* (ref. lost)—*Mexicans* (Tern. x, 78)—*Japan* (ref. lost)—*France* (Greg. bk. vii, ch. 21)—*Peruvians* (Garc. bk. ii. ch. 12)—*Japan* (Alc. i, 63)—*Rome* (Mom. i, 159)—*Salic* (Gui. i, 464)—*Scotland* (Innes, "Mid. Ages," 197)—*England* (Stubbs, i, 211). § 533. *Chippewayans* (School. "I. T." v, 177)—*Shoshones* (Banc. i, 435)—*Haidahs* (Banc. i, 168)—*Sandwich I.* (Ell. "Tour," 400)—*Greece* (Gro. ii, 107, 110, 129)—*Rome* (Maine, "A. L." 372; Mom. ii, 130)—*Basutos* (Arb. 37)—*Abyssinia* (Par. ii, 204-5)—*Sumatra* (Mars. 249)—*Dakotas* (School. "I. T." ii, 185)—*N. Americans* (Kane, 115)—*Dakotas* (Morg. 331)—*Araucanians* (Thomps. i, 405). § 536. *Bushmen* (Lich. ii, 194)—*Chippewayans* Banc. i, 118)—*Arawaks* (Roy. G. S. ii, 231). § 537. *Ahts* (Banc. i, 191)—*Comanches* (School. "I. T." i, 232)—*Brazilians* (Roy. G. S. ii, 195-6)—*Chippewayans* (School. "I. T." v, 177)—*Bedouins* (ref. lost). § 538. *Rechabites, &c.* (Ew. iv, 79-80; Kue. i, 181-2)—*Dakotas* (School. "I. T." ii, 185)—*Comanches* (School. "I. T." ii, 131)—*Iroquois* (Morg. 326)—*Bechuanas* (Burch. ii, 531)—*Damaras* (And. 114-15)—*Kafirs* (Shoot. 16)—*Koosas* (Lich. i, 271)—*New Zealanders* (Thom. i, 96)—*Sumatrans* (Mars. 244-5)—*Mexicans* (Sart. 68)—*Damaras* (And. 147)—*Todas* (Marsh. 206)—*Congo* (Pink. xvi, 108)—*Slavs* (Lav. 185)—*Swiss* (Lav. 82)—*Hebrews* (Mayer, n, 362 note)—*Rome* (Mom. i, 160, 193)—*Teutons* (Stubbs, i, 56). § 539. *Drenthe* (Lav. 282)—*Ardennes* (Lav. 301)—*Lombardy* (Lav. 215)—*France* (Lav. 212)—*Abyssinia* (Bruce, iv, 462)—*Kongo* (Ast. iii, 258.—*Mexico* (Tern. x, 253-4)—*Iceland* (Mall. 289)—*Swiss* (Lav. 83). § 540. *Slavs* (Lav. 189; 194-5)—*Lombardy* (Lav. 216). § 542. *Dakotas* (School. "I. T." iv, 69)—*Abipones* (Dob. ii, 106)—*Patagonians* (Falk. 123)—*Greece* (Gro. ii, 84; 85)—*Germans* (Tac. xv)—*England* (Dyer 3)—*Guaranis* (Wai. iii, 422)—*Rome* (Mom. —). § 543. *Loango* (Pink. xvi, 577)—*Tongans* (Mar. i, 231 note)—*Cashmere* (Drew 68-70)—*Kaffirs* (Shoot. 104)—*Sandwich Islands* (Ell. "Tour," 292)—*Mexico* (Zur. 250-1)—*Yucatan* (Landa § xx)—*Guatemala, &c.* (Zur. 407)—*Madagascar* (Ell. "M." i, 316)—*Fiji* (Sec. 232)—*Tahiti* (Ell. "P. R." ii, 361). § 544. *England* (Stubbs ii, 612-3). § 545. *Quanga and Balonda* (Liv. 296, 307)—*Bhils* (Mal. i, "C. I." 551-2; 185)—*Mexico* (Clav. bk. vii, ch. 37)—*Greece* (Glad. iii, 62; Pot. 90)—*England* (Ling. iii, 7). § 557. *France* (Dar. "Cl. Ag." 537). § 558. *Americans* (Hearne, 151)—*Dahomey* (Bur. i, 220-5; 226; Dalz. 175; Bur. i, 52, note)—*Peru* (Gar. bk. ii, chap. xv; bk. vi, chap. viii; bk. v, chap. xi)—*Egypt* (Shar. i, 188; Brug. i, 51; Shar. i, 182)—*Sparta* (Gro. vol. ii, pt. ii, chap. vi)—*Russia* (Cust. ii, 2; Wal. 289; Cust. —; Bell, ii, 237).

§ 559. *Rome* (Dur. iii, 155-60; iii, 183-7, 9; iii, 173-4; iii,172-3,; iii, 176)
—*Italy* (Sis. [Lard.] 8-9).   § 560. *Greeks* (Gro. ii, 88)—*Japan* (Mit. i, 32-3)—*France* (Corn. xxvii (1873), 72)—*Montenegro* (Boué, ii, 86)—*Dahomey* (For. i, 20)—*Sparta* (Thirl. i, 329)—*Merovingian* (Amp. ii, 305; reg. lost)—*Dahomey* (Bur. ii, 248)—*Japan* (M. and C., 34)—*Egypt* (Wilk. i, 189)—*Persia* (Raw. iii, 242)—*Araucanians* (Thomps. i, 406)—*Fiji* (Ersk. 464)—*Dahomey* (Dalz. 69)—*Egypt* (Brug. i, 53).   §573. *Todas* (Shortt, pt. i, 9)—*Pueblos* (Banc. i, 546)—*Karens* (Gov. Stat. 64; McM. 81)—*Lepchas* (Hook. i, 120-30; Eth. Soc. "Jour." N. S. i, 150-1)—*Santáls* (Hunt. "Ann." —; "Stat." xiv, 330)—*Shervarog* (Shortt, pt. ii, 7; 42)—*Todas* (Shortt, pt. i, 7-9; Hark. 16-17)—*Arafuras* (Kolff. 161-3) -*England* (Hall., chap. viii)—*France* (Lev. ii, 48)—*England* (Free. "Sk." 232; Bage. 281)—*France* (Taine, *passim*)—*England* (Mart. "Intro." 17; Buck. vol. ii, ch. 5; Pike, ii, 574).   §574. *Bodo and D.* (As. S. B. xviii, 745-6)—*Lepchas* (Eth. Soc. "Jour." N. S. i, 152)-*Santál* (Hunt. "Ann." i, 209; As. S.B. xx, 554)—*Jakuns* (Fav. ii, 266-7)—*Bode and D.* (As. S.B. xviii, 745) -*Neilgherry II.* (Ouch. 69)—*Lepchas* (Eth. Soc. "Jour." N. S. i, 150)—*Jakuns* (Fav. ii, 266)—*Arafuras* (Kolff. 161-3)—*Lepchas* (Eth. Soc. "Jour." N. S. i, 150-1; Hook. i, 176)—*Santáls* (Hunt. "Ann." i, 217) - *Hos* (Dalt. 206)—*Todas* (Shortt, pt. i, 1)—*Shervaroy H.* (Shortt, —)—*Jakuns* (Fav. ii, 266)—*Malacca* (Jukes, 219-20)—*Bodo and D.* (As. S.B. xviii, 745)—*Santál* (Hunt. "Ann." i, 209-10)—*Lepchas* (Hook. i, 176, 129)—*Jakuns* (Fav. ii, 266)—*Arafuras* (Kolff. 163-4)—*Lepchas* (Hook. i, 134)—*Santáls* (Hunt. "Ann." 208)—*Bodo and Dhimals* (As. S.B. xviii, 708)—*Santál* (Hunt. i, 217)—*Bodo and Dhimals* (As. S.B. xviii, 744)—*Todas* (Eth. Soc. "Trans." vii, 254).

TITLES OF WORKS REFERRED TO.

(Unless otherwise stated, London is to be understood as the place of publication.)

Acos.—Acosta (Joaq.) *Compendio Histórico del Descubrimiento y Colonizacion de la Nueva Granada.* Paris, 1848.
Ad.—Adams (F. O.) *History of Japan.* 1874-5.
Alc.—Alcock (Sir Rutherford) *The Capital of the Tycoon.* 1863.
All. and T.—Allen (W.) and Thomson (T. R. H.) *Narrative of Expedition to River Niger in* 1841. 1848.
Amp.—Ampère (J. J.) *Histoire littéraire de la France avant le douz. siècle.* Paris, 1839.
A.L.F.—*Anciennes lois françaises*, éd. Jourdain, Isambert, et Decrusy. Paris, 1828, &c.
And.—Andersson (C. John) *Lake Ngami.* 1856.
Arb.—Arbousset (T.) and Daumas (F.) *Narrative of an Exploratory Tour to the North-east of the Cape of Good-Hope.* Trans. Cape Town, 1846.
As.S.B.—Asiatic Society of Bengal, *Journal.* 1855.
Ast.—Astley (T.) *Collection of Voyages and Travels.* 1745-7.
Bage.—Bagehot (Walter) *The English Constitution.* 1872.
Bail.—Bailey in *Journal Ethnological Society.* Vol. 2. 1870.
Bak.—Baker (Sir Samuel W.) *The Albert N'Yanza.* 1866.
Banc.—Bancroft (H. H.) *The Native Races of the Pacific States of North America.* 1875-6

REFERENCES.

Bar.—Barth (H.) *Travels and Discoveries in North and Central Africa.* 1857-58.
Bast.—Bastian (A.) *Der Mensch in der Geschichte.* Leipzig, 1860.
„ „ *Africanische Reisen.* Bremen, 1859.
Bates—Bates (Henry W.) *Naturalist on the River Amazons.* 1873.
Bay.—Bayard (Chev.) *History of.* Trans. by Kindersley. 1848.
Beech.—Beecham (John) *Ashantee and the Gold Coast.* 1841.
Bell—Bell (Robt.) *History of Russia* in Lard. *Cyclopædia.* 1836-8.
Bent·—Bent (J. Theodore) in *Fraser's Magazine* for December, 1880.
Bird—Bird (Miss Isab.) *Unbeaten Tracks in Japan.* 1881.
Black.—Blackstone (Sir W.) *Commentaries.* Ed. by R. Malcolm Kerr. 1876.
Bon.—Bonwick (James) *Daily Life and Origin of the Tasmanians.* 1870.
Boss.—Bossuet (J. B.) *Œuvres choisies.* Paris, 1865.
Boué – Boué (Am.) *La Turquie d'Europe.* 1840.
Bowr.— Bowring (Sir John) *The Kingdom and People of Siam.* 1857.
Boy.—Boyle (F.) *Adventures among the Dyaks of Borneo.* 1865.
Bren.—Brentano (Lujo) *Preliminary Essay on Gilds: English Gilds.* (Early Eng. Text Soc.) 1870.
Broo.—Brooke (Chas.) *Ten Years in Sarāwak.* 1866.
Bru.—Bruce (Jas.) *Travels to Discover the Source of the Nile.* Edinburgh, 1805.
Brug.—Brugsch (Dr. H.) *History of Egypt.* Trans. 1879.
Buck.—Buckle (H. T.) *History of Civilization in England.* 1867.
Burch.—Burchell (W. J.) *Travels in the Interior of Southern Africa.* 1822-4.
Burck.—Burckhart (J. L.) *Notes on the Bedouins and Wahabys.* 1831.
„ „ „ *Travels in Arabia.* 1829.
Bur.—Burton (Sir R. F.) *Mission to Gelele, King of Dahomé.* 1864.
„ „ „ *Abeokuta and the Camaroons Mountains.* 1863.
„ „ „ *Lake Regions of Central Africa.* 1860.
„ „ „ *Pilgrimage to El-Medinah and Mecca.* 1855-6.
Burt.—Burton (John Hill) *History of Scotland.* Edinburgh, 1873.
But —Butler (Major John) *Travels and Adventures in Assam.* 1855.
Cæs.—Cæsar (C. J.) *Commentarii de bello Gallico.* Recog. F. Oehler, Lips. 1863.
Call.— Callaway (Bp. H.) *The Religious System of the Amazulu.* Natal, 1868-70.
Camp.—Campbell (Major-General John) *Wild Tribes of Khondistan.* 1864.
Chér.--Chéruel (A.) *Dict. historique des institutions, mœurs et coutumes de la France.* Paris, 1874.
„ „ „ *Histoire de l'administration monarchique en France.* Paris, 1855.
Cie.—Cieza de Leon (P. de) *Travels.* Trans. by Markham (Hakluyt Society). 1864.
Clam.—Clamageran (J. J.) *Histoire de l'impôt en France.* Paris, 1867-76.
Clav.—Clavigero (Fr. S.) *The History of Mexico.* Translated by Ch. Cullen. 1787.
Coop.—Cooper (T. T.) *Mishmee Hills.* 1873.
Corn.—*Cornhill Magazine.* 1873.
Coul.—Coulanges (F. de) *La Cité Antique.* Paris, 1864.
„ „ in *Revue des deux Mondes.* Vol. xcix. 1872.
C. & W.—Crichton (A.) and Wheaton (H.) *History of Scandinavia* (Edinburgh Cab. Liby). 1838.

POLITICAL INSTITUTIONS.

Crantz—Crantz (David) *History of Greenland*. Trans. 1820.
Cur.—Curtius (E.) *History of Greece*. Trans. 1868-73.
Cust.—Custine (Marquis A. de) *Empire of the Czar*. Trans. 1843.
„   „   *La Russie en 1839*. Paris, 1843.
D'Alb.—D'Albertis (Signor L. M.) In *Transactions of Royal Colonial Institution*, Dec. 1878.
Dalt.—Dalton (Col. E. T.) *Descriptive Ethnology of Bengal*. Calcutta, 1872.
Dalz.—Dalzel (Arch.) *History of Dahomy*. 1793.
Dar.—Dareste de la Chavanne (C.) *Histoire des Classes Agricoles*. Paris, 1858.
„   „   *Histoire de l'Administration en France*. Paris, 1848.
Diaz—Diaz de Castillo (Bernal) *Memoirs* [1508]. Translated by J. Ingram Lockhart. 1844.
Dick.—Dickson (W.) *Japan*. 1869.
Dob.—Dobrizhoffer (Martin) *Account of the Abipones*. Trans. 1822.
Dom.—Domenech (Em.) *Seven Years' Residence in the Great Deserts of North America*. 1860.
Dove—Dove (Rev. T.) In *Tasmanian Journal*. Vol. 1. Hobart Town, 1842.
Drew—Drew (Fred.) *The Jummoo and Kashmere Territories*. 1875.
Duc.—Ducange (Ch. Dufresne, Sieur) *Dissertations sur l'histoire de Saint Louys:* appended to his *Glossarium*, t. vii. Paris, 1850.
Dunh.—Dunham (A. S.) *History of the Germanic Empire* (in Lardner's Cyclopædia). 1837.
„   „   *History of Poland* (in Lardner's Cyclopædia). 1830.
„   „   *History of Spain* „   „   1832.
Dur.—Duran (Fr. D.) *Historia de las Indias de Nueva España*. Mexico, 1867.
Duruy—Duruy (V.) *Histoire des Romains*. Paris, 1876.
„   „   *Histoire de France*. Nouv. éd. Paris, 1860.
Dyer—Dyer (T. F. Thistleton) *British Popular Customs*. 1876.
Ed.—Edwards (B.) *History of the British West Indies*. 1801-19.
Egin.—Eginhardus, *Life of the Emperor Karl the Great*. Trans. 1877.
Ell.—Ellis (Rev. W.) *Tour through Hawaii*. 1826.
„   „   „   *Polynesian Researches*. 1829.
„   „   „   *History of Madagascar*. 1838.
Ersk.—Erskine (Capt. J. E.) *Journal of a Cruise among the Islands of the Western Pacific*. 1853.
Eth. Soc.—Ethnological Soc. *Journal*. 1848-70.
„   „   *Transactions*. 1859-69.
Ew.—Ewald (H.) *The History of Israel*. Trans. Vols. iii & iv. 1878.
Falk.—Falkner (Thos.) *Description of Patagonia*. 1774.
Fav.—Favre (Rev. P.) In *Journal of the Indian Archipelago*. Vol. ii. Singapore, 1848.
Fis.—Fischel (E.) *The English Constitution*. Trans. 1863.
Fitz.—Fitzroy (Admiral Robert) *Narrative of the Surveying Voyages of the Adventure and Beagle*. 1839-40.
For.—Forbes (F. E.) *Dahomey and the Dahomians*. 1851.
For.—Forsyth (Captain J.) *Highlands of Central India*. 1871.
Forst.—Forster (Dr. J. R.) *Observations during a Voyage round the World*. 1778.
Fowl.—Fowler (Geo.) *Lives of the Sovereigns of Russia*. 1858.
Frank.—Franklin (Capt. Sir J.) *Narrative of two Journies to the Shores of Polar Sea*. 1823.

Free.—Freeman (Ed. A.) *The Growth of the English Constitution.* 1876.
„   „   „   *General Sketch of European History.* 1874.
„   „   „   *History of the Norman Conquest of England.*
   Oxford, 1867-76.
Froiss.—Froissart (Sir J.) *Chronicles of England, France, Spain, &c.*
   Trans. by Johnes. 1839.
Gall.—Gallatin (A.) *Notes on the Semi-civilized Nations of Mexico,
   Yucatan, and Central America* (in *Transactions of the American
   Ethnological Society*). Vol. i. New York, 1845.
Gar.—Garcilasso de la Vega, *The Royal Commentaries of the Yncas*
   Translated by Cl. R. Markham. Hakluyt Society, 1869-71.
Gho.—Ghosh (Jogendra Chandra) *Caste in India,* in *Calcutta Review* for
   1880.
Gib.—Gibbon (E.) *Fall of the Roman Empire.* Ed. by H. H. Milman. 1838.
Glad.—Gladstone (W. E.) *Studies on Homer.* Oxford, 1858.
Glas.—Glasfurd in *Selections from the Records of Government of India.*
   (Foreign Department.)
Gom.—Gomara (F. Lopez de) *Historia General de las Indias.* (In *Biblioteca de Autores Españolas,* Tomo xxii.) Madrid, 1852.
Gon.—Goncourt (E. et J. de) *Histoire de la société française pendant
   la Révolution.* Paris, 1854.
Gov. Stat.—*Government Statement on the Moral and Material Progress
   of India for* 1869-70.
Gr.—Grant (J. A.) *A Walk across Africa.* 1864.
Gra.—Grattan (T. C.) *History of the Netherlands.* (In Lardner's Cyclo.)
   1830.
Greg.—Gregory of Tours. *Historiæ Ecclesiasticæ Francorum, libri* x.
   Paris, 1836-8.
Grey—Grey (Sir Geo.) *Journals of two Expeditions of Discovery in
   Australia.* 1841.
Grimm—Grimm (Jacob) *Teutonic Mythology.* Trans. by Stallybrass,
   1880-3.
Gro.—Grote (G.) *History of Greece.* 1846-56.
Guiz.—Guizot (F.) *The History of Civilization.* Trans. (Bohn's Ed.) 1856.
Gutz.—Gutzlaff (Rev. C.) *China Opened.* 1838.
Hall.—Hallam, *Europe in the Middle Ages.* 11th Ed. 1855.
„   „   *Constitutional History.* 1854.
Hark.—Harkness (Capt. Henry) *The Neilgherry Hills.* 1832.
Har.—Harris (Sir W. C.) *Highlands of Æthiopia.* 1844.
Hawk.—Hawkesworth (Dr. J.) *Account of Voyages of Discovery in the
   Southern Hemisphere.* 1773.
Haz.—Hazlitt (W. Carew) *History of the Venetian Republic.* 1860.
Hearne—Hearne (Saml.) *Journey from Prince of Wales's Fort to the
   Northern Ocean.* Dublin, 1796.
Heer.—Heeren (A. H. L.) *Reflections on the Ancient Nations of Africa.*
   Trans. Oxford, 1832.
Herm.—Hermann (C. F.) *Manual of the Political Antiquities of Greece.*
   Trans. Oxford, 1836.
Her.—Herrera (Ant. de) *The General History of the vast Continent and
   Islands of America.* Trans. 1725-6.
Heug.—Heuglin (Th. von) *Reise in das Gebiet des Weissen Nil.* Leipzig,
   1869.
Hinc.—Hincmar, *De Ordine Palatii. Epistola.* Ed. by M. Prou. Paris, 1884.
Hodg.—Hodgson (B. H.) *Kocch, Bódo, and Dhimál Tribes.* Calcutta, 1847.
Hom.—Homer. *The Iliad.* Trans. by A. Lang, W. Leaf, and E. Myers. 1883.
„   „   *The Odyssey.* Trans. by S. H Butcher and A. Lang. 1879.

Hook.—Hooker (Sir J. D.) *Himalayan Journals.* 1854.
Huc—Huc (Prêtre Missionnaire) *Recollections of a Journey through Tartary, Thibet, and China.* Trans. 1852.
  " *The Chinese Empire.* 1855.
Humb."—Humboldt (A. von) *Personal Narrative of Travels to the Equinoctial Regions of America.* Trans. 1852-3. (Bohn.)
Hume—Hume (D.) *History of England.* 1854-5.
Hunt.—Hunter (W. W.) *Annals of Rural Bengal.* 1868.
  "   "   " *Statistical Account of Bengal.* 1875-7.
  "   "   " *Comparative Dictionary of the Languages of India and High Asia.* 1868.
Hutch.—Hutchinson (T. J.) *Buenos Ayres and Argentine Gleanings.* 1865.
Innes.—Innes (Cosmo) *Scotland in the Middle Ages.* Edinb. 1860.
  "   "   " *Lectures on Scotch Legal Antiquities.* Edinb. 1872.
Jer.—Jervis (Rev. W. H.) *History of the Gallican Church to the Revolution.* 1872.
Join.—Joinville (J. de) *Saint Louis.* Trans. by Hutton. 1868.
Jukes—Jukes (J. B.) *Voyage of H.M.S. Fly.* 1847.
Kane—Kane (Paul) *Wanderings of an Artist among Indians of North America.* 1859.
Kel.—Kelly (W.) *Excursion to California.* 1851.
Kem.—Kemble (J. M.) *The Saxons in England.* 1876.
Ken.—Kenrick (Rev. John) *Ancient Egypt under the Pharaohs.* 1850.
Kit.—Kitchen (G. W.) *A History of France.* Oxford, 1873-7.
Kœnigs.—Kœnigswarter (L. J.) *Histoire de l'organisation de la famille en France.* Paris, 1851.
Kol.—Kolben (P.) *Present State of the Cape of Good-Hope.* Trans. 1731.
Kolff—Kolff (D. H.) *Voyages of the Dutch brig Dourga.* Trans. 1840.
Kou.—Koutorga (M.) *Essai sur l'organisation de la Tribu.* Trad. par M. Chopin. Paris, 1839.
Krapf—Krapf (J. L.) *Travels in Eastern Africa.* 1860.
Kue.—Kuenen (A.) *The Religion of Israel.* Trans. by A. H. May, 1874-5.
Lact.—Lactantius, *De Mortibus Persecutorum.* Paris, 1863.
L. & O.—Laird (Macgregor) and Oldfield (R. A. K.) *Expedition into Interior of Africa.* 1837.
Landa—Landa (Diego de) *Relation des Choses de Yucatan* [1566] *Texte Espagnol et traduction française.* Par Brasseur de Bourbourg. Paris, 1864.
Lan.—Lander (Richard) *Records of Capt. Clapperton's last Expedition.* 1830.
Lav.—Laveleye (Emile de) *Primitive Property.* Trans. 1878.
Lay.—Layard (Sir A. H.) *Nineveh and its Remains.* 1849.
Lel.—Lelewel (Joachim) *Histoire de Pologne.* Paris, 1844.
Les.—Leslie (Prof. T. E. C.) in *Fort. Rev. for* 1875.
Lev.—Levasseur (E.) *Histoire des classes ouvrières en France jusqu'à la Révolution.* Paris, 1859.
Lévy.—Lévy (Daniel) *L'Autriche-Hongrie, ses Institutions et ses Nationalités.* Paris, 1871.
L. & C.—Lewis (Capt. M.) and Clark (Capt. W.) *Travels to the Source of the Missouri, &c.* 1814.
Lich.—Lichtenstein (Henry) *Travels in Southern Africa.* 1812-15.
Ling.—Lingard (Rev. Dr. John) *History of England.* 1849.
Liv.—Livingstone (D.) *Popular Account of Missionary Travels, &c. in South Africa.* 1861.

REFERENCES. 23

Lloyd—Lloyd (G. T.) *Thirty-three Years in Tasmania and Victoria.* 1862.
Loub.—Loubère (M. de la) *Du Royaume de Siam en 1687-88.* Amsterdam, 1691.
Low—Low (Hugh) *Sarawak; its Inhabitants and Productions.* 1848.
Lubb.—Lubbock (Sir John) *The Origin of Civilization and the Primitive Condition of Man.* 1882.
Macch.—Macchiavelli (N.) *Works.* Trans. by Farneworth. 1772.
M'Cull.—M'Culloch, *Selections from Records of Government of India.*
McM.—McMahon (Lieut. A. R.) *The Karens of the Golden Chersonese.* 1876.
Macph.—Macpherson, *Report upon the Khonds of Ganjani and Cuttack.* Calcutta, 1842.
Maine—Maine (Sir H. S.) *Early History of Institutions.* 1875.
„   „   „   *Village Communities in the East and West.* 1876.
„   „   „   in *Fortnightly Review* for Nov. 1881.
„   „   „   *Ancient Law.* 1861.
Mal.—Malcolm (Sir J.) *Memoir of Central India.* 1832.
„   „   „   *History of Persia.* 1815.
Mall.—Mallet (P. H.) *Northern Antiquities.* Trans. by Bishop Percy. 1847.
M. & C.—*Manners and Customs of the Japanese.* New York, 1845.
Mar.—Mariner (W.) *Account of the Natives of the Tonga Islands.* 1818.
Mars.—Marsden (W.) *History of Sumatra.* 1811.
Marsh.—Marshall (Lieut.-Col. W. E.) *A Phrenologist among the Todas.* 1873.
Martin, H.—Martin (H.) *Histoire de la France.* Vols. i, iii (Ed. 1855-61), others, 2nd Ed. Paris, 1844.
Martin, M.—Martin (M.) *Description of the Western Islands of Scotland.* 1716.
Mart.—Martineau (Harriet) *History of England during the Thirty Years' Peace.* 1849-50.
„   „   „   *Introduction to the History of the Peace.* 1851.
Mau.—Maury, in *Rev. des Deux Mondes*, tom. cvii, 1873.
May—May, Lord Farnborough (Sir Thos. Erskine) *Democracy in Europe.* 1877.
Mayer—Mayer (S.) *Die Rechte der Israeliten, Athener u. Römer.* Leipzig, 1862.
Maz.—Mazoroz (J. P.) *Histoire des Corporations Françaises d'arts et de métiers.* Paris, 1878.
Méray—Méray (A.) *La vie au temps des trouvères.* Paris, 1873.
Metz—Metz (Rev. F.) *Tribes Inhabiting the Neilgherry Hills.* Mangalore, 1864.
Mich.—Michie (Alex.) *Siberian Overland Route.* 1864.
Mitch.—Mitchell (Sir T. L.) *Three Expeditions into the Interior of Eastern Australia.* 1839.
Mit.—Mitford (A. B.) *Tales of Old Japan.* 1871.
Moff.—Moffat (Robt.) *Missionary Labours and Scenes in Southern Africa.* 1846.
Mom.—Mommsen (Theod.) *History of Rome.* Trans. Dickson. 1862.
Morg.—Morgan (L. H.) *League of the Iroquois.* Rochester, U.S.A., 1851.
Mor.—Morier (Sir R. B. D.) in *Cobden Club Essays on Local Government and Taxation.* 1875.
Mot.—Motley (J. L.) *Rise of the Dutch Republic.* 1855.

Mov.—Movers (F. C.) *Die Phoenizier.* Bonn u. Berlin, 1841-56.
Onde.—Ondegarde (P. de) *Report* in *Narratives of the Rites and Laws of the Yncas.* Translated by Markham. 1873.
Ord.—*Ordonnances des rois de France.* Paris, 1723, &c.
Ouch —Ouchterlony (Col.) *A Geographical and Statistical Memoir of a Survey of the Neilgherry Mountains.* 1847. [Printed with Shortt's Hill Ranges, Pt. 2.]
Pal.—Palgrave (W. G.) *Narrative of a Year's Journey through Central and Eastern Arabia.* 1865.
  "   "   In *Encyclopædia Britannica.* 9th ed. Article *Arabia.*
Pall.—Pallas (P. S.) *Voyages en differentes Provinces de l'Empire de Russie, &c.* Paris, 1788-93.
Park—Park (Mungo) *Travels in the Interior of Africa.* Edinb. 1858.
Par.—Parkyns (Mansfield) *Life in Abyssinia.* 1853.
Patt.—Patterson (Arthur J.) *The Magyars: their Country and Institutions.* 1869.
Pear.—Pearson (C. H.) *History of England during the Early and Middle Ages.* 1867.
Pied.—Piedrahita (L. Fernandez de) *Historia del Nuevo Reyno de Granada.* Amberes, 1688.
Pike—Pike (L. O.) *History of Crime in England.* 1873-6.
Pink.—Pinkerton (J.) *General Collection of Voyages and Travels.* 1808-14.
Pot.—Potter (Bp. John) *Archæologia Græca; or, the Antiquities of Greece.* 1837.
Pres.—Prescott (W. H.) *Conquest of Peru.* 1847.
Prid.—Pridham (Chas.) *Historical, Political, and Statistical Account of Ceylon.* 1849.
Pul.—Puliga (Comtesse de) *Madame de Sévigné.* 1873.
Raf.—Raffles (Sir T. S.) *History of Java.* 1817.
Ram.—*Rambles in Syria.* 1864.
Ranke—Ranke (Leop.) *The Civil Wars and Monarchy in France.* Trans. 1852.
Raw.—Rawlinson (G.) *Five Great Monarchies.* 1862-7.
Rea.—Reade (W. Winwood) *Savage Africa.* 1863.
Rec.—*Records of the Past, being English Translations of the Assyrian and Egyptian Monuments.* 1873-81.
Ree.—Reeves (J.) *History of the English Law.* New ed. 1869.
Ren.—Rennie (Dr. D. F.) *Bhotan and the Story of the Dooar War.* 1866.
Rev. Sib.—*Revelations of Siberia.* 1853.
Rich.—Richter (G.) *Annalen der deutschen Geschichte im Mittelalter.* Halle, 1873.
Roth.—Roth (P.) *Feudalalität und Unterthanenverband.* Weimar, 1863.
Roy. G. S.—Royal Geographical Soc. *Journal.* 1852.
Saha.—Sahagun (Fr. Bernardino de) *Historia General de las Cosas de Nueva España* [1569]. Por C. M. de Bustamente. Mexico, 1829-30.
Saint John—St. John (Sir Spenser) *Life in the Forests of the Far East.* 1863.
Saint Sim.—Saint-Simon (Duc de) *Memoirs.* Abridged by St. John. 1857.
Salv.—Salvianus, *De Gubernatione Dei.* Paris, 1608.
Sang.—Sangermano (Father) *Description of the Burmese Empire.* Rome, 1833.
Santa C.—Santa Cruz in *Narratives of the Rites and Laws of the Yncas.* Translated by Cl. R. Markham. (Hakluyt Society.) 1873.

Sart.—Sartorius (C.) *Mexico.* Edited by Dr. Gaspey. 1858.
Shool.—Schoolcraft (H. R.) *Expedition to the Sources of the Mississippi River.* Philadelphia, 1855.
„ „ *Travels in the Central Portions of the Mississippi Valley.* New York, 1825.
„ „ *Information respecting the Indian Tribes of the United States.* 1853-6.
Seel.—Seeley (Prof. J. R.) *Lectures and Essays.* 1870.
See.—Seeman (B.) *Viti: an Account of a Mission to the Vitian or Fijian Islands.* Cambridge, 1862.
Shar.—Sharpe (Samuel) *History of Egypt.* 1876.
Shoot.—Shooter (Rev. Jos.) *The Kafirs of Natal and the Zulu Country.* London, 1857.
Shortt—Shortt (Dr. J.) *Hill Ranges of Southern India.* Madras, 1870-1.
Sim.—Simon (P.) *Noticias Historiales de las Conquistas de Tierra Firme en el Nuevo Reyno de Granada.* 1624. In Kingsborough's *Antiquities of Mexico*, Vol. viii.
Sis.—Sismondi (J. C. L. de) *History of the Italian Republics* (in Lardner's Cyclopædia). 1832.
„ „ *Histoire des Républiques Italiennes.* Paris, 1826.
Ske.—Skene (W. F.) *Celtic Scotland.* Edinburgh, 1876-80.
Smith—Smith (E. R.) *The Araucanians.* New York, 1855.
Sm.—Smyth (R. Brough) *Aborigines of Victoria.* Melbourne, 1878.
Sohm—Sohm (R.) *Die altdeutsche Reichs- und Gerichtsverfassung.* Bd. i. Weimar, 1871.
Sou.—Southey (Rob.) *History of Brazil.* 1810-19.
Spix.—Spix (J. B. von) and Martius (C. F. P. von) *Travels in Brazil.* Trans. 1824.
Squi.—Squier (E. G.) *Nicaragua.* New York, 1852.
„ „ *Observations on the Geology and Archæology of Peru.* 1870.
Stew.—Stewart (Lieut. R.) in *Journal Asiatic Society, Bengal.* 1855.
Stubbs—Stubbs (Bp. Wm.) *The Constitutional History of England.* Oxford, 1880.
Sturt—Sturt (Capt. Chas.) *Two Expeditions into the Interior of Southern Australia.* 1833.
Sup. Rel.—*Supernatural Religion.* 1874.
Tac.—Tacitus (C. C.) *Germania.* Trans. by John Aikin, 1823.
Taine—Taine (H. A.) *The Ancient Régime.* Trans. 1876.
Tay.—Taylor (Dr. W. C.) *Student's Manual of Ancient History.* 1849.
Ten.—Tennant (Sir J. Emerson) *Ceylon; an Account of the Island, &c.* 1859.
Tern.—Ternaux-Compans (H.) *Recueil de Pièces relatives à la Conquête du Mexique.* In *Voyages, &c. pour servir à l'histoire de la découverte de l'Amerique.* Vol. x. Paris, 1838.
Thie.—Thierry (A.) *Formation and Progress of the Tiers Etat.* Trans. 1859.
Thirl.—Thirlwall (Bp. C.) *History of Greece* (Lardner's Cyclopædia.) 1835-47.
Thomp.—Thompson (Geo.) *Travels and Adventures in Southern Africa.* 1827.
Thomps.—Thompson (G. A.) *Alcedo's Geogr. and Historical Dictionary of America, &c.* 1812.
Thom.—Thomson (Dr. A. S.) *The Story of New Zealand: Past and Present.* 1859.

Thor.—Thorpe (B.) *Diplomatarium Anglicum Œvi Saxonici, a Collection of English Charters, &c.* 1865.
Tie.—Tiele (C. P.) *Outlines of the History of Religion to the Spread of the Universal Religions.* Trans. 1877.
Tits.—Titsingh (I.) *Annales des Empereurs de Japon.* 1834.
Tor.—Torquemada (J. de) *Monarquia Indiana.* Madrid, 1723.
Toz.—Tozer (Rev. H. F.) *Lectures on the Geography of Greece.* 1873.
Turn.—Turner (Sharon) *History of England.* 1839.
Tur.—Turner (Rev. G.) *Nineteen Years in Polynesia.* 1861.
Vieus.—Vieusseux (A.) *History of Switzerland.* 1840.
Wai.—Waitz (Georg) *Deutsche Verfassungsgeschichte*, Kiel, Vols. i and ii, second edition, 1865-70; Vols. iii and iv, 1860-1.
Waitz—Waitz (Dr. Theodor) *Anthropologie der Naturvölker.* Leipzig, 1859, etc.
Wal.—Wallace (D. M.) *Russia.* 1877.
Warn.—Warnkœnig (L. A.) und Stein (L.) *Französische Staats- und Rechtsgeschichte.* Basel, 1846.
Wilkes—Wilkes (Capt. C.) *Narrative of United States Exploring Expedition.* 1845.
Wilk.—Wilkinson (Sir J. G.) *Manners and Customs of the Ancient Egyptians.* New edition, by Samuel Birch. 1878.
Wil.—Williams (Rev. T.) *Fiji and the Fijians.* 1858.
Will.—Williams (S. Wells) *The Middle Kingdom.* 1848.
Wint.—Winterbottom (T.) *Account of the Native Africans in the Neighbourhood of Sierra Leone.* 1803.
Wood—Wood (Lieut. J.) *Journey to the Source of River Oxus.* 1841.
Xim.—Ximenes (F.) *Las Historias del Origen de los Indios de Guatemala.* Viena, 1857.
You.—Young (Ernest) In *Anglo-Saxon Family Law.* Boston, U.S.A. 1876.
Zur.—Zurita (A.) *Rapport sur les différentes classes des Chefs de la Nouvelle Espagne.* In *Voyages, relations, etc. pour servir à l'histoire de la découverte de l'Amerique.* Par H. Ternaux-Campans. Paris, 1840.

REFERENCES.

(For explanation see the first page of References.)

ECCLESIASTICAL INSTITUTIONS.

§ 583. *The deaf* (Kit. 200; Sm. 4)—*Weddas* (Ha·ts. 413)—*Dôr* (Hcug. 195)—*Bongo* (Schw. i, 304-5)—*Zulus* (Gard. 72)—*Latooki* (Bak. i, 247-50). § 584. *Australians* (Smy. i, 107)—*Malagasy* (Rév. 9-11)—*Japanese* (Sat. 87; 79-80)—*India* (Ly. 18)—*Greeks* (Pla. iv; Gro. iii, 187). § 585. *Zulu* (Call. 230-1)—*Andamanese* (J.A.I. xii, 162)—*Waraus* (Brett, 362)—*Chinooks* (U. S. Ex. v, 118)—*Andamanese* (J.A.I. xii, 142)—*Waraus* (Bern. 53)—*Urua* (Cam. ii, 110)—*Zulus* (F.S.A.J. ii, 29)—*Nicaraguans* (Banc. ii, 801)—*Ahts* (Banc. iii, 521)—*Gonds* (His. 19)—*Ukiahs and Sanéls* (Banc. iii, 524)—*Zulus* (Call. 372)—*Shillook* (Schw. i, 91)—*Indians* (School. v, 403)—*Indians* (School. v, 403)—*Chibchas* (Boll. 12)—*China* (Edk. 42)—*E. English* (Kem. ii, 208-9)—*Mongols* (Prej. i, 76)—*Vera Paz* (Banc. ii. 799)—*Mosquitos* (Banc. i, 744)—*Wakhutu* (Thoms. i, 19)—*Africa* (Pinto, i, 124)—*Borneo* (Bock, 78)—*Greeks* (Mau. ii, 33-4)—*Egypt* (Klunz. 103-5)—*Gambia* (Ogilby, 369)—*Blantyre* (MacDon. i, 59—110)—*Dyaks* (St. J. i, 199)—*Nyassa* (Liv. i, 353)—*S. Leone* (Bast. "Mensch," ii, 129)—*Damaras* (Anders. 229)—*Bhils* (T.R.A.S. i, 72)—*Wahehe* (Thoms. i, 237)—*Bongo* (Schw. i, 305)—*Blantyre* (MacDon. i, 62-3)—*Poland* (Mau. ii, 463; 58)—*Apaches* (Banc. iii, 527)—*Nayarit* (Banc. iii, 529)—*Babylonians* (ref. lost)—*Ainos* (Bird, ii, 97; 98)—*Mongols* (How. i, 33)—*England* (Free. i, 768, 521)—*Borneo* (Boy. 229)—*Esquimaux* (Hayes, 199)—*Edinburgh* (Kitto, 199-200)—*Californians* (Banc. iii, 523)—*Mangaia* (ref. lost)—*Hawaii* (Cum. i, 295)—*Natches* (ref. lost)—*Egypt* (ref. lost)—*Beirût* (Jessup, 243)—*Bushmen* (F.S.A.J. ii, 42-3)—*Greece* (Gro. i, 14; Sm., W. ii, 319)—*Amandabele* (Sel. 331)—*Hindoos* (Ly. 19)—*Gauls* (Coul. i, 89; 91)—*Teutons* (Vel. Pat. c. 105)—*Norse* (Das. xviii; Mal. 153)—*Hamóa* (Mar. ii, 112).
§ 586. *Egypt* (Ren. 153; Rec. ii, 11; Ren. 151-2; 153; Bru. i, 70; Rec. iv, 130-1; Mus. "Rév. Sci." 819; Herod. ii, 206; Rec. vi, 144; Bru. i,

84; T.B.A.S. vii, pt. i; Mns. "Rév. Sci." 819; Stu. 94; 150-2; Rcc. viii, 95, 98; Bru. i, 425, 124; Rcc. iv, 58-9; Bru. i, 88; Rcc. viii, 77-8; Ren. 86-7)—Note (Bru. i, 114; chap. iii). § 587. *Hindus* (Wil. 32-4)—*Assyrians* (Rcc. v, 3-4; Smith, 13-14)—*Hebrews* (Chey. 33; Müll. "S. of R." 110)—*Abraham* (Ew. i, 295)—*Hebrew Pantheon* (Sup. Rel. i, 110)—*Bedouins* (Burck. i, 259 et seq.)—*Greeks* (Pot. i, 172)—*Egypt* (Rcc. vi, 101-2)—*Peruvians* (Mol. 17)—*Greece* (Push. i, 213-4)—*Early Romans* (Mom. i, 183)—*Sandwich I.* (Vanc. ii, 149)—*Chaldea* (Rcc. vii, 133)—*America* (School. iii, 317; Brett, 401)—*Egypt* (Rcc. vi, 103)—*Cent. Amer.* (Ovie. bk. xlii, ch. 2)—*Mongols* (How. i, 37)—*Peru* (Anda. 57)—*Mangaia* (Gill, 118)—*Fiji* (Wil. 185)—*Padam* (Dalt. 25)—*Greece* (Gro. iv, 82-5; 95; i, 626). § 589. *Patagonians* (Fitz. ii, 152)—*N. Americans* (Burt. 131)·-*Guiana* (Dalton, i, 87)—*Munducurús* (Bates, 225). § 590. *Zulus* (Call. 157)—*Bouriats* (Mich. 200)—*Kibokwé* (Cam. ii, 188-9)—*Kamtschatkans* (Kotz. ii, 13)—*New Zealand* (Yate, 141)—*Wáralis* (J.R.A.S. vii, 20). § 591. *Uaupés* (Wall. 499)—*Great Cassan* (Ogil. 355-6). § 592. *Egypt* (Ren. 211-12)—*Assyria* (Smith, 16). § 594. *New Britain* (Pow. 197)—*Santáls* (Hun. i, 183)—*Karens* (J.A.S.B. xxxiv, 205). § 595. *Samoans* (Tur. "Samoa," 151)—*Banks Islanders* (J.A.I. x, 286)—*Blantyre Negroes* (MacDon. i, 61). § 596. *New Caledonia* (Tur. "Poly." 427)—*Madagascar* (Ell. "Mad." i, 390)—*India* (Per. 303). § 597. *Samoans* (Tur. "Pol." 239)—*Tahitians* (Ell. "Pol. Res." ii, 208)—*Madagascar* (Dru. 236)—*Ostyaks* (Pri. iii, 336)—*Gonds* (His. 19)—*Chinese* (Gutz. i, 503)—*Sabæans* (Pal. ii, 258)—*Hebrews* (Kue. i, 338-9)—*Aryans* (Maine, 85). § 598. *Egypt* (Ren. 138)—*Aryans* (Dunc. iv, 252, 264-5)—*Jews* (Zim. 495-6)—*Corea* (Ross, 322). § 599. *Japan* (Ada. i, 6)—*Rome* (Hun. "Ex." 746)—*Aryans* (Maine, 55, 78, 64, 79, 55; Hun. "Intro." 149)—*Christendom* (Maine, 79)—*India* (Maine, 56). § 600. *Egypt* (Ren. 134-5; Brug. ii, 40-1)—*Assyria* (Rcc. v, 81, 8). § 601. *China* (Doo. ii, 226)—*Corea* (Ross, 335). § 602. *Asia* (Huc, ii, 55)—*Ethiopians* (Rcc. vi, 73-8)—*Peruvians* (Garci. v, 8)—*New Caledonians* (Tur. "Poly." 526). § 603. *Tanna* (Tur. "Pol." 88)—*Mangaia* (Gill, 293-4)—*New Zealanders* (Thom. i, 114)—*Madagascar* (Ell. "Mad." i, 359)—*Sandwich Islands* (Ell. "Pol. Res." ii, 235)—*Humphreys Island* (Tur. "Samoa," 278)—*Pueblo* (Banc. iii, 173)—*Maya* (Banc. ii, 647)—*Peru* (Pres. 11-12)—*Siam* (Thom. J. 81)—*Javanese* (Craw. iii, 15)—*China* (Mcd. 133)—*Japan* (ref. lost)—*Greeks* (Blac. 45; Gro. ii, 475; Mau. ii, 382-4)—*Romans* (See. 55)—*Scandinavians* (Das. xlvi & lxii)—*Europe* (Fréd. ii, 414, v, 433). § 604. *Blantyre Negroes* (MacDon. i, 65, 64-5, 64)—*Niger* (Bur. 132)—*Samoa* (Tur. "Samoa," 18-19)—*Scandinavians* (Das. xiii)—*Greeks* (Glad. "Homer," iii, 55)—*Hebrews* (Kue. i, 338-9). § 606. *Romans* (Cou¹. "Cité," 233)—*Blantyre Negroes* (MacDon. i, 64)—*New Zealanders* (Ang. i, 247)—*Mexican* (Cla. i, 271)—*Peru* (Garci. bk. ii, ch. 9)—*Khonds* (Macph. 30)—*Tahiti* (Ell. "Pol. Res." ii, 208)—*Ashantee* (Dup. 168)—*Maya* (Banc. ii, 648)—*Egypt* (Bru. i, 46)—*Damaras* (And. 223)—*Dahomans* (Burt. ii, 173)—*Peru* (Mol. 25)—*Chibchas* (Sim. 247-8)—*Karens* (J.A.S.B. xxxiv, 206). § 607. *Ostyaks* (Erm. ii, 44)—*Gonds* (For. 142)—*Kukis* J.A.S.B. xxiv, 630)—*Latooka* (Bak. ii, 4-5)—*Bechuanas* (Hol. i, 324)—*Gonds* (His. 19). § 608. *Damaras* (And. 224)—*Gonds* (His. 19)—*Santáls* (Hun. i, 200-1)—*Peruvians* (Garci. bk. ii, ch. 9). § 610. *Malagasy* (Ell. "Mad." i, 395)—*Egypt* (Bru. i, 15; Wilk. i, 173)—*Rome* (See. 93)—*Mexicans* (Cla. i, 271)—*Peru* (Ciez. 262). § 611. *Egyptians* (Gro. iii, 438)—*Peruvians* (Mol. 54-5)—*Greece* (Cur. i, 323). § 612. *Fiji* (Wil. —)—*Greece* (Cur. i, 369). § 613. *Aryans* (Müll. "Sans. Lit." 533)—*Peruvians* (Garci. bk. iii, ch. 8; Herr. iv, 343). § 614.

Mexico (Brin. 56-7)—*Peru* (Mol. 11). § 615. *Comanches* (School. i, 231)—*New Zealand* (Cook, "Hawk," 389)—*Fiji* (Wil. 185)—*Christians* (Bing. iii, 13; Mos. i, 283). § 617. *Nagas* (J.A.S.B. xxiv, 608; But. 150)—*Comanches* (School. i, 231, 237)—*Eastern Slavs* (Tie. 188)—*Bodo and Dhimáls* (Hodg. 159, 162; J.A.S.B. xviii, 721)—*Arabs* (Tie. 64)—*Greeks* (Glad. "Juv. Mun." 181)—*Tahi'i* (Ell. "Pol. Res." ii, 208)—*Ancient Egypt* (Sha. i, 11)—*Japanese* (Grif. 99-100)—*China* (Gutz. ii, 331; Tic. 29). § 618. *Mexico* (Cla. i, 269, 270; Herr. iii, 220)—*Peru* (Arr. 23)—*Mexico* (Herr. iii, 203)—*Abyssinia* (Bruce, iv, 466; v, 1). § 619. *Egyptians* Tic. 45-6)—*Romans* (Sm. Geo. 105)—*Christian Society* (Guiz. i, 35-6)—*Bodo and D.* (J.A.S.B. xviii, 733)—*Mexico* (Cla. i, 271, &c.)—*Peru* (Garci. bk. ii, ch. 9; Herr. iv, 344)—*Egypt* (Ken. i, 450-2)—*Babylon* (Mau. —)—*Rome* (See. 93)—*Mexico* (Cla. i, 272)—*Europe* (Guiz. ii, 45-6)—*Christian Churches* (Mos. i, 144-6)—*Anglo-Saxon Clergy* (Ling. i, 146). § 620. *Guatemala* (Xim. 177)—*Monachism* (Blun. 487; Hook, 5th ed. 618; Ling. i, 149). § 622. *Ostyaks* (Lath. i, 456). § 623. *Egyptians* (Heer. ii, 114; Herod. ii, 76, note)—*Greeks* (Gro. ii, 324-5; Cur. ii, 2; i, 112; ii, 19)—*Etruscans* (Mom. i, 141)—*Alba* (Mom. i, 43)—*Rome* (See. 89). § 624. *Tahitians* (Ell. "Pol. R." i, 114)—*Chibchas* (Pic. bk. ii, ch. 7)—*Latium* (Mom. i, 44)—*Greeks* (Gro. iv, 91; Curt. i, 116-7; ii, 12)—*Europe* (Hal. 365). § 625. *Zoroaster* (Rob. xxiii-iv). § 626. *Ancient Mexicans* (Diaz, ch. 20?)—*San Salvador* (Pala. 75)—*Chibchas* (Sim. 248-9)—*Karens* (J.A.S.B. xxxiv, 207)—*Rome* (Mom. i, 215)—*Nagas* (J.A.S.B. xxiv, 612)—*Todas* (Mars. 81)—*Damaras* (And. 224)—*Germany* (Pesch. 144)—*Scotland* (Mart. 113)—*Creeks* (School. v, 260)—*Dahomey* (Burt. ii, 150)—*Japan* (Dick. 14)—*Greece* (Gro. iii, 68). § 628. *Ancient Mexicans* (Herr. iii, 213)—*Fijians* (Ersk. 428)—*Assyrians* (Rec. iii, 104)—*Sandwich Islanders* (Cook, "Last Voy." 303)—*Ancient Mexicans* (Saha. bk. viii, ch. 24)—*Yucatanese* (Fan. 308)—*Chibchas* (Herr. v, 90)—*Ancient Mexicans* (Herr. iii, 213)—*Assyria* (Smith, 13)—*Fijians* (Ersk. 440). § 629. *Ancient Mexicans* (Ban. ii, 201)—*Romans* (Coul. "Cité," 218)—*Tahitians* (Ell. "Pol. Res." i, 293; ii, 489). § 630. *Dakotahs* (School. ii, 184)—*Abipones* (Dob. ii, 76)—*Khonds* (Macph. 57)—*Spartans* (Hase, 194)—*Gold Coast* (Cruick. ii, 172)—*Yucatanese* (Herr. iv, 16)—*Primitive Germans* (Stub. i, 34)—*Samoans* (Tur. "Poly." 303)—*New Caledonia* (Tur. "Poly." 427)—*Comanches* (School. ii, 131)—*Egyptian War* ("Daily News," Aug. 7, 1882)—*Eggarahs* (Ail. & T. i, 327)—*Ancient Mexicans* (Cla. i, 271)—*Peruvians* (Pres. 164)—*Guatema'a* (Tor. bk. ix, ch. 6)—*San Salvador* (Pal. 73). § 631. *France* (Roth, 320, 317-3; Leb. vii, 119)—*Church* (Guiz. ii, 58)—*Germany* (Dunh. ii, 121)—*France* (Ord. viii, 24; Guiz. iii, 299)—*Fifteenth century* (Mons. iii, ch. 158)—*Montenegrins* (ref. lost; Den. 83-4)—*Richelieu* (Kitch. iii, 61; Chér. i, 299, 300). § 633. *Polynesians* (Ell. "Pol. Res." ii, 377)—*Assyria* (Lay. ii, 473-4). § 634. *France* (Bed. i, 8; Guiz. i, 36)—*Germany* (Dunh. i, 135)—*England* (Hul. 101)—*Thirteenth century* (Hal. 367). § 635. *Coast Negroes* (Lan. i, 281)—*Yucatan* (Liç. 8)—*Egyptians* (Wilk. i, 186)—*Old English* (Kem. ii, 393)—*Ecclesiastical Courts* (Jer. i, 71). § 636. *Gold Coast* (Cruik. ii, 157)—*Fijian Chiefs* (U.S. Ex. iii, 89; Will. 191)—*Abyssinia* (Harr. iii, 25)—*Marutse* (Holl. ii, 241)—*Dyaks* (Boy. 201)—*Tartars* (Huc, "Christ." i, 232)—*Mexico* (Clav. i, 271)—*Michoacan* (Banc. —)—*Egypt* (Wilk. i, 168)—*Burmah* (Sang. 53). § 638. *Mangaia* (Gill, 293)—*Egyptians* (Herod. "Hist." ii, 43)—*Bhutan* (Bog. 33)—*Egyptians* (Wilk. iii, 354). § 639. *Zulus* (Call. 340)—*Rome* (Mom. i, 158-9)—*Chibchas* (Sim. 248-9)—*Mediæval Europe* (Dun. ii, 63)—*Mandalay* (Fyt. ii, 195)—*Ancient Mexicans* (Zur. 387)—*Peruvians* (Onde. 156)—*Egypt* (Ken. ii, 37)—*Rome* (Mom. ii, 433).

§ **640.** *Zulus* (Call. 378)—*Samoans* (Bodd. 228–31)—*Greeks* (Curt. i, 151)—*Romans* (Mom. ii, 423)—*Japanese* (Dick. 41))—*Nahuan nations* (Ban. ii, 142). § **644.** *Primitive Methodists* (Hook. 7th ed. 497-8). § **646.** *Tahitians* (Ell. "Pol. Res." ii, 478)—*Mexicans* (Herr. i i, 212)—*Chibchas* (Pie. bk. i, ch. 4)—*Belochis* (Burt. "Sind," ii, 169)—*Chibchas* (Pic. bk. i, ch. 2)—*Domras* (see vol. i of this work, 3rd ed. p. 785)—*Friendly Islanders* (ref. lost)—*Caribs* (Heri. 335)—*Brazilian tribes* (J.R.G.S. ii, 198). § **647.** *Polynesia* (Ell. "Pol. Res." ii, 378). § **648.** *Tonga Islands* (Mar. ii, 220)—*Polynesia* (Ell. "Haw." 394)—*New Zealanders* (Thom. i, 103)—*New Hebrides* (J.E.S. iii, 62)—*Timor* (Wall. "Mal. Arch." 196)—*Congoese* (Bast. "Af. R." 78 ; "Mensch." iii, 225). § **649.** *Dakotahs* (School. ii, 195)—*Mangaia* (Gill, 26)—*Peruvians* (Acos. bk. v, ch. 25). § **650.** *Waldenses* (Boo. 18). § **653.** *Egyptians* (Ren. 26 ; Mau. "Rev. Sci.")—*Mexicans* (Tern. i, 86)—*Indo-Aryans* (Raj. i, 423)—*Romans* (Clar. 334)—*Hindus* (Sher. lxxi, 33)—*Thracians* (Gro. iv, 29).

TITLES OF WORKS REFERRED TO.

(If not otherwise specified, London is to be understood as the place of publication.)

Acos.—Acosta (Jos. de) *Historia Natural y Moral de las Indias.* Sevilla, 1590.
Ada.—Adams (F. O.) *History of Japan from the Earliest Period to the Present Time.* 1874–5.
All. & T.—Allen (W.) and Thomson (Dr.) *Narrative of Expedition to River Niger in 1841.* 1848.
Anda.—Andagoya (P. de) *Narrative of the Proceedings of Pedrarius Davila.* Trans. Markham (Hakluyt Socy.). 1865.
And.—Andersson (C. J.) *Lake Ngami.* 1856.
Ang.—Angas (G. F.) *Savage Life and Scenes in Australia and New Zealand.* 1847.
Arr.—Arriaga (P. J. de) *Extirpacion de la Idolatria del Piru.* Lima, 1621.
Bak.—Baker (Sir Samuel) *Albert N Yanza.* 1866.
Banc.—Bancroft (H. H.) *The Native Races of the Pacific States of N. America.* 1875–6.
Bast.—Bastian (A.) *Der Mensch in der Geschichte.* Leipzig, 1860.
" —— *Africanische Reisen.* Bremen, 1859.
Bates—Bates (H. W.) *Naturalist on the River Amazons.* 1873.
Bed.—Bedollierre (E. de la) *Histoire des Mœurs et de la Vie privée des Français.* 1847.
Bern.—Bernau (Rev. J. H.) *Missionary Labours in British Guiana.* 1847.
Bing.—Bingham (Rev. J.) *Works.* 1855.
Bird.—Bird (Miss I. L.) *Unbeaten Tracks in Japan.* 1880.
Blac.—Blackie (Prof. J. S.) *Horæ Hellenicæ.* 1874.
Blun.—Blunt (Rev. J. H.) *Dictionary of Doctrinal and Historical Theology.* 1872.
Bock.—Bock (Carl) *Head-Hunters of Borneo.* 1881.
Bodd.—Boddam-Whetham (J. W.) *Pearls of the Pacific.* 1876.
Bog.—Bogle (G.) *Narrative of Mission to Tibet.* Ed. by C. R. Markham. 1876.
Boll.—Bollaert (W.) *Researches in New Granada, Equador, Peru, and Chili.* 1860.

REFERENCES.

Boo.—Boone (Rev. T. C.) *Book of Churches and Sects.* 1826.
Boy.—Boyle (F.) *Adventures among the Dyaks of Borneo.* 1865.
Brett—Brett (Rev. W. H.) *Indian Tribes of Guiana.* 1868.
Brin.—Brinton (D. G.) *The Myths of the New World.* New York, 1868.
Bruce—Bruce (James) *Travels to Discover the Source of the Nile.* Edinb. 1805.
Bru.—Brugsch-Bey (H.) *History of Egypt.* Trans.
Burck.—Burckhardt (J. L.) *Notes on the Bedouins and Wahábys.* 1831.
Bur.—Burdo (Adolphe) *The Niger and the Benueh.* Trans. 1880.
Burt.—Burton (Capt. R. F.) *Sind Revisited.* 1877.
„ —— *Mission to Gelele, King of Dahomey.* 1864.
„ —— *City of the Saints.* 1861.
But.—Butler (Major John) *Travels and Adventures in Assam.* 1855.
Call.—Callaway (Bp.) *The Religious System of the Amazulu.* Natal, 1868-70.
Cam.—Cameron (Com. V. L.) *Across Africa.* 1877.
Chér.—Chéruel (A.) *Histoire de l'Administration monarchique en France.* Paris, 1855.
Chey.—Cheyne (Rev. T. K.) *The Book of Isaiah chronologically arranged.* 1870.
Cioz.—Cieza de Leon (P. de) *Travels.* Markham's translation (Hakluyt Society) 1864.
Clar.—Clarke (Rev. J. Freeman) *Ten Great Religions.* 1871.
Cla.—Clavigero (Fr. S.) *The History of Mexico.* Trans. 1787.
Cook—Cook (Capt. J.) *Voyage round the World*, in *Hawkesworth's Voyages.* Vol. II.
„ —— *Last Voyage.* 1781.
Coul.—Coulanges (Fustel de) *Hist. des Institutions politiques de l'ancienne France.* Paris, 1874.
„ —— *La Cité Antique.* Trans. Boston, 1874.
Craw.—Crawfurd (John) *History of the Indian Archipelago.* 1820.
Cruik.—Cruikshank (Brodie) *Eighteen Years on the Gold Coast of Africa.* 1853.
Cum.—Cumming (C. F. Gordon) *Fire Fountains; the Kingdom of Hawaii, its volcanoes, etc.* 1883.
Cur.—Curtius (E.) *History of Greece.* Trans. 1868-73.
Dalt.—Dalton (E. T.) *Descriptive Ethnology of Bengal.* Calcutta, 1872.
Dalton—Dalton (H. G.) *History of British Guiana.* 1855.
Das.—Dasent (Sir G. W.) *Story of Burnt Njal.* 1861.
Den.—Denton (Rev. W.) *Montenegro: its people and their History.* 1877.
Diaz.—Diaz del Castillo (B.) *Memoirs.* [1598.] Trans. 1844.
Dick.—Dickson (Walter) *Japan, being a Sketch of the History, Government, and Officers of the Empire.* 1869.
Dob.—Dobrizhoffer (M.) *Account of the Abipones.* Trans. 1822.
Doo.—Doolittle (Rev. Justus) *Social Life of the Chinese.* 1866.
Dru.—Drury (R.) *Madagascar, or Journal during Fifteen years' Captivity on that Island.* 1731.
Dup.—Dupuis (Joseph) *Journal of a Residence in Ashantee.* 1824.
Dunc.—Duncker (Prof. Max) *History of Antiquity.* Trans. 1882.
Dunh.—Dunham (S. A.) *History of the Germanic Empire* (in Lardner's Cyclopædia). 1834.
Edk.—Edkins (Rev. J.) *The Religious Condition of the Chinese.* 1859.
Ell.—Ellis (Rev. W.) *History of Madagascar.* 1838.
„ —— *Polynesian Researches.* 1829.
„ —— *Narrative of a Tour through Hawaii.* 1827.
Erm.—Erman (G. A.), *Travels in Siberia.* Trans. 1848.

Ersk.—Erskine (Capt. J. E.) *Journal of a Cruize among the Islands of the Western Pacific.* 1853.
Ew.—Ewald (H.) *The History of Israel.* Trans. 1867-71.
Fan.—Fancourt (Ch. St. J.) *History of Yucatan.* 1854.
Fitz.—Fitzroy (Admiral R.) *Narrative of the Surveying Voyages of the "Adventure" and "Beagle"* 1839-40.
F.S.A.J.—*Folk-lore (South African) Journal.* Vol. ii.
For.—Forsyth (Capt. J.) *Highlands of Central India.* 1871.
Fred.—Fredegarius. *Chronique:* in Guizot's *Collection des Mémoires relatifs à l'histoire de France.*
Free.—Freeman (E. A.) *History of the Norman Conquest of England.* 2nd edition. 1870-6.
Fyt.—Fytche (Lt.-Gen. A.) *Burma, Past and Present.* 1878.
Garci.—Garcilasso de la Vega. *First Part of the Royal Commentaries of the Yncas.* Trans. 1869-71 (Hakluyt Soc.)
Gard.—Gardiner (Capt. A. F.) *Journey to the Zoolu country in South Africa.* 1836.
Gill.—Gill (Rev. Wm. W.) *Myths and Songs from the South Pacific.* 1876.
Glad. Gladstone (W. E.) *Studies on Homer.* 1858.
  " —— *Juventus Mundi.* 1869.
Gro.—Grote (G.) *History of Greece.* 1846-56.
Grif.—Griffis (W. E.) *The Mikado's Empire.* New York, 1876.
Guiz.—Guizot (F.) *History of Civilization.* Trans. 1856.
Gutz.—Gützlaff (Rev. C.) *China Opened.* 1838.
Hal.—Hallam (Hy.) *Europe in the Middle Ages.* Reprint of 4th ed. 1869.
Harr.—Harris (Sir W. C.) *Highlands of Æthiopia.* 1844.
Harts.—Hartshorne (B. T.) *The Weddas* in *Fortnightly Rev.,* vol. xix, N.S.
Hase.—Hase (H.) *Public and Private Life of the Ancient Greeks.* Trans. 1836.
Hayes—Hayes (I. I.) *Arctic Boat-Journey.* 1860.
Heer.—Heeren (A. H. L.) *Historical Research into the Politics, Intercourse, and Trade of the Carthaginians, Ethiopians, and Egyptians.* Trans. 1832.
Heri.—Heriot (G.) *Travels through the Canadas.* 1807.
Herod.—Herodotus, *History.* Translated by Rev. Geo. Rawlinson. 1858.
Herr.—Herrera (Ant. de) *The General History of the vast Continent and Islands of America* [1601]. Trans. 1725-26.
Heug.—Heuglin (Th. von) *Reise in das Gebiet des Weissen Nil.* 1869.
His.—Hislop (Rev. S.) *Aboriginal Tribes of the Central Provinces.*
Hodg.—Hodgson (B. H.) *Kocch, Bōdo, and Dhimál tribes.* 1847.
Hol.—Holub (Dr. Emil) *Seven Years in South Africa.* Trans. 1881.
Hook—Hook (Dean W. F.) *A Church Dictionary.* 5th ed. 1846, and 7th ed. 1854.
How.—Howorth (H. H.) *Hist. of the Mongols.* 1876, etc.
Huc—Huc (E. R.) *The Chinese Empire.* Trans. 1855.
  " —— *Christianity in China.* Trans. 1857.
Hun.—Hunter (Wm. A.) *Introduction to Roman Law.* 1880.
  " —— *Systematic and Historical Exposition of Roman Law.* 1885.
Hun.—Hunter (W. W.) *Annals of Rural Bengal.* 1868.
J.A.I.—*Journal of the Anthropological Institute.*
J.A.S.B.—*Journal of the Asiatic Society, Bengal.*
J.R.A.S.—*Journal of the Royal Asiatic Society, London.*
J.E.S.—*Journal of the Ethnological Society.*
J.R.G.S.—*Journal of the Royal Geographical Society, London.*
Jer.—Jervis (Rev. W. H.) *History of the Church of France.* 1872.

Jessup.—Jessup (Rev. H. H.) *The Women of the Arabs.* 1874.
Kem.—Kemble (J. M.) *The Saxons in England.* 1849.
Ken.—Kenrick (Rev. J.) *Ancient Egypt under the Pharaohs.* 1860.
Kit.—Kitto (Dr. J.) *The Lost Senses.* 1853.
Kitch.—Kitchin (G. W.) *A History of France.* 1873.
Klunz.—Klunzinger (C. B.) *Upper Egypt.* Trans. 1878.
Kotz.—Kotzebue (Otto von) *New Voyage Round the World.* Trans. 1830.
Kue.—Kuenen (A.) *The Religion of Israel.* Trans. 1874-5.
Lan.—Lander (Richard) *Records of Captain Clapperton's last Expedition, etc.* 1830.
Lath.—Latham (R. G.) *Descriptive Ethnology.* 1859.
Lay.—Layard (Sir A. H.) *Nineveh and its Remains.* 1849.
Leb.—Leber (C.) *Collection des meilleures dissertations relatifs à l'histoire de France.* 1826-42.
Liç.—Liçana (Bernardo de) *Historia de Yucatan.* 1633.
Ling.—Lingard (Rev. Dr. J.) *The History and Antiquities of the Anglo-Saxon Church.* 1845.
Liv.—Livingstone (D.) *Last Journals in Central Africa.* 1874.
Ly.—Lyall (Sir Alfred C.) *Asiatic Studies.* 1882.
MacDon.—MacDonald (Rev. Duff) *Africana: or the Heart of Heathen Africa.* 1882.
Macph.—Macpherson (Lieut. —) *Report upon the Khonds of Gaujani and Cuttack.* Calcutta, 1842.
Maine.—Maine (Sir Henry S.) *Dissertations on Early Law and Custom.* 1883.
Mall.—Mallet (P. H.) *Northern Antiquities.* Trans. 1847.
Mar.—Mariner (W.) *An account of the Natives of the Tonga Islands.* 1818.
Mars.—Marshall (Lieut.-Col. W. E.) *A Phrenologist among the Todas.* 1873.
Mart.—Martin (M.) *A description of the Western Islands of Scotland.* 1716.
Mas.—Maspero (G.) in *Revue Scientifique.* March, 1879.
Mau.—Maury (L. F. A.) *Hist. des Religions de la Grèce Antique.* 1857.
" —— Article in *Revue des Deux Mondes.* Sep. 1867.
Med.—Medhurst (W. H.) *China, its State and prospects.* 1838.
Mich.—Michie (Alex.) *Siberian Overland Route.* 1864.
Mol.—Molina (Ch. de) *An account of the Fables and Rites of the Yncas.* (Hakluyt Society.) 1873.
Mom.—Mommsen (T.) *Hist. of Rome.* Trans. 1868.
Mons.—Monstrelet (E. de) *Chronicles.* Trans. 1840.
Mos.—Mosheim (J. L.) *Institutes of Ecclesiastical History.* Trans. 1758. New ed. 1863.
Müll.—Müller (F. Max) *Introduction to the Science of Religion.* 1882.
" —— *History of Ancient Sanskrit Literature.* 1859.
Ogil.—Ogilby (J.) *Africa.* 1670.
Onde.—Ondegardo (P. de) *Report* in *Narratives of the Rites and Laws of the Yncas.* Trans. Markham. (Hakluyt Soc.) 1873.
Ord.—Ordericus Vitalis. *Ecclesiastical History of England and Normandy.* Trans. 1853-6.
Ovie.—Oviedo y Valdés (G. Fernandez de) *Historia General y Natural de las Indias.* Madrid, 1851-5.
Pala.—Palacio (D. G. de) *Carta al Rey de España,* 1576; *Spanish and English,* by E. G. Squier. New York, 1860.
Pal.—Palgrave (Wm. Gifford) *Narrative of a Year's Journey through Central and Eastern Arabia.* 1865.
Pash.—Pashley (R.) *Travels in Crete.* 1837.
Per.—Percival (Rev. P.) *Land of the Veda.* 1854.

Pesch.—Peschel (Oscar) *The Races of Man and their geographical distribution.* Trans. 1876.
Pie.—Piedrahita (L. F. de) *Historia del Nuevo Reyno de Granada.* Amberes. 1688.
Pinto—Pinto (Major Serpa) *How I Crossed Africa.* 1881.
Pow.—Powell (W.) *Wanderings in a Wild Country.* 1883.
Prej.—Prejevalski (Lieut. Col. N.) *Mongolia.* Trans. 1878.
Pres.—Prescott (W. H.) *History of the Conquest of Peru.*
Pri.—Prichard (J. C.) *Researches into the Physical History of Mankind.* 3rd. ed. 1836-47.
Pla.—Plato. *The Republic.* In *Dialogues.* Trans. by Jowett. Oxford, 1871.
Pot.—Potter (Bp. J.) *Archæologica Græca.* Edinb. 1827.
Ráj.—Rájendralála Mitra. *Indo-Aryans.* 1881.
Rec.—*Records of the Past, being English Translations of the Assyrian and Egyptian Monuments.* 1874-81.
Ren.—Renouf (P. Le P.) *Origin and Growth of Religion as illustrated by the Religion of Ancient Egypt.* Hibbert Lectures. 1880.
Rév.—Réville (Albert) *Histoire du Diable.* Strasb., 1870.
Rob.—Robertson (E. W.) *Historical Essays.* 1872.
Ross—Ross (Rev. John) *History of Corea.* Paisley, 1880.
Roth—Roth (P.) *Feudalität und Unterthanenverband.* Weimar, 1863.
Saha.—Sahagun (Fr. B. de) *Historia General de las Cosas de Nueva España* [1569]. Mexico, 1829-30.
St. J.—St. John (Sir Spenser) *Life in the Forests of the Far East.* 1862.
Sang.—Sangermano (Father) *Description of the Burmese Empire.* Trans. Rome, 1833.
Sat.—Satow (E. M.) *The Revival of Pure Shinto*, in appendix to *Transactions of the Asiatic Society of Japan*, vol. iii, pt. I.
School.—Schoolcraft (H. R.) *Information respecting the History of the Indian Tribes of U. S.* 1853-56.
Schw.—Schweinfurth (Dr. G.) *Heart of Africa.* Trans. 1873.
See.—Seeley (J. R.) *Livy. Bk. I, with Introduction*, etc. Clarendon Press Series. 1871.
Sel.—Selous (F. C.) *A Hunter's Wanderings in Africa.* 1881.
Sha.—Sharpe (Samuel) *History of Egypt.* 3rd ed. 1852.
Sher.—Sherring (Rev. M. A.) *The Natural History of Hindu Caste* in *Calcutta Review.* Vol. lxxi. (1880.)
Sim.—Simon (P.) *Noticias Historiales.* In Kingsborough's *Antiquities of Mexico.* Vol. viii. 1830.
Smith.—Smith (George) *History of Assyria.*
Sm., Geo.—Smith (George) *Religion of Ancient Britain.* 1846.
Sm.—Smith (Rev. Samuel) *Church Work among the Deaf and Dumb.* 1875.
Sm., W.—Smith (Wm.) *Dictionary of Greek and Roman Biography and Mythology.* 1844.
Smy.—Smyth (R. B.) *The Aborigines of Victoria.* Melb. 1878.
Stu.—Stuart (V.) *Nile Gleanings.* 1879.
Stub.—Stubbs (Bp. Wm.) *The Constitutional History of England.* 1880.
Sup. Rel.—*Supernatural Religion.* 2nd. ed. 1874-7.
T.B.A.S.—*Transactions of the Society of Biblical Archæology.*
T.R.A.S.—*Transactions of the Royal Asiatic Society.*
Tern.—Ternaux-Compans (H.) *Recueil des Pièces relatives à la Conquête du Mexique.* Paris, 1838-40.
Thom.—Thomson (Dr. A. S.) *The Story of New Zealand.* 1859.
Thom., J.—Thomson (J.) *The straits of Malacca, Indo-China, and China.* 1875.
Thoms.—Thomson (Joseph) *To the Central African Lakes and back.* 1881.

Tie.—Tiele (C. P.) *Outlines of the History of Ancient Religion to the Spread of the Universal Religions.* Translated by J. E. Carpenter. 1877.
Tor.—Torquemada (J. de) *Monarquia Indiana.* 1723.
Tur.—Turner (Rev. Geo.) *Nineteen Years in Polynesia.* 1861.
„ —— *Samoa a Hundred Years Ago.* 1884.
U. S. Ex.—*United States Exploring Expedition*; by Com. C. Wilkes. Phil. 1845.
Vanc.—Vancouver (Capt. G.) *Voyage of Discovery to the North Pacific Ocean and Round the World.* 1798.
Vell. Pat.—C. Velleius Paterculus. *Historiæ Romanæ libri* ii.
Wall.—Wallace (A. R.) *A Narrative of Travels on the Amazon and Rio Negro.* 1853.
„ —— *The Malay Archipelago.* 1872.
Wilk.—Wilkinson (Sir J. G.) *Manners and Customs of the Ancient Egyptians.* Ed. by Samuel Birch. 1878.
Wil.—Williams (Prof. Monier) *Indian Wisdom.* 1875.
Wil.—Williams (Rev. T.) *Fiji and the Fijians.* 1870.
Xim.—Ximenes (F.) *Las historias del Origen de los Indios de Guatemala.* 1857.
Yate—Yate (Rev. W.) *Account of New Zealand.* 1835.
Zim.—Zimmern (Helen) In *Fraser*, April, 1881.
Zur.—Zurita (Al. de) *Rapports sur les différentes classes de chefs de la Nouvelle Espagne.* Trad. par H. Ternaux-Compans. 1840.

www.ingramcontent.com/pod-product-compliance
Lightning Source LLC
Chambersburg PA
CBHW032138010526
44111CB00035B/613